The Great Pyramid Decoded

Born in 1936, **Peter Lemesurier** read languages at Cambridge and holds the Associate Diploma of the Royal College of Organists. He has been a musician, jet pilot, trimariner, teacher and translator. Over the years, the focus of his interest has moved from prophecy through Ancient Egyptian Mysteries, philosophy, psychology, astrology, mythology and back to prophecy again. He is currently working on Nostradamus.

by the same author

The Armageddon Script
Beyond All Belief
The Cosmic Eye
Gospel of the Stars
The Great Pyramid: Your Personal Guide
The Healing of the Gods
This New Age Business
Nostradamus – The Next 50 Years
Nostradamus: The Final Reckoning

The Great Pyramid Decoded

Peter Lemesurier

ELEMENT
Shaftesbury, Dorset • Rockport, Massachusetts
Brisbane, Queensland

First published in Great Britain in 1977

This revised edition published in Great Britain in 1996 by
Element Books Limited
Shaftesbury, Dorset SP7 8BP

This revised edition published in the USA in 1996 by
Element Books, Inc.
PO Box 830, Rockport, MA 01966

This revised edition published in Australia in 1996 by
Element Books Limited
for Jacaranda Wiley Limited
33 Park Road, Milton, Brisbane 4064

Cover illustration by Mark Topham
Cover design by Bridgewater Books
Text design by Roger Lightfoot
Typeset by WestKey Limited, Falmouth, Cornwall
Printed and bound in the United States of America
by Edwards Brothers, Inc.

British Library Cataloguing in Publication
data available

Library of Congress Cataloging in Publication
data available

ISBN 1-85230-861-3 Hardback UK Edition
ISBN 1-85230-793-5 Paperback US Edition

Contents

APPENDICES:

Table of Illustrations and Diagrams

Acknowledgements

The author is indebted to the following for their kind permission to reproduce copyright material:

The Association for Research and Enlightenment, Inc., of Virginia Beach, Virginia, for the excerpts from the Edgar Cayce readings and for their encouragement; Ernest Benn Ltd., for the diagrams on pages 6, 10, 116, 125, 276, 291, 335, 337, 350 and 357, which are taken in whole or in part from vol. I of *The Great Pyramid: Its Divine Message* by Davidson and Aldersmith; Cambridge University Press for the extract from *Divani Shamsi Tabriz* by Jalal'ud-Din Rumi, translated by R. A. Nicholson; The Hamlyn Publishing Group Ltd., for quotations from *Mythology of the Americas* by Burland, Nicholson and Osborne; The Merseyside County Museums for the picture of Yiacatecuhtli from the *Codex Fejervary-Mayer* on page 305; The Mondadori Press (Milan) for the aerial view of the Great Pyramid on page 11; Penguin Books Ltd., for quotations from the *Bhagavad Gita*, edited and translated by Juan Mascaró (Penguin Classics 1961, copyright © Juan Mascaró 1961); Mario de Sabato, for the extracts from his *Confidences d'un Voyant* (Hachette, 1971); Turnstone Books for the *Djed*-pillar on page 173 (original from *The Sphinx and the Megaliths* by John Ivimy); Roger Viollet (Paris) for the picture of the solar chariot-wheel on page 23; Dr. Adam Rutherford for diagrams taken wholly or in part from his *Pyramidology*, published by the Institute of Pyramidology—namely those on the following pages (Rutherford's page numbers are shown in brackets): 17 (flap), 52 (950), 100 (974), 101 (977), 103 (983), 109 (986), 138 (1065 and 1070), 150 (1088), 177 and 186 (flap), 199 (966), 354 (298) and 379 (1296–7). The late Dr. Rutherford's factual advice and help—generously given despite his vigorous disagreement with many of my working methods and conclusions—has been invaluable throughout, and his meticulous research has been the basis for virtually everything that is valid (and for none of the errors or inconsistencies) in the Pyramid-based sections of this book. The illustrations on pages 19, 66, 75, 80 (upper), 96, 105, 114, 130, 136, 159, and 181 are taken from *The Great Pyramid Passages and Chambers* by J. and M. Edgar, published by Bone & Hulley, and those on pages 4 and 5 from Prof. C. Piazzi-Smyth's *Our Inheritance in the Great Pyramid* of 1864. Except where otherwise stated, biblical quotations are taken from the *New English Bible*, second edition © 1970, by permission of Oxford and Cambridge University Presses.

I tender my sincere thanks to all others who, wittingly or otherwise, have contributed to the book's genesis, to Ray Smith for his redrawing of numerous diagrams, and especially to John Moore, for having acted as midwife.

P. L.

Foreword to the 1996 Edition

IT IS NEARLY two decades now since this book first appeared in print. Little did I suspect at the time that it would achieve world-wide fame or come to be generally regarded as a seminal work on the subject. Nor could I know then how well its findings would stand the test of time, or how accurately its prophetic conclusions would turn out to fit unfolding world events as the years rolled—or, rather, staggered—by.

In the event, the book's solidity (appropriately for its subject, perhaps) has turned out to be positively rock-like. Even some nineteen years later, its data remain virtually unchallenged and its predictions unneedful of revision. Indeed, I have had little to do in this edition beyond listing a number of actual fulfilments of those mooted in the first edition, while adding to chapter 3 a brief summary of some of the more striking theories and investigations that have been reported in print since 1977.

The Great Pyramid, it seems, remains as steady and unchanging a witness as ever, its revelations supported by a host of human prophets stretching from those of the Old Testament, via the great Michel Nostradamus, to such modern luminaries as Edgar Cayce, Jeane Dixon and the astonishing Mario de Sabato. And always the underlying message – whatever the language in which one chooses to express it – is that humanity's relationship with the wider cosmos has become fatally flawed, but that renewed harmony remains eternally possible.

Ultimately, it seems, the answers to the big questions lie in our own hands, so that what goes on to befall humanity reflects directly the extent of our willingness to grasp them. If, however, we fail to make the necessary changes, the universe will inevitably take matters firmly into its own hands. Indeed, on present evidence, this latter outcome seems a good deal more likely.

Not that this necessarily has anything to do with biblical revenge or retribution. It is simply a matter of natural feedback within a universe that, as the Pyramid itself reveals, will always insist on returning to a state of balance.

And so collective doom may very well lie just around the corner for us, while paradise may take just a little longer . . .

P. L.

PART ONE

The House of Hidden Knowledge

'Then . . . there began the building of that now called Gizeh . . . the Hall of the Initiates . . . This, then, receives all the records from the beginnings of that given by the priest . . . to that period when there is to be the change in the earth's position and the return of the Great Initiate to that and other lands for the folding up of those prophecies that are depicted there. All changes that came in the religious thought in the world are shown there, in the variations in which the passage through same is reached, from the base to the top—or to the open tomb and the top. These are signified by both the layer and the colour [and] in what direction the turn is made.

'This, then is the purpose, for the record and the meaning to be interpreted by those that have come and do come as the teachers of the various periods, in the experience of this present position, of the activity of the spheres of the earth. . .'

EDGAR CAYCE, *30th June 1932*

1

A Message from the Dead?

DURING THE LAST hundred years or so of its known four thousand year history, the Great Pyramid of Giza in Egypt—the first, and last, of the Seven Wonders of the ancient world—has attracted more than its fair share of cranks and pyramidomaniacs. Enthusiasts all, their theories have ranged from the fantastic, via the sublime, to the ridiculous. And as a direct result of their labours, any theory which sees the Pyramid as anything more than a finely wrought funerary heap of stones is nowadays liable to be brushed off as 'sheer speculation' without so much as a glance at the evidence presented.

And yet, if the claims of the lunatic fringe are sometimes fantastic, the Pyramid itself is even more so. Few would care seriously to dispute that it is easily the most massive building ever known to have been erected on this planet[1]—having, for example, at least twice the volume and thirty times the mass of New York's Empire State Building. Again, it would be a rash man who undertook to find, even today, a building more accurately aligned to the True cardinal points of the compass, masonry more finely jointed, or facing-stones more immaculately dressed.

The unconvinced may point out that the Pyramid's north-south axis is slightly out of true—by nearly five minutes of arc, or *one-*

[1] Unless, that is, the colossal adobe-brick temple-platform to Quetzalcoatl at Cholula, Mexico (volume well over 200 million cubic feet) or the Great Wall of China are classed as 'buildings'. In the more usual sense of the term the nearest rival in masonry volume to the Great Pyramid as originally constructed (some 91,500,000 cubic feet with theoretical apex) is the neighbouring Second Pyramid (78,400,000 cubic feet), while the largest Mexican pyramid (the much later Pyramid of the Sun at Teotihuacan) has a theoretical volume of less than 50,000,000 cubic feet.

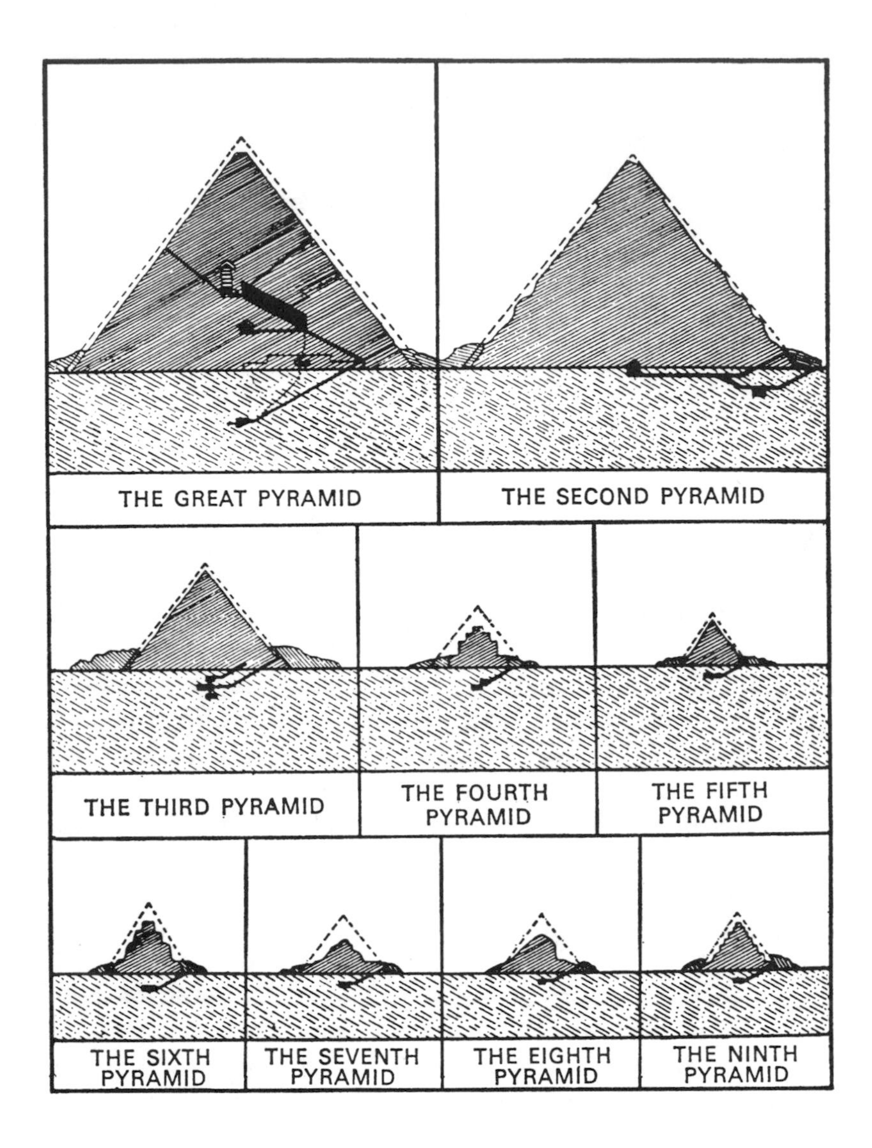

The Pyramids of Giza (reprinted from Prof. C. Piazzi-Smyth's *Our Inheritance in the Great Pyramid*, 1864).

twelfth of a degree. But that would be to ignore the astronomical evidence that the cause even of this minute error is to be found in the gradual movement of the earth's own axis, rather than in any inaccuracy on the part of the building's original surveyors.

Meanwhile, the sceptic may doubt that many of the Pyramid's stones—some of them weighing up to seventy tons—were so finely cut and positioned as to give joints of less than a fiftieth of an inch in thickness; he may scoff at the claim that a fine cement was run

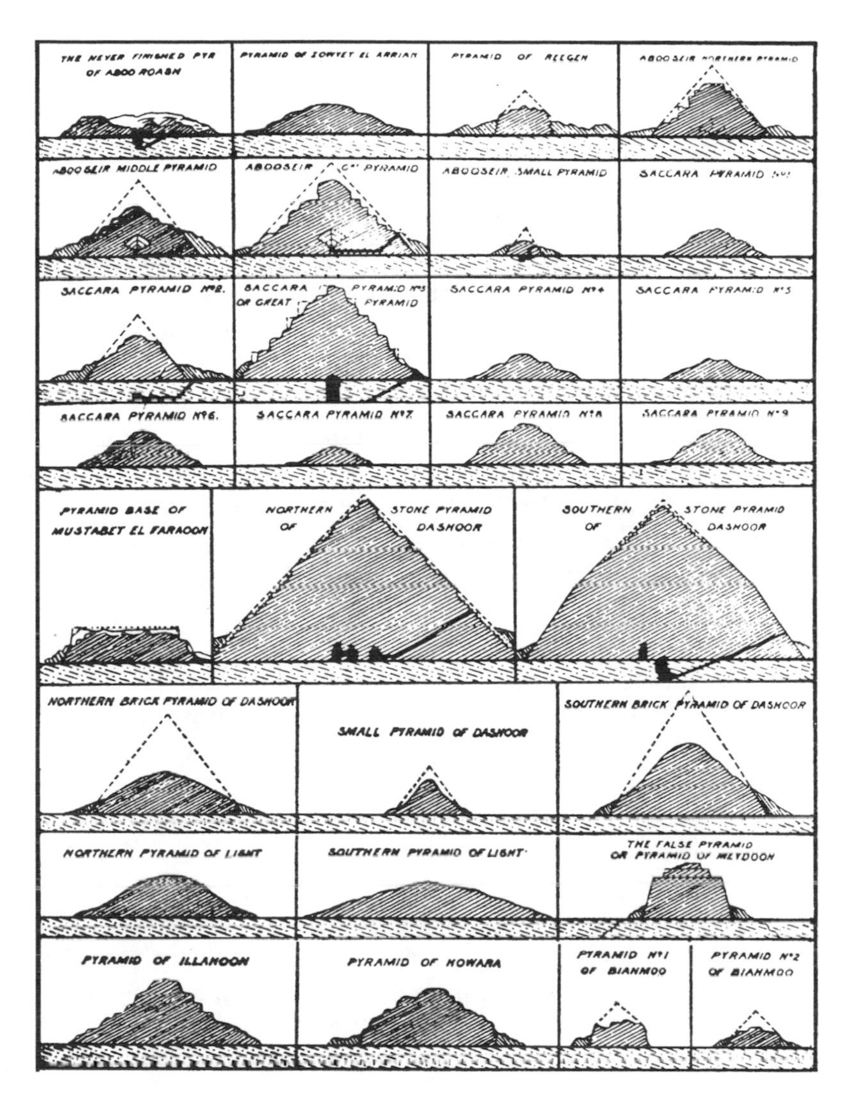

Some of the less well-known Egyptian pyramids (reprinted from Prof. C. Piazzi-Smyth's *Our Inheritance in the Great Pyramid*, 1864).

into these joints so expertly as to give an even coverage of single areas as big as five feet by seven *in the vertical*; he may express profound disbelief when it is pointed out to him that the building's now almost totally despoiled original outer casing of polished limestone (all twenty-one acres of it) was levelled and honed to the standard of accuracy normal in modern optical work. But these, as it happens, are facts which anybody may check who cares to. The stones speak for themselves.

And we are talking, remember, not of the product of twentieth-century engineering, but of a building whose construction dates from the very edge of prehistory.

But how—and why? The logic of the thing seems to defy all analysis.

And so the historians, ably led by the classical Herodotus, have had their field-day. As well they might, bearing in mind that, even to Herodotus, the Pyramid's construction was already as remote in time as Herodotus himself is to us. Knowing precisely nothing about the project's origin, they have naturally fallen back on a process of wild extrapolation from their only slightly less scratchy knowledge of later dynastic times. The Egyptians, it has been established, were obsessed with death and immortality, with the embalming of the dead, with preparations for life in the nether-world. Therefore the Great Pyramid Project represents that same obsession magnified to the *n*th degree. And so the scene described for us is a kind of gothic melodrama unequalled in its sheer antediluvian lunacy. The megalomaniac pharaoh Cheops, brooding over the fate of his own eternal soul, decides to throw his kingdom's entire resources into a colossal real-estate project designed purely to humour his own necromantic illusions of immortality. To satisfy this man's mere superstitious whim, thousands of slaves toil day after day to drag gigantic blocks of masonry up mighty ramps with the aid of nothing better than primitive sleds, levers, ropes and rollers. Overseers drawn straight from the serried ranks of Hollywood extras bark crude orders, wave cruder charts. The whips crack, the ropes creak, the tortured workers groan. For a brief moment in time the seething masses of ignorant, sweating humanity toil with crude tools under the ancient sunlight, and then are swallowed up once more in the primeval mists of antiquity. The silence and the sand return.

And the result? The Great Pyramid—a building so perfect and yet so enormous that its construction would tax the skills and resources even of today's technology almost to breaking-point. Yet that—or something very like it—is offered as the sensible explanation of the project; sensible—when scarcely a hundred years separate this supreme example of the stonemason's craft from the construction of King Zoser's celebrated step-pyramid at Sakkara, said to be the first large stone building ever raised on earth.

The sober truth is, of course, that no historian has yet advanced any explanation of the Great Pyramid's construction that is at all convincing. Nobody alive today knows for certain how the Pyramid was erected, how long it was in the building, how its near-perfect alignments were achieved before the invention of the compass, or how its outer casing was jointed and polished with such unsurpassed accuracy. Nor have historians succeeded in producing any convincing theory as to why such an enormous undertaking, combined with such incredible accuracy, should have been deemed necessary for the construction of a mere tomb and funerary monument to a dead king who in any case apparently never occupied it.[2]

But that is only the beginning of the mystery. It is one thing to align your building exactly with the earth's four cardinal points. It is quite another to site it at the exact centre of the geometrical quadrant formed by the Nile Delta—the ancient kingdom of Lower Egypt. And yet such was the case, as the United States Coast Survey discovered in 1868 and the map on page 8 shows. Meanwhile there was an added bonus. Reference to any equal-area projection of the earth's surface reveals that the chosen site also lies on the longest land-contact meridian on the earth's surface and at the geographical centre of its whole land mass including the Americas and Antarctica. The same could not be said, for example, of Vladivostok or Buenos Aires. But since, of course, the designer could not have known this in view of the limited state of his knowledge, the above facts must be fortuitous.[3] Yet in this case we have a number of further fortuitous facts to take into account. The data set out in Appendices A and B (q.v.) demonstrate, for example, that there are clear mathematical links between the Pyramid's dimensions and the earth's basic geophysical data and orbital astronomy. The basic unit of measurement apparently used by the designer turns out to be an exact ten-millionth of the earth's mean polar radius. The Pyramid's designed base-square has a side measuring just 365·242 of these same units—a figure identical to the number of days in the solar tropical year—and

[2] Compare Herodotus (*Euterpe* 124, 125); Diodorus Siculus (Book I); and see Rutherford pp. 15–22, 1198–1200. The preparation of twin tombs, *and the non-occupation of the northern one*, seems to have been something of a tradition in Khufu's family, being observed in particular by both his parents. By contrast, 'burial' high above ground level (as in the case of the Great Pyramid's King's Chamber) was never the practice of any known pharaonic dynasty. Indeed, it seems almost certain that the semi-legendary Khufu was in fact buried in the great pit-tomb known as Campbell's tomb, some hundred yards west of the Great Sphinx—a tomb which admirably fits Herodotus's description of the royal interment, as received from the Egyptian priesthood.

[3] It is also claimed by some authorities (a) that the height of the Pyramid's summit-platform denotes the earth's mean level of land and sea and (b) that the Pyramid's weight (generally quoted at some 5,955,000 tons) represents a billionth of the weight of the earth. The author has not so far seen any detailed evidence for these propositions, however.

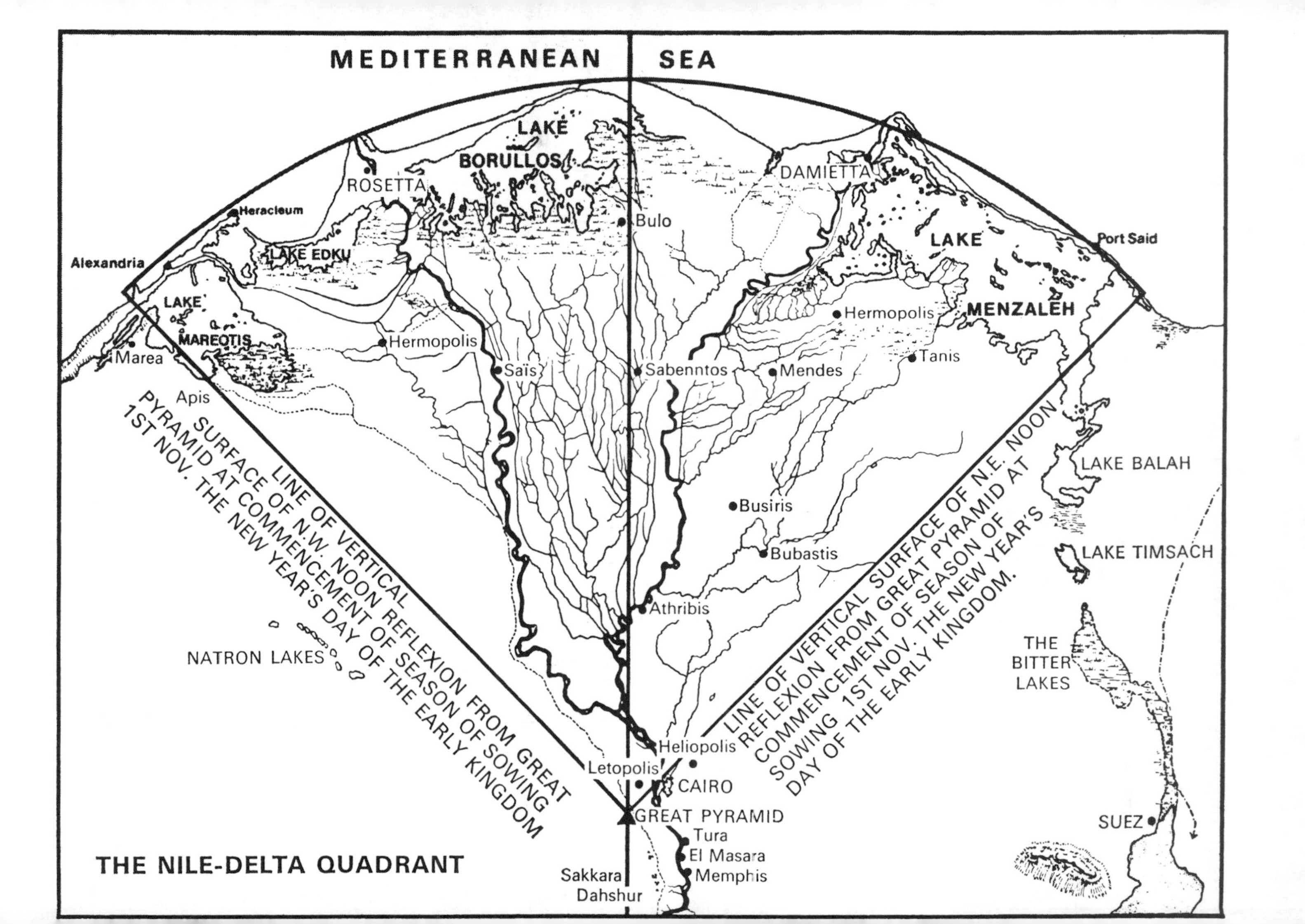

THE NILE-DELTA QUADRANT

the same figure is also to be found in other features of the design. Yet from the slightly indented shape of the base of the core-masonry alternative measurements of 365·256 and 365·259 of these units can be derived—figures which turn out to be the length in days of the sidereal year (the actual time the earth takes to complete a circuit of the sun) and of the anomalistic year (the time taken by the earth to return to the same point in its elliptical orbit, which is itself revolving slowly about the sun). Meanwhile further measurements appear to give exact figures for the eccentricity of that orbit, for the mean distance of the earth from the sun, and for the period of the earth's full precessional cycle (a period of over 25,000 years). If one wished to have an architectural symbol for the planet Earth itself one could scarcely do better than to take the Great Pyramid of Giza.

But that is not all. While it seems that the Pyramid was deliberately left by its builders without its culminating capstone,[4] and while virtually none of its original casing now survives *in situ*, the sockets marking out its designed foundations remain, the angle of slope of its faces can be deduced from such casing-stones as are still preserved, and the architect included in the design of each face a triangular indentation representing a one-fifth scale-model of the Pyramid's cross-section. Consequently we not only know the length of the Pyramid's base-side, but can also calculate accurately its designed height. And it turns out that the ratio of base-perimeter to height is none other than twice the quantity *pi* (π)—in other words, the Pyramid's height is to its base-perimeter as a circle's radius is to its circumference. Using the unit of measurement already referred to (the Sacred Cubit) *every one of the Pyramid's basic external and internal measurements can thus be expressed as a function of the quantities pi and 365·242.*[5] The Pyramid's geometry, in other words, not only combines all the above data into a single, elegant identification of the planet upon which we live; it also gives these quantities durable expression *in terms of each other.*

All of which is no mean feat for an ignorant, superstitious, semi-prehistoric architect to achieve by mere accident. In fact it must by this stage be abundantly clear that there can have been nothing accidental about it at all. The Great Pyramid's measurements reflect not a series of fortuitous coincidences, but an extraordinarily advanced level of knowledge in its designer—a level rivalled only by the technology of its builders. And in view of this we have no

[4] See Appendix H.

[5] The one exception is the quantity 35·76 Primitive Inches and its multiple 286·1 P″. See p. 18. This would suggest a special significance for those quantities.

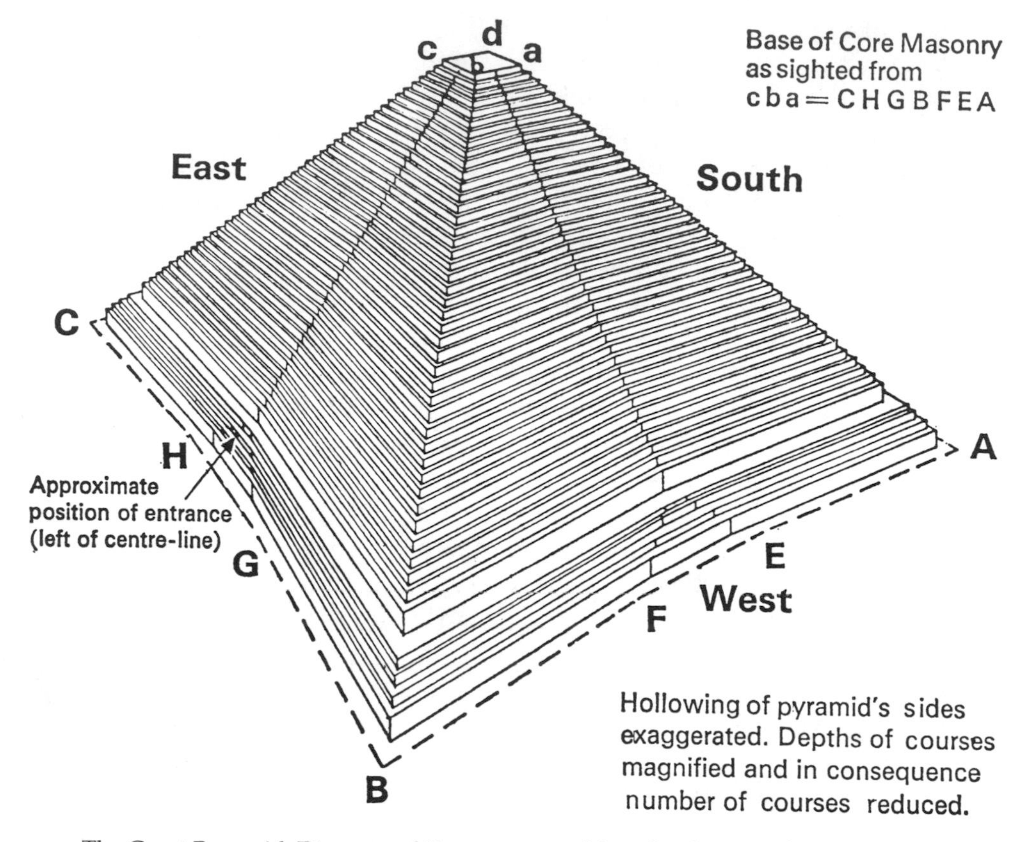

The Great Pyramid: Diagram of Core-masonry. Note the absence of a capstone, the hollowing of the sides and the prominence of the 35th course of masonry (here represented as the fifth course from the bottom). The original base-circuit was well over 1,000 yards. For 3,000 years the whole was sheathed with polished-limestone casing blocks forming a smooth, almost faultless slope at 51°51′14″ to the horizontal. As well as being visible for many miles as a 'golden mountain', brilliant in the sunlight (hence its Egyptian name 'The Light'), the Pyramid cast visible reflections both in the air and on the sand of the Giza plateau, and these may well have been used by the ancient Egyptians to regulate their seasons (see p. 337 and Davidson and Aldersmith's research into the phenomenon).

choice but to look at the questions of how and why afresh. The accepted answers just will no longer do.

There have been attempts to show that the deduction of these data from the Pyramid's measurements is based upon Egyptian measures that did not in fact exist at the time, or at any rate were not used in the building of the Great Pyramid—namely the so-called Primitive Inch (equal to 1·00106 standard inches) and the Sacred Cubit (equal to 25 Primitive Inches). It is widely believed that the eminent and respected Egyptologist Sir Flinders Petrie, in his *Pyramids*

Aerial view of The Great Pyramid (east face in shade). Note boat-pits. A few of the stones of the 203rd course can still be seen on the summit-platform. The apparent entrance is Colonel Vyse's excavation of 1836.

and Temples of Gizeh (1883), succeeded in demolishing the idea that these measures were used in the Pyramid's design. What is less well known, however, is that several clear cases of their use *can be deduced from Petrie's own measurements*, and that the accuracy and independence of these subsequently served for many years as the chief validation for the data already listed. Only during the present century has Petrie's accuracy been improved upon, as a result of advanced surveying techniques backed up by meticulous geometrical calculation—all of which has finally established the validity of the various correlations listed above beyond all question. And this is without even adducing the additional evidence set out in Appendix C to the effect that the ancient Egyptian king-list chronologies were largely fictitious, and had been adjusted in such a way as to enshrine a clear record of the Pyramid's vital dimensions—in those same Primitive Inches whose existence was allegedly demolished by Petrie. *Even if the Great Pyramid no longer existed, it would be possible, by experimental geometrical application of the figures contained in the ancient king-lists, to reconstruct its exterior with a remarkable degree of precision.*

In any case, the historical use or non-use *by the ancient Egyptians* of the Primitive Inch and Sacred Cubit is of largely academic interest. It is, after all, perfectly possible to deduce from an analysis of the Pyramid's dimensions—or even to postulate—units of basic measurement which consistently demonstrate all the correlations already referred to: and these units always turn out to be equal to the Primitive Inch and Sacred Cubit respectively, as the data set out in Appendices A and B make clear. Admittedly, ratios such as those between the three types of astronomical year continue to be expressed by the Pyramid's base-measurements *whatever the units of measurement used*; but as soon as those ratios are expressed in their natural terms (in this case, the number of days in the year) and then re-applied to the Pyramid, the measurement representing one day turns out to be an exact Sacred Cubit in length.

It can thus be conclusively demonstrated, quite independently of any historical considerations, that the postulated basic measurements and listed correlations are all inherent in the Pyramid's design—and consequently we have to assume that the designer who put them there was aware of that fact. Which of course immediately poses problems for the historians.

If, after all, we assume that the inclusion in the Pyramid's design of the data already listed was not fortuitous—and there seems to be little alternative to so doing—then we are virtually admitting that the degree of scientific knowledge exhibited by the Pyramid's designer is completely incompatible with what current historical theory tells us about man's early history. Such is the divide between fact and theory that many writers on the Pyramid have found it necessary to assume that the Pyramid was designed by none other than the Deity—a time-honoured argument for explaining anything one cannot understand, and one which has never so far turned out to be satisfactory in the light of man's subsequent knowledge. A more reasonable explanation would presumably be that human history is not, after all, what we have so often been given to believe—a long, slow, upward gradient leading inexorably from primitive man to the summit of modern society's achievements, with virtually nothing of note apparently happening during the first 1,995,000 years of man's two million or so years of known existence. Instead it would appear to consist of at least two (and possibly many more) such upward gradients, each followed by a sudden collapse and a recommencement, reminiscent of the Hindu and Buddhist 'cycles of existence' or the similar Mayan conception of universal history. Indeed, one of the oldest, most universal and most persistent of legends to be found among virtually all peoples of the world is the apparent folk-memory typified by the story of the biblical Flood—an oral testimony, perhaps, to the cataclysmic destruction of an earlier world whose knowledge and technical achievements were far in advance of anything so far attributed by history to so-called early man, and may indeed have rivalled or even surpassed our own.

Certainly the geological record shows quite clearly that the world's sea-levels rose by over 300 feet between 15,000 and 4,000 B.C., sometimes at a rate of over 30 feet per century.[6] Given, then, that most civilised communities tend to develop along the sea-coasts and on rivers and their estuaries, almost any civilised community that existed during that era must inevitably have suffered some kind of watery catastrophe—unless, that is, its inhabitants had the foresight to 'up sticks' and migrate to some more promising area, somewhat after the style of the biblical Noah. But the more conventional human tendency is simply to build higher sea-defences, and then sit tight and hope. And then one year the sea would at last have broken through, and another lost civilisation would have joined the long lists of world mythology. Could this, then, be what happened to some ancient and advanced civilisation almost all trace of which—apart, perhaps, from the Great Pyramid—has now vanished?

[6] Rhodes W. Fairbridge, 'The Changing Level of the Sea' (*Scientific American*, May 1960, Vol. 202, No. 5).

The idea seems fantastic, but on what other basis can one explain the equally fantastic evidence of the Great Pyramid's technology and the apparent knowledge of its architect? Almost the only alternative would seem to be von Däniken's controversial postulation of men from another world[7]—but this idea would appear to be inherently much more unlikely than the one we have just advanced. Meanwhile the apparent lack of archaeological evidence of earlier highly developed human societies ought not to be regarded as a conclusive argument against the existence of such antediluvian civilisations. We have, after all, no knowledge in experience to indicate what forms these remains would take, where they can be expected to occur, or indeed whether such civilisations can be expected to leave recognisable remains at all; nor can we be sure of the nature or severity of the cataclysms which presumably destroyed them.[8] And there seems to be a very good chance that even such traces as have survived would now be buried beneath many feet of sea-water and mud, very much in the manner of Plato's description of the sunken Atlantis.

Moreover, that lack of archaeological evidence may only be an apparent one. If such civilisations did exist, then it is at least possible that some of their artefacts have been discovered already. But the natural tendency of a traditionally trained archaeologist is to 'fit' his finds into the fashionable preconceived matrix of ideas regarding the shape of prehistory. It is his very ability to do so that has gained him his qualifications as an archaeologist. Consequently, had the earlier civilisation left behind aircraft runways, say, then it is likely that they would be classified as 'ritual causeways'; an earth-and-stone-built chamber for screening off radioactive waste or machinery would of course take on an unintended role as a 'tomb'; an astronomical observatory-cum-computer would immediately start a new life as a 'temple'; and skyward-facing navigational

[7] *Chariots of the Gods?*

[8] Compare the Egyptian priest quoted by Plato (*Timaeus* 22 C–D): 'There have been and there will be many and divers destructions of mankind, of which the greatest are by fire and water, the lesser ones by countless other means. For in truth the story that is told in your country as well as ours how once upon a time Phaeton, son of Helios, yoked his father's chariot, and, because he was unable to drive it along the course driven by his father, burnt up all that was upon the earth and himself perished by a thunderbolt—that story, as it is told, has the fashion of legend, but the truth of it lies in the occurrence of a shifting of the bodies in the heavens which move around the earth, and a destruction of the things on earth by fierce fire, which recurs at long intervals' (translation by R. G. Bury, from *Timaeus, Critias, Menexenus, Epistles*, Heinemann, 1929).

daymarks for airships would become 'magic designs for the appeasement of the gods'. And any artefacts not so easily fitted into the overriding assumption that their makers were members of a primitive hunting society obsessed with considerations of magic and religion would simply be listed as 'unclassified' or 'unidentifiable'.

Meanwhile the total absence of objects made in all but stone and the softer, less useful, metals would be taken as evidence, not that any metal machines used by that former civilisation had long since corroded away, but that the society in question was a primitive, stone- or bronze-age one. In which case one has ample reason to fear for the reputation of our own civilisation among the archaeologists of the year A.D. 20,000.

The cases listed are, of course, highly conjectural, and some scepticism in 'official' circles regarding the possibility of earlier advanced civilisations remains entirely understandable. What clearly *cannot* be accepted, however, is the negative approach—the thesis that the Pyramid's measurements cannot possibly be as described, nor can they have the geophysical and astronomical significance suggested, since such facts (however verifiable) are incompatible with present theory. Clearly, facts must have precedence over theories, not the reverse. And the solid facts are indubitably there, in the land of Egypt, for anybody to examine who cares to—just as they have been for the last four thousand years or more, and will no doubt continue to be for several more centuries to come. As happens when a scientific law is likewise overtaken by the latest research, there seems to be no alternative to overhauling the theories completely.

But to what purpose is all this waste? Why should a member of some primeval but advanced civilisation, apparently realising that he and his contemporaries were on the brink of some cataclysm (if we accept the hypothesis advanced above), decide to draw up plans for the erection of what would appear to be little more than a monument of witness—a colossal expenditure of time and effort undertaken, evidently, for no more useful purpose than to demonstrate to future men that others had been there before? Seen in this light the Pyramid can only be regarded as a rather pointless example of historical boasting.

So it may seem—until one starts to examine other possible motives for such a decision. And then it starts to become apparent that what we have termed historical boasting is in fact a fairly unlikely explanation. The whole thing seems to have been much too carefully planned and premeditated, and to have involved far too much expenditure of time and materials, to permit us to accept so superficial a motivation. It seems likely that the designer had some much more serious purpose in view when he finally drew up the plans for the construction of the mighty edifice.

But *what* serious purpose? Who was he, and what could have been the circumstances that spurred him on? We cannot tell. Perhaps he foresaw the cataclysm which we have already postulated. Or perhaps that cataclysm had already overwhelmed a stricken mankind. In which case we might see the Pyramid's designer as one of a group of advanced emissaries from that former civilisation—colonisers and civilisers of the stamp of Osiris and Thoth, perhaps, and men who, Noah-like, had succeeded in surviving the cataclysm by building mighty boats in which they eventually reached the land of Egypt. And if so, then the remains of the great wooden boats since discovered in the five boat-pits sunk deep in the seashell-laden rock around the Pyramid may have an unexpected significance. For they may commemorate not so much the solar symbolism of the ancient Egyptian funeral rites—itself possibly a later symbolic veneer—as the primeval voyage of those ancient founding fathers under whose lofty auspices the adjacent Pyramid's enormous pile was eventually to be raised. And bear in mind, incidentally, that the great Thoth himself (under his Greek name of Hermes Trismegistus) was credited specifically by the Egyptian priest-historian Manetho[9] with having produced 36,525 books of ancient wisdom[10]—*a figure identical to the number of Primitive Inches in the Great Pyramid's designed perimeter*.

Be that as it may, the question still remains, what serious purpose is a doomed civilisation likely to have in trying to communicate with its successors? The most obvious aim would seem to be the transmission of some kind of *warning or advice* designed to help the later civilisation avoid the pitfalls which trapped its predecessor, or perhaps even to steer it into a path consistent with man's true capabilities and destiny. The knowledge of this, we may perhaps infer, the earlier civilisation had gained but had been unable or unwilling to fulfil, and so wished to pass it on for the benefit of posterity. The idea seems fanciful, perhaps, and yet such a function would appear to be consistent with the salvationist function apparently assigned to the Pyramid by the Saïte recension of the Egyptian *Book of the Dead*; and there are strange echoes of the same idea even in the ancient Judæo-Christian scriptures.[11] In fact, if the designer

[9] As quoted by the neo-Platonist Iamblichus.

[10] The so-called Hermetic writings.

[11] See chapter 7.

believed in some form of resurrection or reincarnation—as is, perhaps, not unlikely—we may even see in the Pyramid's construction a degree of self-interest—a means of carrying over into a future incarnation the knowledge he had gained and would not otherwise remember. Or, at very least, a measure of concern for the ultimate reborn fate of his own contemporaries.

The external features and dimensions of the Pyramid, then, seen in this light, would have the function of validating the designer's main message: they could be interpreted as his credentials—the evidence that he 'knows what he is talking about'. But in this case, where is the message itself?

Logically, if the guarantee is inscribed on the outside of the package, we would expect to find the guaranteed commodity inside it. The message, if there is one, should be *inside* the Pyramid. And yet inside the Pyramid, apart from a few quarry markings in the originally sealed upper chambers, not a single inscription appears to have been left by the builders. There is nothing except a system of rising, descending and horizontal passages—plus what appears

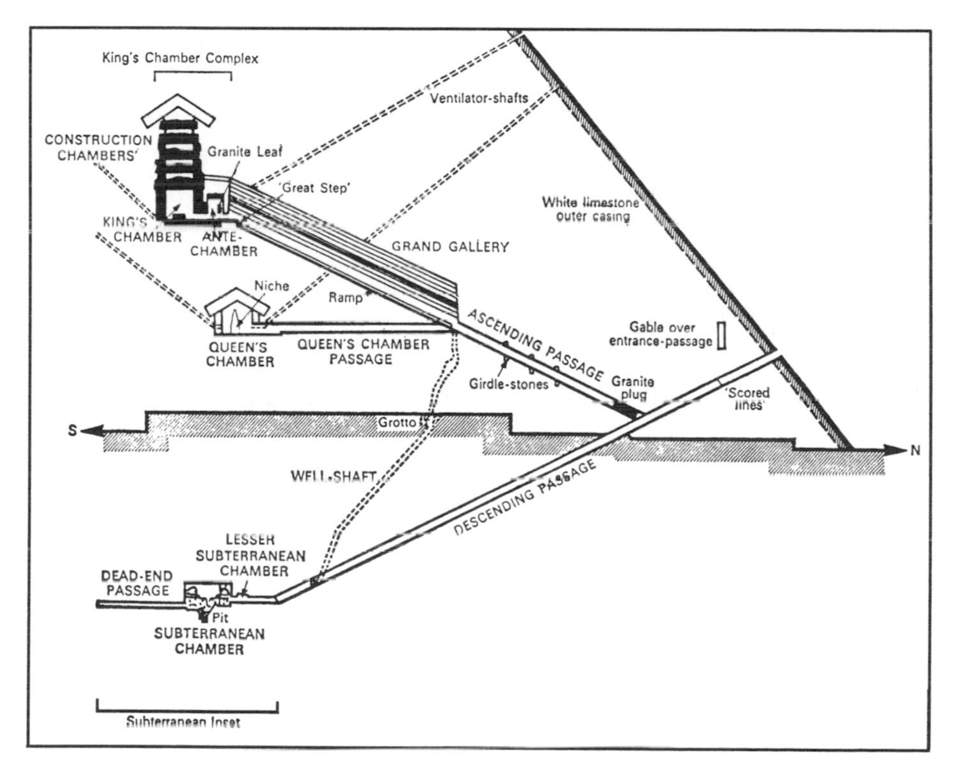

The passages and chambers of the Great Pyramid of Giza, looking west: black represents granite.

to be a dummy well-shaft—linking a number of basically rectangular chambers (see diagram page 17). All we can record about these features therefore, is their dimensions, their angles of slope, their relative positions and the types of rock in which they are constructed. If, then, there is a message at all, it is to be found in these features, and not inscribed in any known language. But then these features are at least expressible in terms of mathematics—which is the only truly universal language—so that it is perhaps *in terms of mathematics* that they need to be interpreted, just as do the exterior dimensions. Indeed, the choice of a mathematical code would actually seem to be a wise one on the designer's part, for he could not be sure how long either his own language or the script in which it was normally written would survive—whereas any civilisation advanced enough to read his credentials in the Pyramid's exterior dimensions must necessarily have sufficient understanding of basic mathematics to read its interior geometry too.

If, however, there is a link between the designer's credentials and the message he wished to convey, then we should expect to find some mathematical link between the Pyramid's external and internal features—other than the basic units of measurement, that is. And sure enough, such measurements are to be found.

For example, the inner square-perimeter of the Pyramid's base falls 286·1 P″ short of the 36524·2 P″ of the outer square perimeter (see diagram page 354). But this is also the distance by which the axis of the passage-system lies to the east of the axis of the Pyramid itself—which is why the ninth-century Caliph Al Mamoun missed the entrance passage when making the first forced entry along the centre line, only to discover his error by a stroke of good fortune when the 'Hidden Lintel' of the Ascending Passage became dislodged by the vibration of his workmen's hammering. Be it noted, incidentally, that he found that passage blocked from top to bottom and, having cleared the blockage, discovered that the King's Chamber contained not treasure but an empty, lidless, uninscribed sarcophagus. Meanwhile that same distance which fooled the Caliph is also found in the height of the Grand Gallery's roof above that of the Ascending Passage which leads into it, and in the floorline distance from the lower Well-Shaft opening to the bottom of the Descending Passage (eastern side).[12]

[12] It is also one-seventh of the perimeter of the original summit-platform, as defined by the heights of the pyramid's four air-shaft outlets (see p. 183 and note 63 p. 126), and one-eighth of the base-perimeter of the full-design capstone (see Appendix H).

By the same token, the concavity of each of the Pyramid's four sides measures 35·76 P″, as is pointed out on page 355. But that same distance is found *inside* the Pyramid in the height of the Great Step, the height of the Subterranean Passage and the distance from the north wall of the Grand Gallery to the axis of the Well-Shaft.

Since, then, the figure 35·76 and its multiple 286·1 are almost alone among the Pyramid's vital statistics in not being direct functions either of the value *pi* or of the quantity 365·242, we are forced to conclude that these measurements were deliberately chosen and deliberately applied, both outside and inside the Pyramid. We can at this stage only guess at what absolute significance—if any—they may have, but one of their functions at least is plain—to lead the observer from the information written in the building's exterior geometry to a set of further information presumably written in the geometry of the interior chambers and passageways.

But at this point we are faced with a further indication of a possible link between the world outside the Pyramid and the message possibly enshrined in its interior. The designed angle of descent and ascent of its sloping passages is 26°18′9·7″—an angle significant for having once been the exact elevation of the Pole Star from the latitude of the Pyramid. In fact during the third millennium B.C. the then Pole Star, Alpha Draconis, actually shone directly down the Descending Passage at its lower culmination—and this was the very millennium which apparently saw the construction of the Pyramid itself. This angle, then—and thus, perhaps, the passage which forms it also—clearly has both an astronomical and a chronological significance. But could it perhaps have others too? If that same angle is laid off from the Pyramid's east-west axis in a north-easterly direction, for example, the rhumb-line so produced marks the bearing of the summer sunrise from the latitude of the Pyramid.[13] Moreover it has long been known that the same line marks off several

[13] Specifically, that of sunrise on 6th June and 7th July, assuming that the horizon-altitude is zero and that sunrise is taken as the moment when the sun's lower limb sits tangent on the horizon (the definition usual among the ancient megalith-builders). With the sun's apparent motion corrected for the third millennium B.C., the same bearing would formerly have been reached on 5th June and 8th July.

It is interesting to note that the bearing misses by 1·7° the obvious target—the contemporary *mid*summer sunrise on E27°43′N—just as the same angle, applied as an altitude, similarly misses the north celestial pole by 3°41′ (it *had* to, since the former Pole Star, like its present counterpart, was not sited exactly at the celestial pole).

It is possible that the sunrise bearing marked at some date the exact rising of *Venus* at the summer solstice—the 'morning star' accorded a specifically Messianic

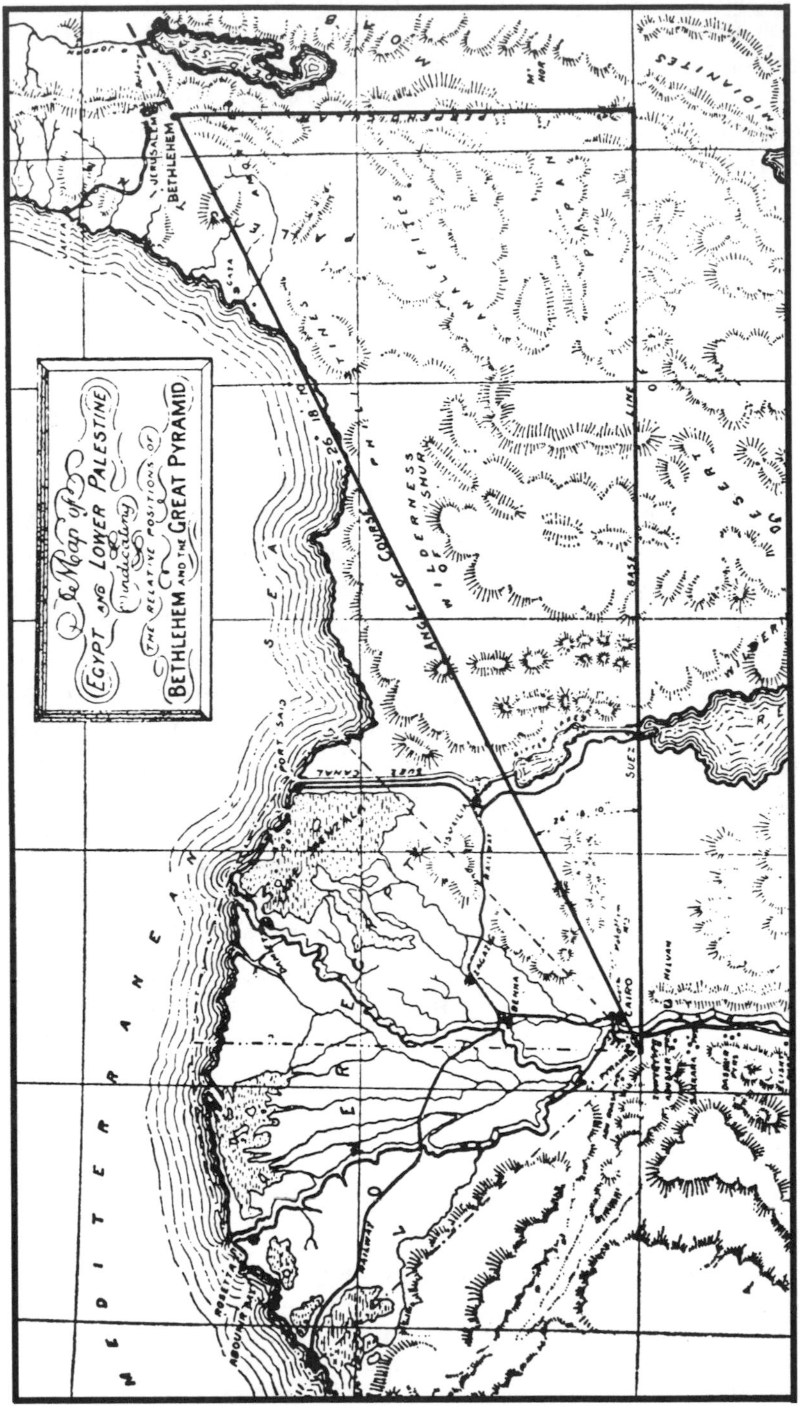

Geographical application of the Pyramid's passage-angle.

oddly significant geographical sites, and in particular, as is shown opposite, passes directly through the Jewish town of Bethlehem.

The suggestion that this fact could be other than pure coincidence must seem laughable. And yet we have already encountered so many other supposed coincidences—many of them just as remarkable in their own way—as to be forced to the conclusion that they were not coincidences at all, but evidence of a remarkable level of knowledge. Are we then to assume that this one fact, alone among all those adduced so far, is purely fortuitous? We could, of course. But the alternative is intriguing, to say the least.

We should remember, to start with, that the capstones of the more illustrious Egyptian pyramids were often gilded to represent the sun—indeed, the slopes of the pyramids were often associated with the rays descending from it. If the builders of the Great Pyramid deliberately omitted to add the capstone to their great symbol for planet Earth, then that omission could conceivably symbolise an incomplete world as yet 'in darkness',[14] and the deliberately reduced dimensions of the original Pyramid as built add weight to the possibility. There is some evidence in the ancient texts that the eventual addition of the capstone (and thus the completion of the Pyramid to its full design) was seen by the initiates as symbolising the return of Light to the world in the Messianic person of the resurrected Osiris. This capstone = sunrise = Messiah notion is also evident in parts of the Judæo-Christian scriptures, and not least in Jesus of Nazareth's overtly Messianic assertion, 'The stone which the builders rejected has become the chief corner-stone.'[15] The

[13] *(continued)* significance in several world religions and mythologies. If so—and the calculation would be a complex one—the datings for this event might well prove to be significant in some way.

Equally complex calculations would be involved in determining whether a single shift of the earth's axis could provide a direct hypothetical correlation between the minute error of some 5′ observable in the building's orientation, the 1·7° midsummer-sunrise error, the missing of the celestial pole by 3°41′ (despite the explanation already advanced) and the fact that the site lies only 1′9″ of latitude (or just over a mile) south of the thirtieth parallel of northern latitude. An alternative explanation for the last mentioned point might be that the Pyramid's site was intended to mark a third of the *surface-distance* from equator to pole, rather than a third of the angle between them (i.e. 30°)—or even, possibly, a compromise point half-way between the two positions (thus effectively defining the degree of flattening at the terrestrial poles).

[14] See Appendix H, and compare the similarly truncated pyramid on the reverse side of the seal of the United States, with its explicit reference to 'a new order of ages'.

[15] Mark 12:10, Matt. 21:42, and Psalm 118:22 (in a passage astonishingly reminiscent of the Osirian ritual of the Egyptian *Book of the Dead*).

appositeness of the remark is quite astonishing, and naturally poses the question whether he himself had been initiated into the Egyptian temple mysteries, *including, perhaps, the knowledge of the Great Pyramid's message itself.* And if Jesus, then why not also his parents and the apparently Essenic sect which seems to have produced him? Could his birth in Bethlehem, in other words, have been a deliberate response to the apparent pyramidal link between the Messianic sunrise of a new age and the geographical rhumb-line already described? Could the knowledge of that link have been the real reason for the parents' apparently hasty journey at the time of the birth? The whole notion seems highly conjectural, perhaps, and yet the Mosaic sections in chapter 9 seem to show that just such a process may have been at work many centuries previously at the time of the Israelite Exodus from Egypt. And if so, then there is no inherent reason why similar ideas could not have been at work among the later Nazarenes.[16]

But if the above was *not* the case, and the coincidence of the Bethlehem-line was *not* accidental, then there can be only one other possible explanation: *the Pyramid's architect could see into the future.* Such, indeed, was the tradition among the ancient Egyptians, as Arabic manuscripts still affirm today: the Pyramid, it was said, contained a record of all that was past and of all things to come. The evidence of chapter 3 (q.v.) suggests that, incredibly, even this notion may have some basis in truth. And in this case, it would certainly need to—for the famous birth in Bethlehem took place at least 2,600 years after the Pyramid's construction.

Of all the astonishing facts surrounding the Great Pyramid of Giza, however, it is this supposed prophetic element which modern man is likely to find, *a priori,* hardest to accept. Unless, like the early pyramidologists, we postulate that the Pyramid was designed by the Divine Architect himself—or, like the ancient Egyptians, by his son[17]—how can we possibly accept that any man could foresee the future, especially with the accuracy often claimed for the Pyramid? After all, even those who claim to be able to perform such occult

[16] The result of the rhumb-line's missing the midsummer sunrise is that it marks *two* summer sunrises during the year instead of one. The possibility ought not to be entirely ignored, therefore, that the line might point symbolically to *two separate* Messianic appearances. Meanwhile, noting that the two dates in question define a period of some 33 days, compare the specifically Messianic code-meaning suggested for the distance 33·5 P″ on page 39. Could the similarity, one wonders, be significant in terms of the Pyramid's code—a deliberate assertion by the designer of a Messianic significance common to passage-distance and sunrise-angle?

[17] Imhotep, son of the god Ptah, the Architect of the Universe.

The cyclic nature of time as symbolised in the Hindu sun-temple at Konarak (Orissa). The 'solar chariot' of the temple is borne on twelve wheels such as that shown, each symbolic of a cycle of zodiacal time, or some 26,000 years. Tradition has it that the present age belongs to the eleventh of these twelve cycles. The eight spokes of each wheel could be interpreted (see p. 28) in terms of human reincarnation through the various cycles.

feats generally profess a certain amount of difficulty in forecasting exact dates and times, as even Jesus of Nazareth himself seems to have admitted (if Matthew 24:36 is to be relied on). Only astrology, it seems, could possibly pinpoint dates as precisely, and as far in advance, as is claimed for the Pyramid (see, for example, chapter 6)—and yet most of those who claim to know about 'the occult' expressly deny that astrological influences have any *absolute* control over men's lives, and suggest that, at best, they result in certain underlying tendencies.

In the event, however, our best plan would be to avoid putting theories before facts. If the Pyramid *has* a message, whether prophetic or otherwise, then let us try to read it. If it then turns out that the message is demonstrably both prophetic and accurate, we shall be bound to accept it as fact—and yet another age-old theory will have to be abandoned and replaced by another. And perhaps the new theory will take a surprising form, for we may be forced to conclude that the Pyramid's message is not so much a prophecy as a *memory*—a 'memory of the future', to quote the original German title of von Däniken's *Chariots of the Gods?* We may come to see the Pyramid's prophetic message (always assuming, of course, that it has one) as saying not so much 'This is what will happen' as 'This is what has happened before'—or even 'This is what *always* happens' (a form of statistical prediction already used in our own day in such fields as long-distance weather-forecasting).

And at this point we shall have come back full circle to King Solomon's alleged view of human destiny: 'For everything its season, and for every activity under heaven its time: a time to be born and a time to die... Whatever is has been already, and whatever is to come has been already, and God summons each event back in its turn.' (Eccl. 3:1, 2, 15.) A cyclic view of history, no less, and one to which Hindus, Buddhists and the ancient peoples of Central America have likewise immemorially subscribed. 'All moons, all years, all days, all winds reach their completion and pass away. So does all blood reach its place of quiet, as it reaches its power and its throne. Measured was the time in which they could praise the splendour of the Trinity. Measured was the time in which they could know the sun's benevolence. Measured was the time in which the grid of the stars would look down upon them; and through it, keeping watch over their safety, the gods trapped within the stars would contemplate them.' (Quoted from the *Chilam Balam of Chumayel*, an ancient Maya text.)

2

Decoding the Pyramid

MANY VOLUMES have already been written on the internal and external measurements and features of the Great Pyramid and on their symbolic significance—or lack of it. Readers will find them set out in meticulous detail in *The Great Pyramid Passages and Chambers* by J. and M. Edgar (Bone & Hulley, 1923), *The Great Pyramid* by D. Davidson and H. Aldersmith (Williams & Norgate, 1925), Dr. Adam Rutherford's five-volume *Pyramidology* (Institute of Pyramidology, 1957) and the celebrated *Pyramids and Temples of Gizeh* of Sir W. M. Flinders Petrie (1883).[1]

Rutherford, in particular, takes advantage of the fact that the quantity π, which is basic to the whole design, can be calculated to a theoretically infinite degree of exactitude; and consequently he actually quotes most of his measurements to an almost incredible ten-thousandth of an inch. Even Petrie, back in the nineteenth century, was already going to almost absurd lengths to ensure the greatest possible accuracy of measurement, and spent many months on the site with his surveyor's instruments and measuring rods, not to mention batteries of thermometers to correct his figures for the expansion and contraction inevitably associated with any temperature variations which might occur. In fact the almost fanatical efforts of later researchers to improve still further on Petrie's figures have led to the point where it can probably be claimed with every justification that the Great Pyramid has now become the most accurately and comprehensively surveyed building in the entire world.

[1] Compare also Peter Tompkins's comprehensive and profusely illustrated survey of the known attempts to fathom the building's meaning in his *Secrets of the Great Pyramid* (Allen Lane, 1973) and Kingsland's *The Great Pyramid in Fact and in Theory* (Rider, 1932).

As for its alleged message, all the works mentioned have much to say on the topic. Rutherford, with time on his side, is naturally by far the most convincing, as well as the most readable, and his coverage is easily the most comprehensive; but many potential readers may find it disturbing that he, like most of his predecessors, sees fit to interpret the Pyramid's message exclusively as a vindication of the Judæo-Christian scriptures—and of a fairly fundamentalist reading of them at that. Yet it seems almost inconceivable that a monument of such resplendent universality as the Great Pyramid could have so restricted a message as this approach would suggest. So unlikely does the thesis appear that there are many critics who would nowadays frankly suspect the Pyramidologists of 'cooking' the figures to fit their own preconceived religious notions, instead of allowing the Pyramid to speak freely for itself, however unorthodox the results.

Fortunately, however, we have a control—and an illustrious one at that. For the Egyptologist Flinders Petrie, back in 1881, had a special reason for going to the extraordinary lengths we have already described in attempting to establish the Pyramid's exact dimensions. His overriding object at the time was to disprove the claims of his Pyramidologist father (William Petrie) that the Pyramid had a message at all, let alone a Christian one. The great edifice was simply a royal tomb, and Petrie's intellectual endeavours to prove the thesis were fired by emotional factors of filial rebellion every bit as compelling in their own way as the religious leanings of the Pyramidologists. Hence his long and patient efforts, his minutely exact measurements, his carefully recorded findings. Of these at least there could be no suspicion, for they had in no way been influenced by any hint of religious fervour. Quite the reverse, in fact. And in the event even the Primitive Inch and the Sacred Cubit were to be cast to the four winds. In short, claimed Petrie, they had never existed—or at least had not been used in the design or construction of the Great Pyramid.

Henceforward, then, thanks to Petrie, no Pyramidologist would ever again dare to falsify the facts. But in the event, no Pyramidologist was ever to *need* to. For Petrie, quoting the Pyramid's statistics in a mixture of British feet and inches and Egyptian Royal Cubits, was blissfully unaware of two extraordinary facts. First, his own data actually revealed several clear instances in the Pyramid of the use of a Primitive Inch and Sacred Cubit of approximately the lengths claimed by the Pyramidologists. And if his figures were then reconverted into these same measures, Petrie's own data actually validated up to the very hilt the Pyramidologists' earlier mathematical claims concerning the Pyramid's dimensions. If, therefore, the Pyramidologists' interpretation of the Pyramid's message were subsequently to prove to be in any way inaccurate or one-sided, then the fault must lie not in the figures themselves but in the Pyramidologists' interpretation of them—at worst, a case of simple human over-subjectivity.

To have one's basic premises unwittingly and irrevocably vindicated by the efforts of one's fiercest opponent is a stroke of good fortune such as rarely befalls the theorist. But that that confirmation should come from the eminent Petrie himself was a turn of events that even the most optimistic Pyramidologist could scarcely have hoped for.

And so the Great Pyramid's major dimensions were at last firmly established beyond all reasonable question, and it needed only the advent of the later researchers already listed, with their more advanced techniques and more favourable site conditions, to bring the data up to their ultimate pitch of present-day refinement. Of the sources mentioned, it is Rutherford who is in the best position to capitalise on and generally sum up the work of the earlier researchers, as well as making his own by no means meagre contribution to the available data through his research on the spot. In fact his figures for the Pyramid as designed are the most up-to-date available, and probably merit the title of definitive. It is these figures which form the basis of the present volume. It now only remains for us, therefore, to examine them with a view to discovering whether they are based on a consistent code and—if they are—to attempt to decode the Pyramid's presumed message strictly for what it is.

Now it is obvious that if an architect proposes to employ 100,000 men on free rations over a period of some twenty years (if Herodotus is to be believed)[2] or six hundred years (which von Däniken, on almost equally slender grounds, judges more likely), merely in order to leave a message in stone, then he must have something supremely important to say. Consequently he will first prepare a comprehensive and total blueprint based on an absolutely clear and consistent code. To do otherwise would be unthinkable. Thus, if a given measurement or arithmetical factor is significant, then it will be used significantly, and not merely gratuitously thrown in at odd moments

[2] As various commentators have noted, this would have involved the laying, *at the very least,* of one block of stone in every two minutes of daylight—a speed totally incompatible with any known technology, let alone with the supreme standards of care and accuracy exhibited by the Pyramid's masonry.

for the sake of mathematical aesthetics. This in turn means that what is done with such features is likely to be as important as the features themselves. Even processes such as addition, subtraction, multiplication and division are each likely to be significant in their own right and to have a definite and consistent meaning when applied. It seems inherently probable that *no* feature of the design (as opposed to the builder's methods of realising it) will be fortuitous. If a given measurement or angle is applied at a given point—even to the height of a passage, say—then it seems reasonable to assume that that measurement or angle was *chosen*, and for a reason, and our interpretation will therefore need to reflect it.

Already, in his *Pyramidology*, Adam Rutherford points out a number of measurements which appear to be of symbolic relevance in this way. The measurement 286·1 P″, for example, is, as we have already pointed out, the exact distance by which the Pyramid's original built perimeter fell short of that of the full design (indicated by the foundation sockets). If, therefore, the eventual completion of the monument to its full design was seen by the ancient initiates as symbolising the return of *light* to a darkened world, then it could be that that *light* is intended to be symbolised by the distance 286·1 P″. Consequently we need in due course to test the various occurrences of that quantity against the possible meaning *light* or *enlightenment*.

The quantity 35·76 P″, on the other hand, to which we have also drawn attention, is exactly one-eighth of the quantity 286·1 P″. If therefore we can establish a meaning for the number 8, we may have some sort of guide to the meaning of the distance 35·76 P″. Now as Rutherford points out, the number 8 has always been associated, in Christian numerology at least, with the notion of resurrection. In Far Eastern religious contexts, however, its connotation is more often that of continuing physical reincarnation (see the Hindu solar chariot wheel on page 23, for example). It therefore seems reasonable at this stage to attach to the quantity 8 an experimental blanket meaning: *rebirth*. Assuming, then—not unnaturally—that the process of division corresponds to the notion *through* or *by* (i.e. the notion of an agent), the meaning of the distance 35·76 P″ would appear to be that of 286·1 / 8—or *enlightenment through rebirth*.

Meanwhile, as Rutherford points out, the distance 35·76 P″ always occurs further on than the distance 286·1 P″ within the Pyramid's passage-system. 286·1 P″, in other words, always seems to lead to 35·76 P″. Could this fact be interpreted, then, as signifying that the *attainment of enlightenment* automatically leads to *enlightenment through rebirth*—i.e. a more enlightened next incarnation? If so, then the notion would certainly square with the teachings of Hinduism and Buddhism on the subject of reincarnation and the law of *karma*—though not, of course, to present-day Christian doctrine. And meanwhile this would tend to give the distance 35·76 P″ the meaning *an enlightened incarnation.*

At the same time we should note that the base-perimeter of the missing capstone would have measured exactly 8 × 286·1 P″ (see Appendix H). If therefore we attach the natural significance to the process of multiplication—namely that represented by the word *of*—this would suggest a significance for the capstone, when finally installed, of *the rebirth of enlightenment*. This was precisely the significance apparently attached by the ancient initiates to the Pyramid's final completion.

Again, Rutherford notes that the number 5 is the Pyramid number *par excellence.* After all, the full design has five points and five sides (i.e. four faces and a base), each of its faces contains a one-fifth scale inset triangle, and a factor of five constantly recurs in its interior measurements—not least in the Sacred Cubit itself, of 5^2P″. Nevertheless, it should be noted that the Pyramid's exterior will be truly five-based only when it is finally completed to that full design, and when the capstone—itself a five-sided pyramid—is finally added. Thus the significance of the number 5 seems to be intimately linked up with the function of the Pyramid itself, and in particular with the *birth of enlightenment* apparently associated with the final addition of its capstone. The tentative reading *initiation, initiate* or *bringer of enlightenment* for the number 5 would therefore seem to be justified initially. Meanwhile the symbolism of the five-pointed and five-sided capstone suggests that it is the specific function of this *light-bringer* to bring an imperfect world to perfection through the power of that enlightenment. The parallel with the ancient Messianic concept is clear, whether represented by the Egyptian Osiris, the Hindu Vishnu, the Mayan Quetzalcoatl, the Buddha Maitreya, the Zoroastrian Shaushyant or their more familiar Judæo-Christian counterpart. Consequently, in addition to its general sense of *initiate,* the number 5 may also be expected to occur with reference to any particularly important Messianic figure—to whom we could then presumably refer as *The Great Initiate* or *the One-who-is-to-come.* But at the same time we should expect some independent code-indication of that initiate's 'special' importance.

Thus, the fact that the Grand Gallery's roof is composed of 40 (8 × 5) clearly marked stone slabs could conceivably be interpreted as meaning that the Gallery in question is connected in some way with *the rebirth of an initiate.*

Meanwhile this same roof is some 1836 P″ long—or 153 × 12 P″. The number 12 is associated throughout the Judæo-Christian scriptures with the notion of *mankind*—not least in the case of the twelve tribes of Israel and the twelve apostles whose task was to take the gospel to them.[3] At the same time the number 153 has a long *Christian* association, at least, with the notion of *the enlightened ones*—apparently connected with the 153 fishes reported as being caught by the disciples towards the end of John's gospel.[4] If, therefore, we accept the possibility that both associations may represent the survival of some much older numerological tradition, then the length of the Grand Gallery's roof could conceivably represent *the enlightened ones of mankind*—or even *the enlightening of mankind* itself.

The fact that the Grand Gallery leads directly to the Great Step—itself 35·76 P″ high—thus suggests a connection between the *rebirth of the initiate* (8 × 5) and the *enlightened incarnation* (35·76 P″) of the *enlightened ones of mankind* (153 × 12 P″). And the fact that the top of the Great Step and the floor of the King's Chamber Complex beyond it are themselves exactly 153 courses of masonry below the summit-platform (designed to support the supposedly Messianic capstone) would further suggest that this level is to be regarded as exclusive to the *reborn enlightened ones* under the future *One-who-is-to-come.* The additional fact that the two low sections of the King's Chamber Passage together measure 153·057 P″ in length finally makes it virtually certain that these three occurrences of the number 153 are not accidental.

Meanwhile we come to the up/down, right/left symbolism also noted by Rutherford. It is his thesis that 'up' and 'right' are intended to signify progress towards enlightenment, while 'down' and 'left' represent its opposite, i.e. human degradation and all that is negative

[3] Note, too, that 12 was at one time the number of the known moons of Jupiter, itself named after God the Father. Could Jupiter somehow be connected with the 'Messianic plan'? See p. 270.

[4] 153 is in fact the sum of the numbers 1 to 17. It is also the product of 9 and 17. Meanwhile the individual figures of the combined numbers 9 and 17 add up to 17, those of the number 153 add up to 9, and those of the combined numbers 1 and 17 also add up to 9. So, indeed, do those of 2 and 16, 3 and 15, 4 and 14 . . . and each succeeding 'pair' of numbers until the central number of the series is reached—and that number *itself* is 9, whose pyramidal meaning seems to be: *utter perfection.* It may be to this series of 'mathematical coincidences' that the number 153 owes its 'special' significance. Oddly enough, however, the number 17 itself seems to have no independent significance in the Pyramid's passageways.

and 'sinister' (which is derived from the Latin for 'left'). The fact that the entrance is in the north side of the Pyramid, and that left therefore corresponds to east and right to west, adds weight to this view, since west was traditionally the glorious 'direction of the funeral destinies' in the symbolism of the later tomb-builders, while east was that of rebirth, as it is in most of the world's religio-symbolic systems—and rebirth, at least under the terms of Hinduism and Buddhism, is a process to be avoided if possible.

Application of this thesis seems to bear out its validity. As we have seen, for example, the axis of the entire passage system is 286·1 P″ to the left (east) of the Pyramid's own axis—which would suggest that it represents the path of those who have lost their enlightenment. Only in the King's Chamber and the Subterranean Chamber is it possible to return westwards far enough to reach the Pyramid's axis again—a fact which would suggest that these chambers represent specific opportunities to regain that enlightenment. Perhaps they represent the possibility of escaping from mortality and the physical world altogether—a notion reinforced by the fact that the King's Chamber contains (and apparently always contained) only an empty lidless sarcophagus, presumably indicative of escape from death or mortality. But the approaches to the King's Chamber, as we have seen, are preceded by the Great Step—an *upward* leap of 35·76 P″, which itself is preceded, at the bottom of the *upward*-sloping Grand Gallery (with its 1836 P″-long roof of 40 slabs), by an *upward* leap in the roof of 286·1 P″. That escape from mortality, in other words—if such is its significance—is closely bound up with progress towards enlightenment, with a subsequent enlightened incarnation and with the rebirth of the initiate(s).

But at this point we need to establish just who it is we are talking about. Who is it that is undergoing these various incarnations? Who is it that is losing or gaining enlightenment, winning or forfeiting immortality? Who, in short, is the deceased?

Not, we may be sure, the mere pharaoh Cheops or Khufu. For if the king *was* ever buried in or under the Pyramid, then it was in some obscure place (as the Egyptian priesthood always maintained), and not in any part of the known system of passages and chambers whose symbolic significance we are here discussing. Nor does any one man appear to have been the subject of the exercise. To the ancient Egyptian initiates, the Great Pyramid's passages and chambers seem to have symbolised the various stages of a solemn initiation into the mysteries of the spirit world. And that initiation seems to have applied as much to the souls of the dead in general as to the postulant solemnly craving initiation during his lifetime into the sacred mysteries of the priesthood. The Pyramid's passage system, in short, was a kind of route-map for the soul, a training for the deadly game of snakes-and-ladders which would comprise its subsequent passage through the underworld towards eventual rebirth.

Not that we have any direct evidence that the passage system of the Great Pyramid was specifically so designed. True, there is some evidence that the Pyramid—either physically or in symbol—may at one stage have played an initiatory rôle in connection with the ancient mysteries. But the major evidence for such a connection is to be found in the Saïte recension of the Egyptian *Book of the Dead*, where the soul's progress through the underworld is portrayed in terms of a system of halls and passageways so vividly reminiscent of the Great Pyramid's as to force the author, W. Marsham Adams, to the natural conclusion that they are one and the same.[5]

Adams's list of passages and chambers is given below with his admittedly somewhat free translations from the *Book of the Dead* on the left and the normal terms on the right. This arrangement permits a comparison which offers some intriguing hints as to the possible symbolic significance of the various internal features of the Pyramid. Compare the diagram on page 17.

Pyramidal identifications of terms from the Book of the Dead

The Descent	The Descending Passage
The Double Hall of Truth	The Ascending Passage and Grand Gallery
The Door of Ascent	Entrance to Ascending Passage
The Hall of Truth in Darkness	The Ascending Passage
The Hall of Truth in Light	The Grand Gallery
The Crossing of the Pure Roads of Life	Intersection of upper passageways
The Well of Life	The Well-Shaft
The Royal Arch of the Solstice	Entrance to King's Chamber Passage
The Passage of the Veil	The King's Chamber Passage
The Chamber of the Triple Veil	The Antechamber

[5] Legend has it that the original of the so-called *Book of the Dead* was written by Thoth, the great founding-father of Egypt. But if that book was identical in essence with the Great Pyramid's spiritual message, might we not see here at least a hint that Thoth himself may have been the Pyramid's designer, and that the building may thus represent in some way his testament in stone? The overt mathematical link between the number of his writings and the dimensions of the Pyramid (see p. 16) certainly seems to add weight to this possibility, while Hermes, his Greek counterpart, was once *god of the stone-heap*.

The Chamber of Resurrection The Chamber of the Grand Orient The Chamber of the Open Tomb	The King's Chamber
The Path of the Coming Forth of the Regenerated Soul	The Queen's Chamber Passage
The Chamber of Regeneration The Chamber of Rebirth The Chamber of the Moon	The Queen's Chamber
The Chamber of Ordeal The Chamber of Central Fire	The Subterranean Chamber
The Secret Places of the Hidden God	The Construction Chambers

For the ancient Egyptian initiates, then, the Great Pyramid's interior passage system apparently symbolised the post-mortem trials of the soul in the underworld. It was 'the way of the dead'. But it seems highly unlikely that this reading could have been the one originally intended by the building's even more ancient architect. For a start, parts of the passage-symbolism seem to refer at one and the same time to the enlightened ones of mankind (the Grand Gallery roof's 153 × 12 P") *and* to their unenlightened counterparts (the eastward displacement of the entire passage-system by 286·1 P"). But if both enlightened and unenlightened are so signified, then there can be only one reasonable conclusion—the reference must be to *the whole of mankind*.

Again, the architect states quite explicitly in the Pyramid's geometry that the building symbolises the planet earth itself. Therefore, if the passageways do symbolise the progress of the souls of men—however one understands the term—then that progress is clearly depicted as taking place not in some mythical underworld *but here on earth*. The fact that the system seems to represent various opportunities for progress towards, or retreat from, enlightenment tends to confirm this conclusion—for if there is one thing that most religions are unanimous about, it is that the soul's ultimate fate is determined primarily by its efforts in the physical sphere, here on earth.

If our argument thus far is valid, then, what is the architect's view of the state of the human soul? It is, alas, a gloomy one. The Pyramid, let us remember, is overtly designed as a tomb, complete with sarcophagi. And the soul condemned to imprisonment in its passageways is none other than the soul of man. And for the first time the building's gigantic size begins to seem a little less preposterous, for its symbolic occupant is not a mere forgotten pharaoh, but something much more all-embracing. The Pyramid's passageways do indeed symbolise 'the way of the dead'—*but we, all of us, are the dead in question.*

The idea sounds a startling one, perhaps. Yet the architect spared no effort to rub it in. Not only did he disguise the whole monument as a tomb—a reference explicit enough in itself, and quite sufficient to fool generations of later historians—he also took great care to site the entrance in the nineteenth course of masonry, which itself is almost exactly 38 P″ (2 × 19 P″) high. The number 38, as Rutherford again points out, has biblical associations (albeit tenuous ones) with the notions of death and sickness—as does the number 19. And certainly inspection of the Pyramid's interior geometry tends to suggest that the number 19 in particular may here be intended to have a deathly significance—perhaps by association with the moon's nineteen-year eclipse cycle, which results in the periodic 'death' of the life-giving sun.

Man, then, it seems, is the 'deceased'. The 'life' that he has lost is presumably the same immortality that can, in symbol, be regained once again via the open coffer of the King's Chamber. But just as his spiritual death is apparently associated with the loss of enlightenment signified by the eastward displacement of the whole passage system, so the regaining of immortality seems to be conditional upon the gaining of enlightenment signified in the height of the Grand Gallery—through which alone the 'Chamber of Resurrection' can be reached. In the final analysis, in other words, it is ultimately some kind of 'knowledge' which alone can 'set man free'.[6]

The Pyramid's passage system somehow represents a blueprint for man's 'soul-evolution'. That would seem to be a reasonable initial hypothesis. But if this is the case, then the whole thing is largely meaningless unless we can regard the whole system as one that *each man's soul* has to traverse—as indeed the ancient initiates themselves seem to have seen it. As a mere representation of 'the average soul', or of the spiritual state of 'man in general', the whole system would be of little more than statistical and academic interest. Each soul, in other words, must be seen as entering by the entrance and as not 'escaping' again until it has attained immortality in one or other of the relevant chambers: each soul apparently has a number of choices to make: and each soul must remain in the 'earth-planes' during the whole of the period signified. If, then, this period exceeds that of a normal lifetime (as seems likely—see pages 36–7), one is bound to conclude that each soul reincarnates many times in the course of its struggle towards eventual immortality. And consequently, since the

[6] Compare John 8:32: 'You shall know the truth, and the truth will set you free.'

main passageways are all two Royal Cubits wide (2 RC = 41·21 P"), this passage-width would appear to symbolise *the reincarnating human soul*. But in this case, the fact that the greatest width of the Grand Gallery is *four* Royal Cubits would suggest that the Gallery somehow symbolises the way of reincarnation of *two souls simultaneously*, and this fact must be reconciled with the overall interpretation eventually arrived at (see pages 102–4).

Meanwhile a turn to the left of 2 RC would presumably indicate actual *rebirth to mortality*, as would any kind of drop associated with the same measurement (e.g. the drop of the roof at the entrance to the 'Passage of the Veil', which is 2 RC high).

A further symbolic factor identified by Rutherford is what he calls the Death Factor—represented by the vertical height of the entrance (37·995 P"—or almost exactly 2 × 19 P"), a measurement which, as he puts it, 'permeates the entire downward passage-system'. But since this measurement is simply the vertical component of any line drawn from floor to roof, parallel to the Pyramid's angle of slope (51°51'14·3"), inside the Descending Passage, it would seem logical likewise to associate its horizontal component (29·8412 P") with the same notion of *death*. One would therefore expect to find examples of the measurement's use with this significance.

In this way, it is feasible, little by little, to work out a whole range of possible symbolic factors in the Great Pyramid's internal measurements—factors which seem to be related to each other via surviving traditional features of the ancient science of numerology (whose most famous exponent, the Greek Pythagoras, apparently learnt this self-same craft from the contemporary Egyptian priesthood). Thus, in the Pyramid, as in other numerological applications, the number 2 would, for example, signify *production* or *productive*; 3, *perfect* or *utter*; 7, *spiritual perfection*; 10, *millennium* or *eternity*; and 12, *mankind*. The number 6, despite a strong traditional association with the notion of physical fulfilment, seems in the Pyramid to signify *preparation*—and thus, in the spiritual sense at least, *incompleteness* or even *imperfection*, just as in the Hebrew tradition the six preparatory days of physical creation are seen to be spiritually incomplete without the seventh. Meanwhile the functions of addition, subtraction, multiplication (= *of*) and division (*through* or *by*) all seem to have their own direct significance.

Consequently a fairly literal and straightforward interpretation of the Pyramid's various symbolic features seems possible, given certain conditions. They are that our initial assumptions are valid, that some sort of code is basic to the Pyramid's design, and that we can successfully reconstruct it. One would also, incidentally, expect self-consistency to be a feature of the code: functions such as $6 = 2 \times 3$—presumably *preparation is productive of the perfect*—would appear to be a good means of testing any such code against itself, once constructed.

Finally, what of any time scale involved? The evidence that the ancient king-list chronologies (see Appendix C) embody a record of the Pyramid's vital dimensions on a scale of one year to one Primitive Inch immediately suggests the possibility that a similar time scale may apply within the passageways—and this idea can claim some support from the apparent link between the sum of the Pyramid's base-diagonals *in Primitive Inches* and the period of the earth's precessional cycle *in years*. The tentative hypothesis that each Primitive Inch along the floors of the passageways normally represents a year therefore seems worth pursuing. The length of the 'year' in question is immutably fixed by the Pyramid itself at 365·242 days—so that even the Pyramid's 'chronograph' (assuming there is one) would be a direct function of the earth's orbital astronomy as defined by the building's exterior dimensions (i.e. the 'polar inch' and the 365·242-day equinoctial year).

None the less, the Pyramid's designer appears to have taken precautions against the decipherment of the Pyramid's message at too early a date. As Rutherford observes, we seem to be faced with scale-changes at certain points, and these scale-changes appear to be indicated by the various 'steps' in the passageways. The thesis is a possible one, certainly, though Rutherford is less clear about the precise relationship between each step and its new scale. Presumably, if scale-changes are signified by steps, then the dimensions of any given step must signify in some way the precise relationship between the old scale and the new. Nevertheless, identification of the correct conversion formula can only be achieved on the basis of *subsequent* historical events—and the major steps seem to come comparatively late in the Pyramid's chronograph. However, our investigation will suggest that the last of the Pyramid's steps was negotiated by mankind some forty years ago, so that the construction and validation of a general scale-change law can now at last be attempted.

As for the starting-date, recent research on stellar alignments has shown that the Scored Lines in the entrance passage are astronomically associated with the date 2141 B.C. (vernal equinox).[7] And certainly, if one then works backwards towards the entrance at one

[7] See page 54, and compare Rutherford I on this point.

Primitive Inch per year, this places the entrance itself within the known dates of the Pharaoh Cheops, or Khufu, whose Pyramid it is alleged to be.

So far, so good. We have suggested the basis for a symbolic reading of the Pyramid's internal features, and it now remains for us to put it to the test. Stage one in the Pyramid's interpretation must be a clear statement of the proposed 'reconstructed' code and its symbolism, as arrived at along the lines suggested. Stage two must then be a simple and consistent read-out, such as any suitably programmed computer could perform. If the result then makes sense, we can dismiss coincidence as a possible explanation, and must conclude that the message as reconstructed is that intended by the Pyramid's architect. Any predictive elements it may contain may then be regarded as potential evidence of its validity.

The following is an attempted reconstruction of the Pyramid's code, based mainly on a critical and exhaustively tested analysis of the data supplied by Rutherford.

Hypothetical Code-Reconstruction

The Pyramid = the planet Earth.
The Capstone (5 points, 5 sides) = the One-who-is-to-come or Great Initiate; the birth of enlightenment.
The Passages and Chambers = the progress of the soul through the earth-planes.

Directional symbolism

Southwards = progress of the soul through time.
Northwards = return to physical existence.
Downwards, Leftwards, Eastwards = descent into evil, spiritual degeneration, negation, rebirth into mortality.[8]
Upwards, Rightwards, Westwards = progress towards enlightenment and immortality.[8]

[8] Interpretations consistent with the known symbolism of the much later tomb of Tutankhamun—see *Tutankhamen*, by C. Desroches Noblecourt, p. 246.

Passage features

Sloping passages at 26° 18′9·7″ = evolutionary progress of the soul through time at one Primitive Inch per year.
Horizontal passages = 'insets' signifying man's attainment of particular evolutionary levels (see *Levels*).
Vertical steps at passage entrances = change of time scale (applicable until next step) based on relationship between height of step and nearest base-unit of previous scale: (a) steps up = more time per inch; (b) steps down = less time per inch.
Vertical steps in body of passage = indication of base-scale affecting whole of passage in question.
Non-vertical steps = change of scale arrived at trigonometrically by projection of old, 'sloping' scale on to new horizontal floor.
Corbelled vaults = 'telescopic' structures, to be thought of as 'closed' or 'collapsed' originally, but opened up to full height by force from within (compare the 'Chinese nests' of shrines surrounding the sarcophagi of later tombs).
Horizontal slots half-way up wall = provision for 'sliding floor', indicating that that which reaches the level indicated does not need to redescend until the end of the feature in question.
Vertical slots on opposite sides of passage = provision for portcullis-type closure of the passage after it has fulfilled its purpose.
Blockages = that which needs to be removed before further progress can be achieved.
Limestone = the way of the physical world, space and time.
Granite = the workings of the spiritual world, the Divine, eternity.
Flat but roughly finished surfaces = earth, the world, the earthly.
Irregular floor-surfaces = without specific time-scale, or time-scale approximate.
Irregularly shaped tops of granite inserts (as though broken off) = that which is 'sent down from above', a message or messenger from the 'spirit-planes'.
Chambers = eras of final decision.
Gabled chambers (7-sided, 10-cornered) = attainment of enlightenment, escape, eternity.
Flat-roofed chambers (8-pointed, 6-sided) = the gateway, through rebirth, to higher planes (i.e. no finality).
Open coffins = escape from mortality and the physical; translation to higher planes.
Far passage-ends = conclusion of the Pyramid's message for the path in question.
Ventilation-shafts = escape from mortality and the physical world.
Scored lines = start of message.

Levels

Level of Queen's Chamber floor = Plane of Life or potential enlightenment.
Level of Subterranean Chamber roof = Plane of Death or unenlightened mortality.

Geometrical symbolism (all inch-measurements quoted in Primitive Inches)

1 *n* (1/100 of a Royal Cubit) = ·206066″ = one year on granite floors (as suggested by the geometrical symbolism of the King's Chamber Complex).[9]
1″ (1·00106 standard inches) = one year (sloping passages only).
20·6066″ = 1 Royal Cubit = (a) horizontally, 100 years (100 *n*); (b) vertically, death or birth.
25″ (5 × 5) = the Great Initiate or the Messianic ideal.
29·84″ = death, mortality (horizontal component of the *pi*-angle when inserted into the sloping passages).
33·5″ = the Messianic or avataric presence. (See note 16 page 22.)
35·76″ (286·1″/8) = effect on the soul's *karma* of gain or loss of enlightenment, as manifested at the next rebirth; an enlightened/unenlightened reincarnation.
37·995″ = death, mortality (vertical component of the *pi*-angle when inserted into the sloping passages).
41·21″ = 2 Royal Cubits = (a) horizontally, subjection to mortality and human reincarnation, the 'shoulder-room' of a reincarnating human soul; (b) vertically, 'the veil', the actual double passage from life to death and back to life or vice versa—i.e. 1 Royal Cubit twice.
67·59″ = height of 'man come of age'.
286·1″ (35·76″ × 8) = gain or loss of enlightenment.
365·242″ = a time, an age.
1881·24″ = the evolutionary distance between blind mortality and enlightenment, or between enlightenment and final escape from mortality (distance between Plane of Death and Plane of Life). (See *Levels*, and pages 180–2.)
5448·736″ = the incomplete or imperfect world.
26°18′9·7″ = the Messianic blueprint for human evolution (Bethlehem-angle).
51°51′14·3″ = the Divine, the spirit (the Pyramid's angle of slope, or *pi*-angle).
Square or rectangle = the physical.
Circle = the spiritual, celestial or eternal.

[9] The '*n*' is known to have been used by the ancient Egyptians as the normal subdivision of the Royal Cubit, which was itself the hundredth part of the side of the square-aroura (see Appendix A).

Circle superimposed on square of equal area = the bringing of the physical into conformity with the spiritual.
Geometric projections through stone = symbolic cross-references.

Arithmetical symbolism

1 = unity, the One. (All the other code-numbers are direct functions of this basic unity—a fact of considerable symbolic aptness.)
2 = production, generation, productive.
3 = perfect, utter, complete.
π (3·1412) = eternity, the eternal, the Divine, the spiritual (see 'Circle').
4 = physical, terrestrial (see 'Square').
5 = initiation, an initiate or Messianic leader; the Great Initiate (when accompanied by the measurement 33·5″—q.v.).
6 = preparation; and thus, spiritually, incompleteness or imperfection.
7 = eternal or spiritual perfection.
8 = rebirth.
9 = utter perfection (3^2).
10 = eternity, a millennium, a Messianic age.
11 = physical reality, realisation, achievement.
12 = all men, mankind, true man.
19 = death, mortality.
25 = the Great Initiate or One-who-is-to-come, the Messianic ideal (5^2).
99 = culmination.
100 = ultimate reward or retribution.
153 = the enlightened, enlightenment.
1,000 = as 10.
Addition = added to, and, with, gives rise to.
Subtraction = without, detracts from.
Multiplication = of, times.
Division = by, through (agent).
Square = *par excellence*, to an infinite degree, complete, utter.
Square-root = the essence of, the inception of, the seed of.

Validation-test for self-consistency

6 = 2 × 3 Preparation is productive of completeness—good.
8 = 2 × 4 Rebirth produces a/the terrestrial—acceptable.
10 = 2 × 5 The millennium produces the initiate—good.
12 = 2 × 6 Mankind is productive of imperfection—good.
12 = 3 × 4 Man is the perfection of the terrestrial—i.e. the culmination of earthly evolution—good.
286·1 = 8 × 35·76 Enlightenment results in the rebirth of an enlightened one—acceptable.
8 + 12 = 2 × 10 Rebirth added to all men produces the millennium—acceptable.

$25 - 6 = 19$ Incompleteness, if applied to the Messianic ideal (?) is death—good.
$4 \times 4 = 2 \times 8$ Utter earthliness produces rebirth—good.
$4 \times 5 = 2 \times 10$ The terrestrial Great Initiate—i.e. the rebirth of the One-who-is-to-come (?)—produces the millennium—good.
$8 \times 5 = 10 \times 4$ The rebirth of the initiate or One-who-is-to-come is synonymous with the earthly millennium—good.

In view of the reasonable consistency of the above, it would at least seem worthwhile experimentally to attempt the application of the code, as reconstituted, to the Pyramid and its passage system. (The main details of the latter are listed in Appendix F.)

Factorisations

Some measurements are capable of a variety of factorisations. Thus 12 could be factorised as 1×12, 12×1, 2×6, 6×2, 3×4 or 4×3: while 40 could likewise represent either 4×10, 10×4, 5×8 or 8×5. And even this assumes that the code does not envisage the breaking-down of any measurement into more than two factors.

Taken in isolation, then, any measurement may suggest a number of alternative interpretations, to which equal weight must be attached initially. Once we observe that measurement in the context of the other features around it, however, we may find that a particular factorisation seems symbolically much more apt than its alternatives—and in this case we can identify that factorisation as the correct one *in that context*.

It should be noted that this is exactly what happens in the case of those more familiar symbols which are our everyday words. Taken in isolation, most words are almost meaningless, so many are the possible ways of interpreting them. The word 'round', for example, has at least five distinct basic grammatical uses—without even going into the variations of meaning that the word can have within those grammatical categories. Research into linguistics demonstrates that, in one way or another, we rely almost totally on context to pinpoint the meanings of the words we use—so that it would be surprising if similar considerations did not also apply to the language of the Pyramid's code.

Tolerances

The entire Pyramid was constructed—as has already been pointed out—with extraordinary precision. But it would be ridiculous to expect that precision to be literally infinite. We are bound to consider the question of what tolerances the architect intended, and to what tolerances the builders actually worked.

Examination of the Pyramid's interior suggests that certain clear priorities were observed in the matter of accuracy. Floorline-length, i.e. chronological dating, seems in all cases to have had priority over other considerations, and here tolerances of as little as a thousandth of an inch seem to have been observed. Moreover, despite four millennia of earth movements and distortion (see diagrams page 125, for example), trigonometrical calculation makes it possible to reconstruct the original design measurements exactly, even today. Other aspects of the geometric and arithmetical codes seem to have been subordinated to this basic consideration, however, so that a less fine degree of accuracy appears to have been observed with respect to some of the other symbolic features and cross-references, especially where any conflict arose with chronological requirements.[10] It is also true to say that, of all the passage features, the floorline measurements have probably been the least affected by earth movements and the passage of time, as several of the Pyramid's surveyors have noted.

There is some evidence in the following pages to suggest that references to the hypothetical code were quite commonly built into the design *correct to whole numbers of inches only*—if the number of apparently significant near-misses is anything to go by. The method employed seems to have involved correcting *downwards* rather than to the nearest whole number—a convention for which Davidson and Aldersmith's work on the king-list chronologies (see Appendix C) provides some supporting evidence. What mattered in the case of the geometric code, in particular, was not so much approximation as the number of complete inches—except where the shortfall in question fell within the relevant design tolerances.

Thus, not only do we find cases of 286·1 P″ apparently corrected to 286 P″, but 67·59 P″ is sometimes treated as though it were 67 P″ and 29·84 P″ as though it were 29 P″. Cases of this type are most common in mathematical functions which involve mixtures of geometric and arithmetical factors—e.g. symbolic arithmetical multiples of geometric quantities—and in measurements apparently

[10] See, for example, the length of the lower sections of the King's Chamber Passage (together 153·057 P″ long), which seems to be intended to denote both a period of time *and* an arithmetical reference to the enlightened (153). This suggests that chronologically the distance had to be exact: but symbolically the significant feature was the completion of the 153rd inch.

comprising direct references to the arithmetical code (see note 10 page 42). In effect, then, the arithmetical code should be regarded as containing whole-number versions of all the quantities in the geometric code.

Significance of masonry courses

The Pyramid's 203 courses of masonry between base and summit-platform have a wide variety of thicknesses (details are given in Appendix G). Since it would clearly have been simpler architecturally to settle for a single, standard thickness, it seems reasonable to assume that the various thicknesses were not chosen arbitrarily, but were assigned to particular courses for cogent symbolic reasons. Both course number and thickness, in other words, may be mathematically significant in terms of the Pyramid's code.

Certain courses in particular seem to be directly related to specific levels within the passage system, while others are significant with respect to the Pyramid's exterior geometry. On the other hand, careful measurement reveals that

(a) even allowing for subsidence, there were originally fractional variations in thickness within each course of core-masonry.
(b) the internal passage levels (as opposed to the five external orifices) did not in most cases correspond exactly with tops or bottoms of core-masonry courses.
(c) the original courses of casing-stones likewise did not always correspond exactly with tops or bottoms of core-masonry courses.

It would thus seem that the core-masonry was never intended to be completed to the same standards of accuracy as either the external casing or the internal passage features. Indeed, there is no obvious reason why it should have been, bearing in mind that its function was merely to act as 'filling'.

Taking (b) and (c) together, however—and still remembering the Pyramid's large and obviously deliberate variations in thickness from course to course—it would seem to be a reasonable hypothesis that the main passage levels may originally have corresponded to the tops or bottoms of given courses of casing-stones, and that the levels of the latter were *at least approximately* those of their counterparts in the core-masonry. From the surviving structure, in other words, we should still be able to determine the architectural levels of the main passage features to the nearest course, and at the same time to establish an average or notional value for the thickness of that course reduced to whole Primitive Inches (see pages 41–3 above)—in so far as subsidence and earthquake distortion allow.

To take a case in point, the level of the Queen's Chamber floor—provisionally identified above as the so-called Plane of Life—comes geometrically 846·0654 P″ above the Pyramid's base. Architecturally, therefore, this level may be related to the 24th course of masonry (whose bottom comes an average of 820·4 P″ above the Pyramid's base) and with the 25th course (whose bottom comes 852·7 P″, and whose top some 885 P″, above the same base). Clearly the nearest architectural point of reference is, in this case, the join between the two courses. Course 24 is some 32 P″ thick, and course 25 just over 33 P″.

Consequently we can presumably identify the level of the Queen's Chamber floor as one notionally based on the 24th course and giving rise to the 25th—based, that is, on physical imperfection and/or preparation (6 × 4), productive of (true?) man (2 × 12), and giving rise to the Messianic ideal or the Great Initiate (5^2). Meanwhile the thickness of the 24th course (32 P″) identifies its level as one of physical rebirth (8 × 4)—a conclusion apparently corroborated by the ancient Egyptian designation of the Queen's Chamber specifically as the 'Chamber of Rebirth'—but giving rise to the 'perfect achievement' or 'achievement of the perfect' (11 × 3).

The main passage levels are listed below, together with details of the courses to which they most closely correspond. In all cases the word 'notional' applies to the approximative process of deduction outlined above.

Passage-features etc.	*Notional architectural level*	*Average thickness of course*[11]
Base course	Course 1	58 P″ (2 × 29)
Approximate mid-point of Granite Plug and top of Grotto in Well-Shaft	Base of course 7	41 P″ (2 Royal Cubits)
Level of Queen's Chamber floor (Plane of Life)	Top of course 24 (6 × 4) Base of course 25 (5^2)	32 P″ (8 × 4) 33 P″ (3 × 11)
Top of Queen's Chamber's north and south walls	Top of course 30 (6 × 5, or 3 × 10)	28 P″ (7 × 4)

[11] Reduced to whole Primitive Inches, and assuming a constructional tolerance of at least 0·1 P″.

Top of Aroura-parallelogram in Pyramid's cross-section (see p. 350)	Axis of course 35 (7 × 5)	50 P″ (10 × 5)
Course above 35th	Course 36 (6^2, or 3 × 12)	41 P″ (2 Royal Cubits)
King's Chamber floor	Top of course 50 (10 × 5)	28 P″ (7 × 4)
King's Chamber roof	Base of course 60 (6 × 10, or 5 × 12)	28 P″ (7 × 4)
Core-masonry outlet of Queen's Chamber air-shafts	Course 90 (9 × 10)	38 P″ (2 × 19)
Casing outlets of Queen's Chamber air-shafts	Course 91 (7 × 13)	35 P″ (7 × 5)
—	Course 100 (10^2)	35 P″ (7 × 5)
Core-masonry outlet of King's Chamber north air-shaft	Course 101 (100 + 1?)	33 P″ (11 × 3)
Core-masonry outlet of King's Chamber south air-shaft	Course 102 (100 + 2?)	28 P″ (7 × 4)
Casing outlet of King's Chamber north air-shaft	Course 103 (100 + 3?)	29 P″
Casing outlet of King's Chamber south air-shaft	Course 104 (100 + 4? or 8 × 13?)	26 P″ (2 × 13?)
Summit-platform	Top of course 203 (7 × 29)	21 P″ (3 × 7)

Approaches to decoding

In attempting the task of decoding the Pyramid's presumed message, the author has two main courses of action open to him, each with its attendant risks. He can either describe and analyse only those features which seem to be direct functions of the code as postulated—in which case he is liable to be accused of ignoring all the features which do not fit his thesis. Or he can list and 'try for size' every conceivable feature of the passage system, however accidental, doubtful or vague its mathematical and symbolic relevance may appear to be. And in this case he is of course liable to be accused of twisting everything to suit his theories.

Of these two courses I have, in the event, tended towards the second, on the grounds that the intelligent reader is perfectly capable of deciding for himself what is relevant and ignoring what is not, once he has been given all the facts. Having done his own pruning, he can then judge for himself how far the conclusions reached are warranted by the remaining evidence. Indeed, I have seen fit to indicate with a question-mark certain features whose apparent significance I myself feel to be either accidental or at best doubtful.

3

The Pyramid Speaks

IN ATTEMPTING to analyse the Pyramid's message in terms of the code set out in the last chapter, our best plan would appear to be to tackle the task from the point of view of the ancient would-be initiate of the Pyramid's mysteries. Accordingly, we shall approach the Pyramid's north side and then enter it as he would have done, examining each feature in turn as we come to it. A general description of that feature will then be followed by a simple reading in terms of the reconstructed code. Throughout, all inch-measurements will be given in the same Primitive Inches originally used by the architect (equal to 1·00106 standard inches), and in all cases we shall assume that the passing of time is represented by our progress along the centre-line of the various passage floors.

General Observations

Our would-be initiate's first distant sight of the Pyramid, if he had approached the Giza Plateau one summer's noon, would have been of a brilliant white structure of enormous size sending reflected beams of sunlight out over the desert and the Nile Delta (see page 337). He might therefore have been struck immediately by the notion that here might be an artefact from some 'superior world'—one whose function was to spread not only physical light but also, perhaps, spiritual and intellectual light across the face of the earth.

Coming nearer, the ancient visitor would no doubt have been astonished to find that the Pyramid, resplendent though it was, had been left by its builders without its culminating capstone (see Appendix H)—a stone which on other pyramids was often gilded to represent the sun. Moreover this omission could only have been intentional, since the outer casing of the known pyramids was normally finished from the top downwards. These facts might consequently have led him to suspect that this enormous structure was somehow meant to be symbolic of the 'unfinished house of man's destiny', and that its missing capstone represented some crowning achievement which was destined to occur only with the sunrise of a new age.

Assuming that the ancient visitor was equipped with a knowledge of the Pyramid's code, he could next have verified his hypothesis. As it stood, the unfinished Pyramid had eight corners and six sides (including base and summit-platform): it therefore symbolised the rebirth of imperfection, or preparatory rebirth—more familiar as the ominous biblical term 'the resurrection of damnation', and apparently indicative merely of human rebirth to mortality or to physical existence. The addition of the final capstone, however, would transform it into a five-sided, five-pointed structure—symbolic of a birth of enlightenment or Messianic initiative. With this completion, moreover, the four separate faces of the former edifice would at last become united at the summit—a symbol, perhaps, of an eventual end to human discord and division. And the capstone which would accomplish this transformation would itself have five points and five sides, and would thus represent the actual Messianic initiate who would bring all this about—the 'sun' which would bring about the dawn of a new and glorious age of peace.[1]

Again, on examining the exterior geometry of the Pyramid, the visitor would have been intrigued to notice that its measurements actually fell short of the clearly marked foundation sockets by amounts directly related to the displacement of the entrance eastwards from the Pyramid's centre-line. Realising that both shortfalls therefore represented some kind of anomaly, i.e. some kind of departure from the ideal design, the initiate might have seen them as signifying in some way the imperfection of the present human condition when compared to the ideal, enlightened one.

Confirmation of this notion would immediately have been forthcoming from the measurements themselves. For the full-design Pyramid's alignments and geometrical characteristics were clearly intended by its designer to be direct functions of the earth's basic geophysical data and of its orbital astronomy. In short, the mighty edifice was intended, when finally completed, to represent the earth

[1] The notion of the pyramid is still used, even today, as the symbol for a hierarchy headed by a single leader.

itself, in harmony with its cosmic environment. Yet our ancient visitor could have seen for himself that it was not so completed—indeed, it had been deliberately built small, its base-perimeter being reduced by a total of 286·1 P″. In view of this he could have read the symbols to mean that only when *light* was restored to the world would our planet achieve its true cosmic harmony.

Meanwhile the Pyramid's patent symbolisation of the earth, plus the further fact that its whole interior design was ostensibly that of a tomb, might have suggested to him that the Pyramid's passages were meant to symbolise the way of those who are entombed (i.e. imprisoned) in the physical world—and thus also, perhaps, that the era symbolised by the final placement of the capstone would represent a time when those entombed might at last rise from the 'dead' to claim a living spiritual inheritance. He might also have admired, in passing, the designer's ingenuity in supplying the inquisitive, via his tomb-symbolism, with a ready made explanation of the monument without at the same time revealing to them his literally vital information until such time as they became evolved enough to understand and profit by it.

Confirmation for these notions would have been readily to hand. Referring to his code, the visitor could have interpreted the fact that the entrance lay in the 19th course of masonry and was almost exactly 38 Primitive Inches high (i.e. 2 × 19″)—as well as having its axis 286·1″ to the *east* (or left) of that of the Pyramid itself—as signifying that the passages stood for the way of the spiritually dead. Consultation of the Egyptian sacred writings, and especially of the Saïte version of the *Book of the Dead*, would have added further confirmation that the passageways were to be thought of as symbolising the evolution of the souls of the 'dead'—even though the context traditionally envisaged seems to have been the *post-*mortem one.

Finally, before entering the Pyramid, the ancient visitor might have glanced up and noticed that the prominent 35th course of casing-stones was much thicker than those above and below it (see diagram page 10 and Appendix G)—no less than 50 Primitive Inches, in fact, as closer examination would have revealed. Meanwhile the next course above it was 2 Royal Cubits thick. Bearing the code in mind, therefore, he could have deduced that the 35th course—with its axis exactly 25 or 5^2 inches from the top and bottom of the course, and set at one-fifth of the Pyramid's total height—represented numerically the spiritual perfection of the initiate (7 × 5), as well as possibly having some connection with gain or loss of enlightenment through rebirth (35·76″ = 286·1″/8). At the same time he could have interpreted its thickness of 50″ as symbolising the Messianic age of the initiate(s) (10×5) and/or the appearance of the Great Initiate himself (2×5^2).

Meanwhile the next course up, to which the 35th course as it were gave rise, was 2RC thick, and therefore had some connection with mortality. Moreover the fact that it was the 36th course of masonry would have suggested that it symbolised not only the perfection of man (3×12) but also, at the same time, utter incompleteness (6^2). And its number too was sufficiently close to 35·76 to suggest a connection with the gain or loss of enlightenment through rebirth.

Our visitor could thus have tabulated the information available to him from these two courses of masonry as follows:

Lower Course		*Symbolism*
(1) Number of lower course from base	35 (7×5)	*The spiritual perfection of the initiate(s)*
(2) Thickness of course	50″ (10×5 or 2×5^2)	*The Millennium of the initiates/productive of the Great Initiate*
(3) Distance of course-axis from base of Pyramid[2]	1162·6″	
(4) Distance of course-axis from base of Pyramid	(365·242″ × 10)/π	*A millennial age through the spiritual or Divine*
(5) Distance of course-axis from base of Pyramid	Pyramid's full height/5	*A world perfected through initiation*
(6) Straight length of full-design Pyramid's side at level of course-axis	2 × 365·242″ × 10	*Productive of a millennial age*
(7) Distance from axis to top or bottom of course	25″ (5^2″)	*The Great Initiate/the Messianic ideal*
Upper Course		
(8) Number of upper course from base	36 (3×12, or 6^2)	*Perfect man/utter imperfection*
(9) Thickness of upper course	41·21″ (2 RC)	*Mortality*
(10) Position of upper course	mounted on lower course	*Arises out of/is based upon*
(11) Both course-numbers may also refer to the geometric quantity 35·76″	286·1″/8	*Enlightenment through rebirth*

[2] See Appendix A.

From these data, the visitor could then have derived a reading such as:

The spiritual perfection of the initiates (1) *will produce the Great Initiate himself* (2) (7) *and a Messianic age* (2) (6) *of world-perfection* (5) *brought about by spiritual means* (4).[3]

This Messianic age or millennium will in turn give rise (10) *either to the perfection of men* (8) *or to their utter imperfection* (8), *as they gain or lose enlightenment through their successive rebirths* (9) (11).

This code-statement, in other words, writ large above the entrance to the Pyramid's passageways, could perhaps be regarded as a kind of summary of the more detailed message likely to be decipherable within—a possible conceptual framework for the whole of our subsequent investigation. Meanwhile this investigation in turn should help to establish whether or not that conceptual framework is a valid one.

Starting with the entrance we shall now begin our investigation, setting out our information in the way already exemplified above.

The non-technical reader may well prefer to skip the listed data and proceed directly to the suggested READING for each section.

The Entrance

(1) Level	19th course of masonry	*Death*
(2) Thickness of course	37·995″ (2 × 19″)	*Productive of death*
(3) Position of passage-axis	286·1″ to left (east) of Pyramid's axis	*Having lost enlightenment*
(4) Angle of outer slope of casing-stones	51°51′14·3″ to the horizontal	*Divine/spiritual*
(5) Width of entrance	41·21″ (2 RC)	*Mortality/rebirth*

READING: *This is the path of the dead* (1) (2)—*the path of those reincarnating mortals* (5) *who have lost their enlightenment* (3) *and forsaken their spiritual nature* (4).

There seems to be at least a suggestion here that man was originally a wholly spiritual creature.

[3] 'Neither by force of arms nor by brute strength, but by my spirit! says the Lord of Hosts.' This verse (6) from the prophetic fourth chapter of Zechariah, comes in a passage which actually seems to refer to the Pyramid's missing capstone (see p. 218–19), and certainly accords with the reading given above.

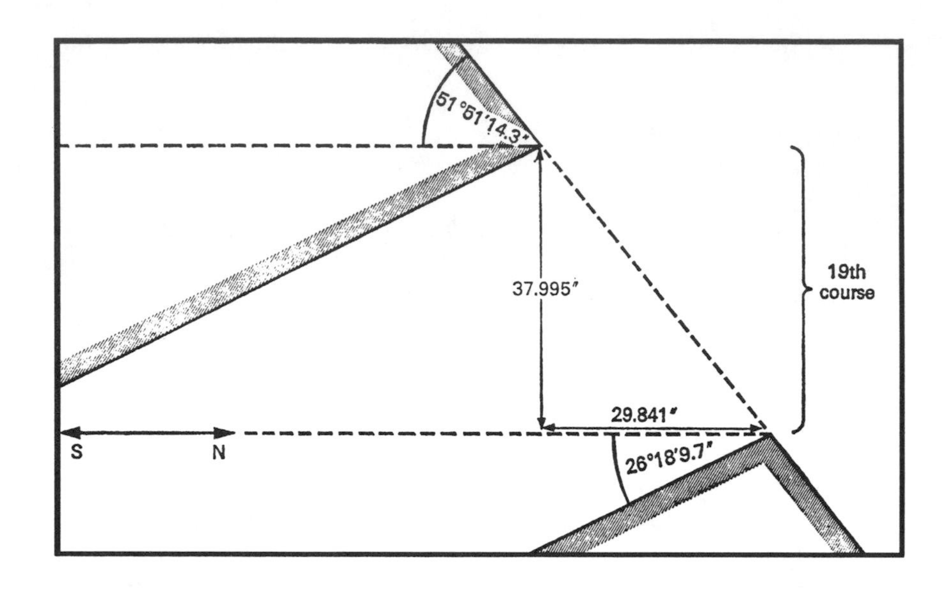

Symbolic features of the Entrance.

The Descending Passage (The Descent)

The main Descending Passage features are as follows:

(1) Direction	due southwards	*Through time*
(2) Slope	downwards	*Degeneration*
(3) Angle of descent	26°18'9·7"	*Human evolution*
(4) Width	41·21" (2 RC)	*Rebirth/mortality*
(5) Vertical component Vof *pi*-angle when inserted in passage	37·995"	*Death*
(6) Horizontal component of *pi*-angle when inserted in passage	29·841"	*Death*
(7) Cross-section	rectangular	*Physical/terrestrial*
(8) Original star-alignment	Alpha Draconis (lower culmination)	*Death/Hell(?)*

READING: *The souls of evolving humanity* (3), *reincarnating* (4) *in the physical world* (7), *will forfeit more and more of their original spirituality* (2) *as time goes on* (1). *For the imprisonment of spirit in mortal flesh is death* (5) (6) (8).

It should be pointed out that Alpha Draconis, the true Pole Star during the third millennium B.C., was the Dragon Star, and had an

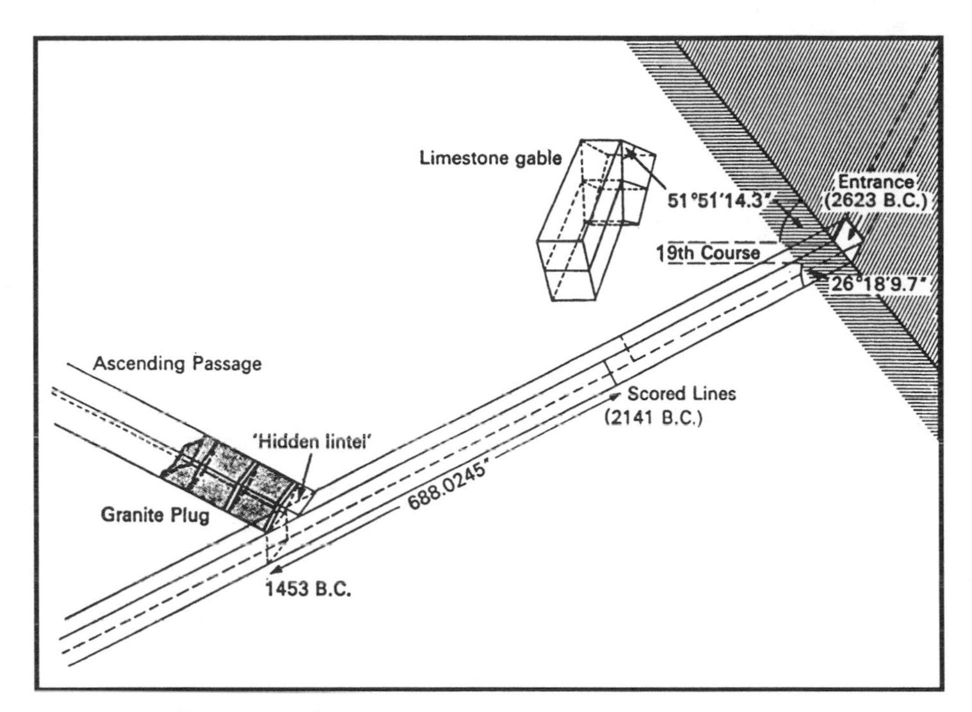

Projection of the Descending Passage.

almost universal diabolical or death-laden significance among the ancients. Yet it has also been associated with Lucifer who, like man (it is said), underwent a 'fall from grace'. Could *man himself*, one wonders, be the fallen angel of the ancient legend? And in this case what strange truth might lie in the speculations of those who assert that man's true origins were extra-terrestrial, and are to be found in the celestial region of the Pole Star? Could the Pyramid's Descending Passage merely represent a continuation of that fall out of the northern sky, via dimensions unknown, into earthly mortality? And may we see here a connection with the fact that *homo sapiens* apparently first appeared—quite suddenly, it seems in the northern hemisphere?

The Scored Lines

(1) These two straight lines, perpendicular to the floor of the Descending Passage and extending from floor to roof, are carefully scored in both walls of the passage at a point roughly beneath the great limestone gable shown in the diagrams above and on page 181 and 481·7457" down the floor from the entrance.

(2) Recent astronomical research has revealed[4] that these lines were in alignment with the star Alcyone of the Pleiades, in the constellation of Taurus the Bull, at noon of the spring equinox (21st March) of 2141 B.C.—a somewhat later date than that originally calculated for this alignment by the astronomers Herschel and Piazzi-Smyth. It is interesting to note, in this connection, that bull-sacrifice has since time immemorial been associated in the Middle East with atonement for sins and with divine salvation, while the Pleiades were firmly linked in the ancient Egyptian tradition with the goddess Hathor, the 'goddess of the Foundation' and instigator of the primeval 'deluge'.

(3) The suggestion that the Alcyone-alignment was intended by the Pyramid's designer is supported by the fact that the feature which corresponds to the Scored Lines in the Trial Passages is a flat surface which could have been used as a pelorus for star-sighting (see diagram page 199).

(4) In 2141 B.C. the axis of the Descending Passage was exactly aligned with the Dragon Star—the then Pole Star—at its lower culmination. The Dragon and its star is associated traditionally (as pointed out above) with the forces of evil and thus with spiritual death. The fact that the star in question 'looked' straight down the Descending Passage to the beginning of the Subterranean Passage (q.v.) seems to confirm the association.

(5) Of no other date in the current precessional cycle can (2) and (4) above *both* be said; nor does any other possible star-alignment suggest itself.

READING: *This is the start of the prophecy and of the Messianic blueprint for human evolution. Start counting from here (2141 B.C.).*

On this basis the entrance, counting backwards at 1" per year (see code), comes at a point representing the summer solstice of 2623 B.C. This may conceivably be connected with the date of the Pyramid's construction: but it almost certainly does not refer to the inception of its planning, which may have dated from some hundreds or even thousands of years previously. The date does, however, appear to tally with the reign of the pharaoh Cheops, with whom the Pyramid's construction is generally associated.

[4] See Rutherford I.

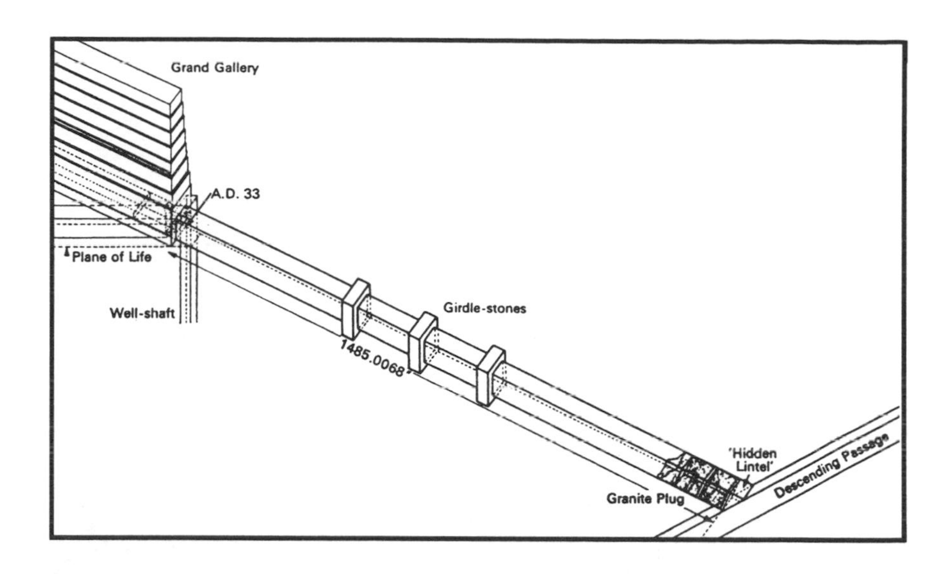

Projection of the Ascending Passage.

The Beginning of the Ascending Passage (The Door of Ascent)

(1) Position	688·0245″ after Scored Lines[5]	*(Time-measurement)*
(2) Symbolic dating	688·0245 years after spring equinox of 2141 B.C.	*(Time-measurement)*
	= 30th March 1453 B.C. (notionally 10.34 a.m.)	*(Dating)*
(3) Nature of entrance	originally blocked by a limestone slab (apparently cognate with the *Book of the Dead's* Hidden Lintel)	*Physical impediment to progress—removal necessary*
(4) Features beyond slab	passage is further blocked at lower end.	*Further impediment*
(5) Nature of blockage	three built-in blocks of red granite—the top two joined as one (the so-called Granite Plug)	*Double (?) spiritual impediment to progress —removal (acceptance) necessary*
(6) Peculiarities of Granite Plug	top end of top block is roughly shaped or has been broken off	*Sent down from a higher plane*[6]
(7) Length of Granite Plug	originally about 10 RC (2 RC × 5)	*Millennium (?)/Death or rebirth of initiates (?)*
(8) Height of Granite Plug	exactly as for Descending and Ascending passages	*'Built-in': thus, planned from the beginning (?)*

[5] Compare diagram p. 53 for the precise application of this measurement.

[6] Curiously enough, the symbolism at this point also seems to suggest the reading 'sent backwards through time'.

(9) Width of Granite Plug	2 RC at upper end, but slightly tapered (3″) at lower end for tight fit in similarly-tapered passageway (therefore tapered width = some 38″ or 2 × 19″)	*Productive of death*
(10) Passage beyond Plug	originally filled with further limestone blocks	*Further physical obstacles*
(11) Direction of Passage	southwards	*Through time*
(12) Slope of Passage	upwards	*Evolutionary progress*
(13) Continuation of Descending Passage past entrance	all features remain unchanged (see The Descending Passage).	*Rebirth/mortality/ death*

A preliminary reading of these features makes it clear that entry to the Ascending Passage symbolically involves 'taking down' first the limestone slab of the Hidden Lintel, and then the three huge blocks of the Granite Plug. To clear the way, the latter would then have to be slid down into the lower part of the Descending Passage, which they would of course effectively block.

In symbol, then, a physical initiative needs to be followed at this point by the acceptance of a heavy *spiritual* burden, which in turn would effectively block the downward, death-laden path, thus sparing humanity the events portrayed in the Subterranean Inset. That spiritual burden seems to be connected numerically with the assumption of mortality by an initiate or initiates and/or with the Millennium (7).

In the event, however, the Pyramid's design shows that the blocks of the Granite Plug *cannot* be 'taken down'. They are built into the Ascending Passage as a permanent feature, and are further held in place by a tapering of the passage walls. Being granite, they are also virtually impenetrable. Progress up the passage therefore involves 'burrowing' into the softer limestone masonry surrounding the granite blocks (much as Caliph Al Mamoun's workmen did in the ninth century of our era), followed by further 'burrowing' through the limestone blocks which originally filled the remainder of the passage.

In symbol, then, the spiritual conditions for entry into the Ascending Passage are seen as constituting too great a burden for man, who is instead destined to 'storm' the path by purely physical means.

READING: *688 years after the prophecy's beginning (i.e. on 30th March 1453 B.C.)* (1) (2), *men will start an attempt at spiritual improvement* (12) *through events in the physical sphere* (3).

But they will find their way blocked (4) *unless they also fully accept the spiritual conditions laid down 'from above'* (5) (6)—*conditions which will prepare the ground for the rebirth of the initiates* (7) *and for the eventual Millennium* (7).

If accepted fully, the spiritual conditions in question would finally block the way of spiritual death for man (8) (9) (13)—*but they would also place too great a burden on him at this time. Consequently man will prefer to circumvent the new dispensation and substitute for it the painful and laborious observance of purely physical restraints* (10). *The way of death will therefore be left wide open still* (13).

The historical period indicated seems to have been characterised, on a fairly worldwide basis, by the imposition of fresh ritual restraints on everyday life and by a movement towards monotheism. In India, for example, the Hindu Vedas—which exhibit both tendencies—were being written down at about this time.

In the Middle-Eastern context similar processes are known to have been at work among both Egyptians and Jews. Indeed, Rutherford adduces a great deal of evidence to show that 1453 B.C. was also the probable starting-date of the Israelite Exodus from Egypt under the semi-mythical Moses. In fact, if 1453 B.C. was the year in question, then the astronomical conditions relating to the Jewish Passover (the first full moon after the spring equinox) do indeed fix the morning of 30th March as the only possible date of departure.

In this case it is an odd coincidence that the red granite of the Granite Plug is virtually identical with that of Mount Horeb—the mountain on two blocks of whose stone Moses is alleged to have received the Divine Law for his 'kingdom of priests and holy nation'. And their mission, as apparently laid down at the time, was eventually to rescue all mankind from its present all-enveloping darkness—a Messianic role, no less.

Whether the biblical story is based on actual events or purely on inside knowledge of the Pyramid's symbolism—or whether the one led to, or simply predicted, the other—one thing seems plain. The Pyramid's prophecy seems to have been borne out at this point by a fairly widespread series of developments in the evolution of world religious thought.

The Ascending Passage (The Hall of Truth in Darkness)

(1) Direction	southwards	*Through time*
(2) Slope	upwards	*Evolutionary progress*
(3) Angle of ascent	26° 18′ 9·7″	*Human evolution*
(4) Width	41·21″ (2 RC)	*Rebirth/mortality*

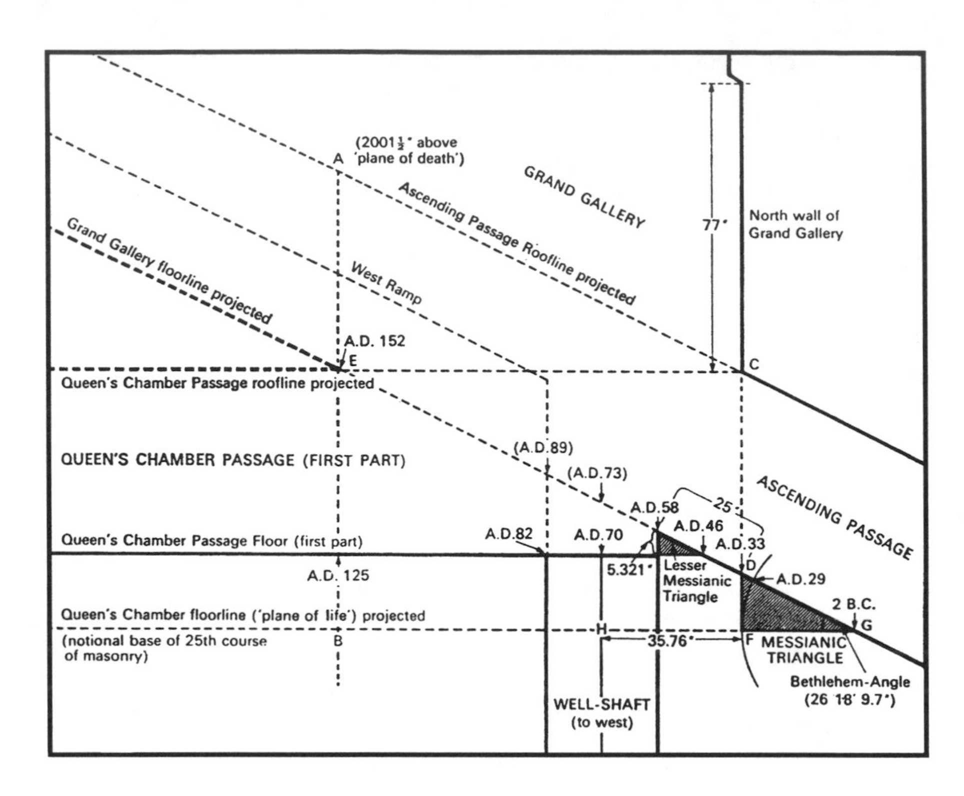

The Crossing of the Pure Roads of Life (looking west).

(5) Cross-section	rectangular	*Physical/terrestrial*
(6) Further features	passage is bored through three distinctive limestone 'girdle-stones'.	
(7) Peculiarities of 'girdle-stones'	Nos. 1 and 3 are preceded by 'markers' on the *west* wall, and No. 2 by a 'marker' on the *east*.	*Three specific physical obstacles, Nos. 1 and 3 'favourable', No. 2 'unfavourable'.*
(8) Apparent datings	Stone No. 1: 797–765 B.C. Stone No. 2: 592–559 B.C. Stone No. 3: 384–352 B.C.	*(Datings)*
(9) Length of passage-floor	1485·0068″	*(Time-measurement)*
(10) Dating of top end of passage	1485·0068 years after 30th March, 1453 B.C. = 1st April A.D. 33[7] (notionally 10.04 p.m.)	*(Dating)*

[7] Note that since, under our present system of year-numbering, the year 1 B.C. was followed immediately by the year A.D. 1 (there was never a year nought in the system as originally devised), calculations across the B.C./A.D. boundary always produce a dating which is apparently 'one year short'.

(11) Features at top end of passage	roof height jumps by 286·1″	*Attainment of enlightenment*
(12) Further features at top end of passage	floorline continues into the Grand Gallery (The Hall of Truth in Light)	*Entry into an enlightened age (?)*

READING: *The souls of evolving humanity* (3), *reincarnating* (4) *in the physical world* (5), *will continue their painstaking efforts at spiritual improvement* (2) *as time goes on* (1).

During three periods in history (797–765 B.C., *592–559* B.C. *and 384–352* B.C.*) their efforts will be guided and focused by notable physical events* (6) (8); *and the second of these periods will be of an especially exacting nature* (7).

In the spring of the 1485th year after the commencement of the upward path (1st April A.D. *33)* (10), *man will receive sudden enlightenment* (11), *and thereafter a path of possible enlightenment will open itself to humanity* (12).

The Ascending Passage seems to beg specific identification with the historical development of Judaism—though other, parallel, evolutionary paths may equally well be alluded to. The historical significance of the three girdle-stones is none too easy to ascertain, however. Certainly the Jewish captivity in Babylon commenced around 590 B.C., though it apparently continued until around 534 B.C. None the less, this could conceivably correspond to the 'trying times' apparently foretold by the Pyramid's chronograph. What is striking is the fact that the same period immediately preceded the sudden dawning in the East of Buddhism, Confucianism and Taoism, and the rise in the West of the immensely influential Pythagoreans. The datings of the third girdle-stone would seem to correspond to the era of Plato, Aristotle and Alexander the Great.

The dating for the top end of the Ascending Passage corresponds closely to the founding in the Middle East of a new offshoot of Judaism—namely Christianity—while in the East the same period saw the foundation of the Buddhist doctrine of 'salvation by faith' through the merit of a 'saviour' or *bodhisattva*.

The Messianic Triangle

(1) The level of the Queen's Chamber Floor (Plane of Life), if projected, cuts the Ascending Passage floor at G on diagram opposite.	*Attainment of 'Life' (?)*
(2) The Ascending Passage's roof-level increases initially by some 77″ (11 × 7) at C	

(entrance to Grand Gallery, or Hall of Truth in Light), and subsequently by 286·1" at a point 25" (5^2) beyond C (compare diagram page 100).		*Achievement of spiritual perfection* *Messianic enlightenment*
(3) The Grand Gallery's north wall intersects the Ascending Passage floor at point D on diagram . . .		*Dating for (2)*
(4) . . . and the Queen's Chamber floor-level at F.		*South side of triangle GDF*
(5) The floor-width between G and D remains 41-21" (2 RC) and the passage's cross-section remains rectangular.		*Rebirth/mortality* *Physical/terrestrial*
(6) Angle DGF	26°18' 9·7"	*Human evolution*
(7) The path from the Plane of Life (at G) to the actual inception of enlightenment (at D) is a direct function of triangle GFD.		*Triangle GFD is a 'wedge' giving man a 'leg up'*
(8) But FD	14·85" (code-equivalent: 2 × 7) = approximately 29·8412"/2	*Productive of spiritual perfection* *(?)*
(9) GF	30·043" (code-equivalent: 6 × 5)	*Preparation of Messianic leader or initiate*
(10) GD	33·5116" (code-equivalent: 33·5")	*Messianic presence*
(11) Meanwhile point D represents	1st April, A.D. 33	(*See text below*)
(12) Thus, point G represents	27th September, 2 B.C.[8] (notionally 7.20 a.m.)	(*See text below*)
(13) Level of base (BHFG)	notional base of 25th (5^2) course of masonry	*Birth of life-giving (see* (1)*) Great Initiate or Messianic ideal*

READING: *The rebirth on earth* (5) *of the Life-giving* (1) *Messianic presence* (10) *will date from 27th September, 2* B.C. (12). *A man yoking his life to the establishment of the Messianic ideal* (13) *will attain spiritual perfection* (2) (8) *on the completion of his preparation as the Great Initiate* (9). *This event will occur in* A.D. *33 (1st April)* (11). *From that moment a path of enlightenment, based on the following of that same Messianic ideal, will start to open itself to mankind* (2).

[8] Rutherford also points out that distance GF, if marked off on hypotenuse GD, gives the date 14th October A.D. 29.

The triangle GFD, in other words, seems to represent a Messianic figure who was to be born in 2 B.C. and who was to achieve the full stature of the Great Initiate in A.D. 33—whence the suggested term 'Messianic Triangle'.

There is considerable evidence (presented at length by Rutherford) that these dates were in fact those of the Jewish religious leader known to us as 'Jesus of Nazareth'[9] a conclusion which must tend to suggest that the enlightenment of the Grand Gallery may be connected directly with the application of his teachings.

Whether present-day Christian teaching bears any real comparison with those original teachings by Joshua the Nazarene is a moot point. Indeed, if the Pyramid's message is as universal as we have supposed, then it seems unlikely that the 'enlightenment' of the Grand Gallery is available solely to any one religious sect or tradition. In which case the original Christian teachings would need to be seen as a further development of *world*-religion, and not merely of Judaism. Jesus of Nazareth, in other words, would need to be seen as standing on the shoulders of other teachers such as Krishna and the Buddha, and not merely of Moses and the Hebrew prophets. True, the geometry of the Great Pyramid shows the Messianic Triangle as growing directly out of the Ascending Passage tradition, of which it seems to represent the natural summit. But Judaism is only one of the various traditions to which the Ascending Passage could refer. The original Christian teachings, in short, could in 'Pyramidal' terms be seen as the crowning additive to *any* religion, and in this case their founder could truly be regarded as the Great Initiate, the even more advanced successor of all the world's earlier avatars.

Meanwhile our identification of the Messianic Triangle and the subsequent enlightenment with Jesus of Nazareth and his teachings, seems to be corroborated not only by the explicit references to him as 'the light' in John's gospel—a title formerly shared, in that case, with the Pyramid itself—but also by the more explicit geographical reference of the Pyramid's Bethlehem-angle. Indeed, there seems to be an even more intriguing biblical link. For the three sides of the Messianic Triangle are direct functions of (a) the Grand Gallery's '*way* of the enlightened' (hypotenuse), (b) the Grand Gallery's north wall, whose height signifies the attainment of enlightenment or *truth* (perpendicular), and (c) what we have termed the Plane of *Life* or potential enlightenment (base). Thus, Jesus' own reported claim, 'I

[9] Rutherford further suggests that the date 14th October A.D. 29, as mentioned above, could represent the date of Jesus' baptism in Jordan.

am the way; I am the truth and I am life' (John 14:6), could be seen as *a direct reference to the Great Pyramid's definition of himself in the Messianic Triangle.*

At the same time it should be noted that the geometry of the Messianic Triangle does not speak with any certitude of the *death* of the figure represented. And it will be recalled that Jesus too seems to have been sufficiently unconvinced of the necessity of his death to pray that, if possible, the cup might pass from him. Could there, one wonders, be some link here with the strange and ancient tradition, shared by Muslims and Templars alike, to the effect that the 'real' Jesus was not crucified at all—or with Schonfield's more recent conjecture that he could somehow have survived the cross, as, in a different sense, his followers subsequently claimed?[10]

As for the biblical references to forty days of posthumous appearances to the living, these likewise have no symbolic equivalent at this point in the Pyramid's passageways. But then their origin could conceivably lie in a symbolic reference to the hoped-for 'second coming', i.e. the 'rebirth of the One-who-is-to-come' (8 × 5).

The Crossing of the Pure Roads of Life

At this 'nerve-centre' of the Pyramid a number of closely interlocking features are apparent, as the diagram (p. 58) shows. They are as follows:

(i) Features of the continuation of the Ascending Passage floor:

(1) Direction	due southwards	*Through time*
(2) Slope	upwards	*Evolutionary progress*
(3) Angle of ascent	26° 18′ 9·7″	*Human evolution*
(4) Width	41·21″ (2 RC)	*Rebirth/mortality*
(5) Route	through Grand Gallery	*(Path of 'Light'?)*

[10] The nearest 'deathly' reference seems, in fact, to be in the first overlap of the Grand Gallery's 286·1″-high north wall above the south end of the Ascending Passage. This section appears to measure some 38″ (2 × 19) in height, a fact which would tend to suggest the attainment at this point of a level of enlightenment necessitating physical death. If this were seen as a specific reference to the *Great Initiate's* death, however, then the 'pyramidal' dating for that event would appear to be later than that suggested above—conceivably as late as A.D. 38 or 39—while 1st April A.D. 33 would presumably then refer merely to the completion of his preparation for the Messianic rôle. The identification with the Great Initiate is not specific, however, and it is therefore very much open to question whether these facts have any relevance to the dating of the crucifixion.

(ii) Features of the Grand Gallery roof:

(6) Height	286·1″ above Ascending Passage roof	*Attainment of enlightenment*
(7) All other features	as (1) to (5) above.	*(See* (1) *to* (5)*)*

(iii) Features of the Queen's Chamber Passage entrance:

(8) Direction	due southwards	*Through time*
(9) Slope	horizontal	*Static level of attainment*
(10) Level	1 RC above Plane of Life (or notional base of 25th course)	*Birth/death based on Messianic ideal*
(11) Width	41·21″ (2 RC)	*Rebirth/mortality*
(12) Height	46·99″	*(?)*
(13) Nature	a natural architectural continuation of the Ascending Passage	*Lower or inferior alternative to* (5)
(14) Entrance	via *downward* step through torn-up floor of Grand Gallery.	*A decline from* (5)

GENERAL READING: *Any human souls who attain enlightenment at this time* (5) (6) *will continue to evolve spiritually* (2) (3) *through their rebirths* (4) (7).

Those who fail to achieve enlightenment at this time, however, (14) *will find the upward path of evolution too steep* (9) (13), *and will instead proceed* (13) *into a path through time* (8) *characterised by physical rebirth and re-death* (11). *Yet their clinging to mortality will actually be based in some way on the life-giving Messianic knowledge itself* (10).

In the light of this reading, a more detailed examination of the various features of the Crossing of the Pure Roads of Life suggests itself, as follows:

(i) Features of the Ascending Passage (entry into Grand Gallery):

(1) Direction	due southwards	*Through time*
(2) Slope	upwards	*Evolutionary progress*
(3) Angle of ascent	26° 18′ 9·7″	*Human evolution*
(4) Width	41·21″ (2 RC)	*Rebirth/mortality*
(5) Cross-section	rectangular	*Physical/terrestrial*
(6) Roof-height at Grand Gallery entrance	begins to shoot up by 286·1″ at 1st April A.D. 33, culminating 25″ (5^2) further on.	*Attainment of enlightenment through Messianic ideal or leader*

(7) Other features at same point	only floor continues unchanged.	*Continued upward evolution*
(8) Length of original floor-continuation	25″ (1 SC)	*Messianic ideal*
(9) Terminal feature	downward step of 5·321″	*Decline/scale-change*
(10) Next feature	beginning of floor of first part of Queen's Chamber Passage	*Start of 'inferior path'*
(11) Level of floor in first part of Queen's Chamber Passage	1 RC above Plane of Life (or notional base of 25th course of masonry)	*Birth/death based on the Messianic ideal (?)*

READING: *The souls of those who have taken the path of upward evolution* (3), *reincarnating* (4) *in the physical world* (5), *will continue to advance spiritually* (2) *until 25 years* (8) *after the dawning of the Messianic enlightenment* (6). *Then (from 1st April* A.D. *58) a sudden decline will occur* (9), *causing a clinging to mortality despite the knowledge of the Messianic teachings* (10) (11).

(ii) Initial features of the Grand Gallery floor:

(1) Direction	due southwards	*Through time*
(2) Slope	upwards	*Evolutionary progress*
(3) Angle of ascent	26° 18′ 9·7″	*Human evolution*
(4) Width	41·21″ (2 RC)	*Rebirth/mortality*
(5) Length of first portion	25″ (5^2) after north wall	*Messianic ideal/(dating)*
(6) Height of north wall	286·1″ above Ascending Passage roof	*Gaining of enlightenment*
(7) Dating of north wall	1st April, A.D. 33	*(See page 59)*
(8) Next feature	downward step and gap corresponding to entrance of Queen's Chamber Passage	*Decline leading to inferior path*
(9) Notional point of recommencement	point E on diagram page 58 (intersection with Queen's Chamber Passage roof-line)	*'Parting of the ways'?*
(10) Distance GE	152·54″ (arithmetical code-equivalent = 19 × 8 (or 8 × 19)	*Mortal rebirth, 'bridgeable' only by the enlightened* (153)
(11) Thus, dating of point of recommencement	April A.D. 152	*(Dating—153 years after 2* B.C.*)*
(12) Distance DE	152·54″—33·51″ (GD) = 119·03″ (code-equivalent = 4 × 29·84″ or 19 + 100)	*Removal of Messianic presence*
(13)		*Physical death*

(14) Nature of point E	final divide between Grand Gallery and Queen's Chamber Passage	*'Parting of the ways'*

READING: *The souls of reincarnating humanity* (4) *will continue to evolve* (3) *and to make progress* (2), *enlightened by the Messianic teachings* (5) (6). *But from the spring of* A.D. *58* (5) *their path will be undermined* (8), *and only those who have by then gained full enlightenment* (11?) *will have enough spiritual momentum to continue on the upward path* (9) *after the Messianic presence has been removed* (12). *They will have to pass through physical death* (13), *but this will bring its reward* (13).

From the spring of A.D. *152* (11) *the path of the truly enlightened* (6) (11?) *will separate from the path of those who will cling to mortality despite the Messianic teachings* (14).

(iii) Features of the Well-Shaft (top end):

(1) North-south position	immediately after point denoting A.D. 58 and beginning of Grand Gallery's 'roof of enlightenment' (see diagram page 100)	*Pertaining to the enlightened (?) or associated with realisation of the Messianic ideal (25")*
(2) North-south width	31" measured on slope	(*Time-measurement*)
(3) North-south width	26·7021" horizontally[11] (code-equivalent: 2 × 13")	*Productive of (13?)*
(4) North-south position of axis	35·76" south of Grand Gallery's north wall (FH on diagram p. 58)	*Enlightenment through rebirth*
(5) East-west position of axis	89·61" to *west* of passage-axis (code-equivalent: 3 × 29·84")	*Utter death*
(6) Level of well-shaft's roof	level of Queen's Chamber Passage floor (first part)	*Level of rebirth (?)*
(7) Level of well-shaft's lip	level of Queen's Chamber Passage floor (second part) —1 RC below (6)	*Shaft indicates loss of life (?) associated with the Messianic ideal*
(8) Level of well-shaft's lip	Plane of Life (notional base of 25th course)	
(9) Hence, time-scale for north-south transit of mouth of shaft	as for first part of Queen's Chamber Passage (q.v.)	
(10) Thus, dating for N. edge	A.D. 58 (spring)	(*Dating*)
(11) Thus, dating for axis	A.D. 70 (5th June)	(*Dating of central event*)

[11] Also the approximate interior width of the King's Chamber coffer.

(12) Thus, dating for S. edge	A.D. 82 (9th August)	*(Dating)*
(13) Height of entrance tunnel	1 RC	*Death/birth*
(14) Height of downward step into entrance tunnel	1 RC	*Death*
(15) Cross-section of upper part of well-shaft	square	*Physical/terrestrial*
(16) Nature of entrance	a low tunnel to the right (west) immediately after the downward step marking A.D. 58 . . .	*Progress towards enlightenment by following the Messianic ideal (25"),*
(17)	. . . and leading directly to the lip of the square limestone shaft	*leading to physical death*

READING: *In the year* A.D. *58* (1) *a direct path towards spiritual life* (5) (8) (17) *will lead those prepared to take it directly to an enlightened future incarnation* (4).

But the price of entering this soul-path (3) *will be readiness to undergo physical death* (5) (13) (14) (15)[12] *during a period lasting from* A.D. *58 to 82* (10) (12), *and which will revolve around events in the summer of* A.D. *70* (11) (17).

In the course of the last three readings we have been gradually getting a picture of a time of crisis foretold as occurring between the years A.D. 58 and 82. There are hints that a distortion of the Messianic teaching at this time will undermine that teaching, and allow the majority of its adherents to enter a spiritually static path of continuing mortality and reincarnation (the Queen's Chamber Passage), while the few who are truly enlightened will be able to continue their upward path only by undergoing sudden death and destruc-

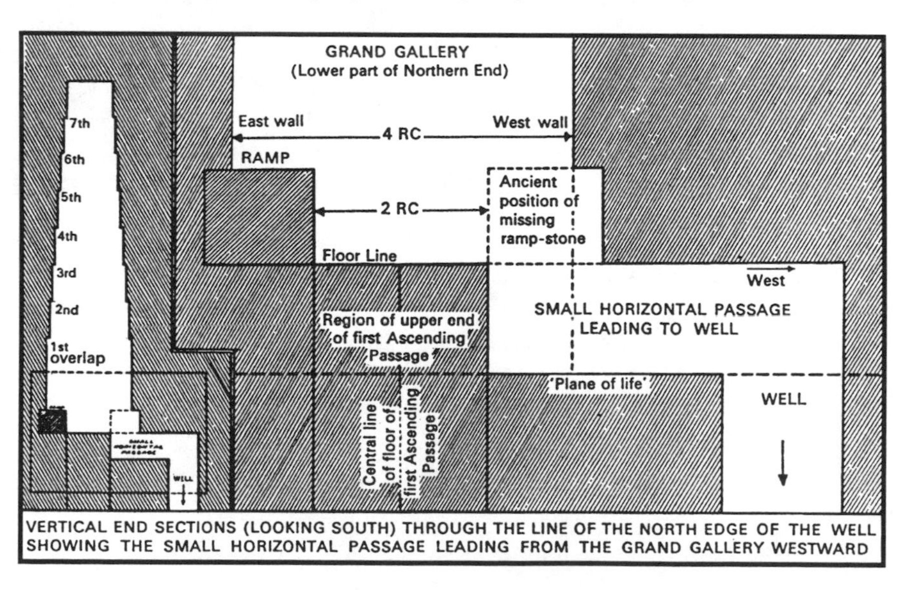

VERTICAL END SECTIONS (LOOKING SOUTH) THROUGH THE LINE OF THE NORTH EDGE OF THE WELL SHOWING THE SMALL HORIZONTAL PASSAGE LEADING FROM THE GRAND GALLERY WESTWARD

tion (i.e. by traversing the entrance to the Well-Shaft).[12] *Only* to those few truly enlightened ones will the spiritual path represented by the Grand Gallery be available.[13]

It is striking to note that the years in question designate the very period during which the original Nazarenes—the exclusively Jewish sect led by Jesus of Nazareth—were to be virtually wiped from the face of the earth. Already by A.D. 58—the year in which Paul the Pharisee was himself arrested—a measure of persecution had set in under the emperor Nero. But with the Messianically-inspired Jewish uprising of A.D. 66 the full might of Rome was to be turned on the defiant Jewish nation. And the culminating act was the massacre of thousands of Jews and the Roman sacking of Jerusalem itself *in the summer of* A.D. *70,* in the course of a bloody war which concluded with the collapse of the final fortress of Masada when its entire garrison committed suicide in A.D. 73.

The effect of this national calamity was to scatter most of the few Nazarenes who still survived to the four winds, and thereafter neither they nor the few aged survivors still lingering in Judaea were able—despite a brief revival early in the second century—to prevent the sect's gradual decay and eventual disappearance from the stage of world-history. The best they could do as the moment of crisis approached was to set down in writing the precious teachings and dispatch the documents for preservation to less troubled parts of the Empire—a selfless act to which, indirectly, we owe the biblical gospels.

The year A.D. 70, then, may well be taken as signifying the beginning of the end of the Nazarenes—the original Jewish Christians—and their teachings. It clearly does not mark, however, the disappearance of Messianic Judaism in general—which was to produce two further rebellions, in A.D. 115 and A.D. 132, among the Palestinian Jews. Nor does the date A.D. 70 mark the death of *non*-Jewish Christianity—the version of the Nazarene teachings originally propounded by Paul the Pharisee. Indeed, *its* battles were to continue unabated after A.D. 82 under the reign of Domitian, and by a process of diplomacy, patience, fortitude and compromise the movement eventually managed to survive right up to the present day.

[12] It is clear from (5) (14) and (17) above that 'stumbling into the Well-Shaft' is intended to have a totally death-laden significance. Its function as a 'Well of Life' clearly applies only to those souls symbolically ascending it from below.

[13] Compare the remarkably similar idea at Matt. 10:39—'By gaining his life, a man will lose it; by losing his life for my sake, he will gain it.'

Consequently the layout of the Pyramid's Crossing of the Pure Roads of Life makes it clear that the Grand Gallery, with its floorline symbolically 'broken off' between A.D. 58 and 152, specifically represents the path of *the original Nazarenes only*—presumably those who received their enlightenment directly at the hands of Jesus of Nazareth before their eventual extermination. And interestingly enough this fact is reflected directly in Jesus' own recorded insistence that his personal mission was to 'the lost sheep of the house of Israel', and not to the rest of mankind (Matt. 10:6, 15:24). The path of the rest of Ascending Passage humanity—including perhaps, that of the surviving non-Jewish, or Pauline, form of Christianity—must be represented, if at all, by the Queen's Chamber Passage, since only this passage has an unbroken floorline from the year A.D. 58 onwards. And the dating of either A.D. 125 (QCP floorline) or A.D. 152 (Grand Gallery floorline) for the notional 'point of separation' (see page 58) seems, in the event, an apt one: for it was at around this time that Western Christianity made its final break, not only with the remnants of Jewish Christianity (see the evidence of St. Jerome) but also with the orthodox Jewish tradition which had given birth to both.

Meanwhile we should perhaps note, once again, that it would be inconceivable for the Pyramid's message apparently to concern itself with the internal bickerings of a single religion such as Christianity unless that religion represented the logical development and culmination of all the other world-religions that had gone before—or unless, of course, those bickerings were paralleled in the other religions.

At this point, however, we can test our conclusions by examining the Lesser Messianic Triangle (see page 58)—a triangle formed by the downward step at the point marking A.D. 58, the first part of the Grand Gallery floor, and the floorline of the first part of the Queen's Chamber Passage. If, after all, the main Messianic Triangle represents the historical Jesus of Nazareth, then its equally unmistakable lesser counterpart must surely likewise foreshadow some historical, semi-Messianic successor. And if we can positively identify that figure, then we should be able to achieve a clearer historical identification of the path to which that figure so clearly leads—namely the Queen's Chamber Passage.

The details of the Lesser Messianic Triangle are:

(1) Plane of hypotenuse	plane of Grand Gallery's path of the enlightened	*An enlightened one (?)*
(2) Plane of south side	plane of Well-Shaft's north wall, and of top	*Enlightenment leading to death (?)*

	section of Grand Gallery's north wall (diagram p. 100).	
(3) Plane of base	1 RC above Plane of Life (notional base of 25th—5^2—course)	*Life/mortality based on the Messianic ideal*
(4) Plane of base	level of first part of Queen's Chamber Passage floor	*Rebirth (?)*
(5) Plane of base	cuts in two the 25″ (5^2) of floor-line following the Messianic Triangle proper	*Destroys the Messianic ideal*
(6) Nature of south side	*downward* step into Queen's Chamber Passage at the moment when the full enlightenment of the Grand Gallery and the death denoted by the Well-Shaft finally begin (see p. 100)	*Falling away at the time of enlightenment/failure to measure up to the standard of the enlightened*
(7) Height	5·32068″ (code-equivalent: 5)	*An Initiate or Messianic leader (?)*
(8) Length of hypotenuse	12·009″ (code-equivalent: 12)	*Mankind*
(9) Thus, floor-dating of beginning of hypotenuse	A.D. 46 (28th March)	*(Dating)*
(10) Floor-dating of end of hypotenuse	A.D. 58 (1st April)	*(Dating)*
(11) Date produced by marking off base-measurement on hypotenuse from point marking A.D. 46[14]	A.D. 57 (January)	*(Intermediate dating)*
(12) Position of Queen's Chamber Passage floor	contiguous with base of triangle and *directly below* Grand Gallery	*Inferior path produced by figure denoted by triangle*

READING: *In* A.D. *46* (9) *a further Messianic leader* (7) *will attempt to open to all men* (8) *the Messianic path of enlightenment* (1). *In* A.D. *57 his work will reach its climax* (11), *but in* A.D. *58* (10) *it will be broken off* (6) *at the beginning of a period of death for the enlightened* (2).

This leader will found an inferior offshoot (12) *of the true Messianic path* (5), *an offshoot which will produce a spiritually static path of continued mortality, but which will none the less be based on the Messianic teachings* (3) (4) (6).

History appears to offer us a clear identification of this leader.

[14] This is a procedure adopted by Rutherford—apparently to good effect—in respect of the *main* Messianic Triangle (see notes 8 and 9 on pp. 60–1).

For A.D. 46 marked the beginning of the first missionary journey of Paul the Pharisee to the 'gentiles', dedicated to throwing open the Messianic teachings to the whole world; A.D. 57 was the year of his eventual return to Jerusalem at the end of his world-shaking mission; and the spring of A.D. 58 marked his final arrest there, prior to his last journey and death in Rome. The identification with Paul could, of course, be coincidental—but if so, it would merely be one more in the Pyramid's long chain of similar 'coincidences', and thus, paradoxically, a piece of further evidence tending to suggest that they are, in reality, not coincidences at all.

Meanwhile this identification inevitably leads us to a further conclusion. If the Lesser Messianic Triangle represents Paul the Pharisee, then the contiguous Queen's Chamber Passage—whatever else it may stand for—must represent the particular teaching which Paul propounded, namely the ancestral form of Christianity as we know it today.

But in that case the Pyramid has a further revelation in store for us. For its prophecy is quite insistent that Paul's 'Christianity' was destined to be an *offshoot*, or deviation, from the teachings of Jesus of Nazareth. Indeed, as the above data and reading clearly show, it was to represent a spiritually static, inferior path based on some kind of diminution of the Nazarene teachings (compare feature (5)). Moreover the Pyramid's symbolism shows that initial downward step as occurring at the very moment when enlightenment is apparently attained by those prepared to undergo death (6)—a symbolic link apparently indicating a refusal, at the critical moment, to 'swallow the Messianic hook' in its entirety.

That 'gentile' Christianity had its origin in the apostle Paul is neither a new nor a particularly controversial suggestion: but that it may represent a deviation or 'falling away' from the teachings of Jesus of Nazareth—despite the latter's own reported misgivings at Matt. 16:5–12 about the danger to his own 'bread' of 'the leaven of the Pharisees'—is a notion which is still relatively in its infancy at the time of writing. Indeed, it seems likely that it will be many years yet before acceptance by Christians of the idea becomes at all general and before the original message of Jesus himself—whatever it was—is rediscovered and re-accepted by mankind at large. After all, the *volte-face* involved is a considerable one, since Christians have long assumed, lacking convincing evidence to the contrary, that what Paul taught and what his followers wrote *was* the original message of Jesus. It has taken the research of scholars such as Dr. Hugh Schonfield (of *Passover Plot* fame) to point out the dubiousness of this assumption.

None the less it should be remembered that what one might term the 'Pauline heresy' is apparently seen by the Pyramid as a *necessary* development, as subsequent history tends to confirm. Had a compromise version of Christianity not appeared when it did, it seems inherently likely that the teaching would have disappeared along with the Nazarenes themselves, instead of being preserved in its present, albeit imperfect, form. The Queen's Chamber Passage may represent an inferior path, but at least it does provide a floor to walk on.

It is ultimately for the theologians to argue the doctrinal rights and wrongs of Paul's case. Or then again, perhaps not—after all, it is the theologians who are responsible for the present doctrinal situation. At all events, our own duty at this point is merely to record the fact that the Pyramid's chronograph appears to portray the future Paul as a drastic 'revisionist', and indicates that the movement he was to found would take the form of a serious departure from the original Messianic teachings. Let us simply leave the matter there, without further comment.

Before we leave the Crossing of the Pure Roads of Life, it is worth noting that the Grand Gallery floor and Queen's Chamber Passage roof were not produced by the builders to their geometric point of intersection (E in the diagram overleaf). Perhaps for constructional reasons they were in fact cut somewhat short to terminate in a vertical surface some 38 P" high (WZ in the diagram). At the same time provision was made for a bridging-slab to seal the entrance to the lower passage, supported by five unevenly spaced crossbeams. The fact that the positions of all these features must have been deliberately chosen suggests that it may be worthwhile to explore their possible symbolic and chronographical significance.

It is possible, for example, to attach notional datings to the various features of the passage-intersection and 'cut-off' both on the Grand Gallery time-scale and on that of the Queen's Chamber Passage. They are as follows:

	Queen's Chamber Passage	*Grand Gallery*
Geometric point of intersection	A.D. 125	A.D. 152 (April)
Cut-off	A.D. 220	A.D. 256 (September)
Upper sill of cut-off		A.D. 296 (October)

It should be remembered, however, that our dating-theory throughout has been based exclusively on *floor*-measurements. Since, therefore, the above features pertain to the floor of the Grand Gallery, rather than to that of the Queen's Chamber Passage, it would seem

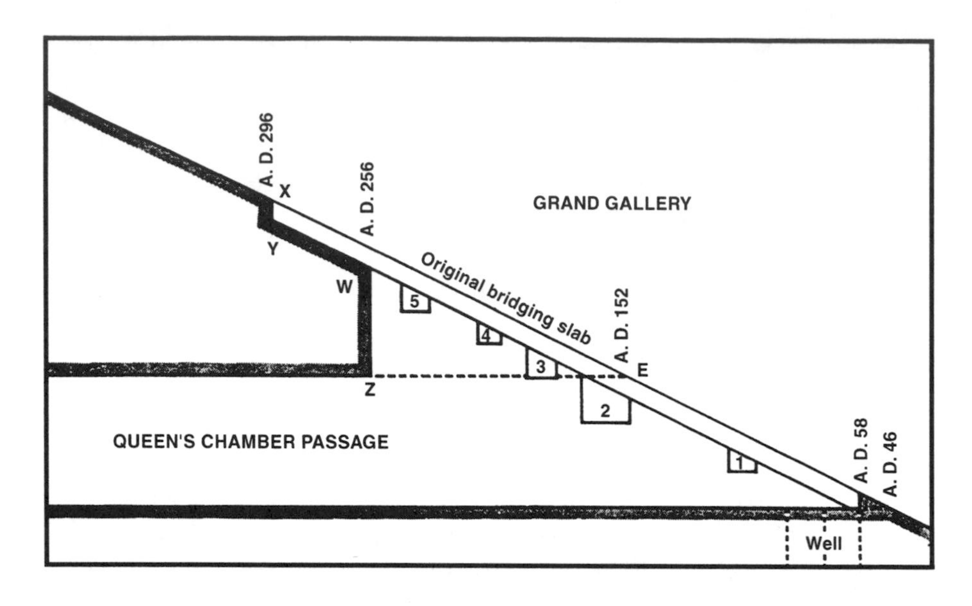

The 'cut-off', showing original bridging-slab and approximate positions of supporting beams

that the Grand Gallery's floor-datings are more likely to be the significant ones.

Meanwhile a number of symbolic measurements also seem to be present. Petrie's measurements suggest that distance EX on the diagram measures just over 144″, apparently indicating that the Grand Gallery floor symbolises the path of the 'elect' (12 × 12 = the 'men of men'). The length of the 'sill' (WY) is 40″ (8 × 5) and its height (XY) 8″—apparently suggestive of the notion that the path of the elect depends in some way upon 'the ultimate rebirth of the initiates' (8 × 8 × 5). And the height of the main part of the cut-off (as we saw above) is some 38″ (2 × 19), which would seem to indicate that the final, decisive breakthrough from the Gallery into the lower passage is in some way 'productive of death'. Finally, the symbolism of the bridging-slab clearly shows that the upper path is to be regarded as accessible only while the slab remains in place: as soon as it is torn up to gain entry to the lower path, the Grand Gallery is, in symbol, closed. None the less the two paths remain in contact until the vertical surface of the cut-off finalises the division between them.

On this basis we can therefore construct a more detailed reading than was hitherto possible for the intersection of the Grand Gallery and Queen's Chamber Passage.

READING: *The inferior path through time and mortality founded between* A.D. *46 and 58 will part company from the upward path of the elect by the year* A.D. *152. A period of indecision or vacillation will none the less ensue. Even as late as* A.D. *296, attempts will still be made to reconcile the two divergent paths. But by the year* A.D. *256 the mortal direction of the lower path will already have become firmly established, and all remaining prospects of entering the upper path will, by the same token, be fatally undermined.*

Bearing in mind our earlier conclusion that the Queen's Chamber Passage represents, among other things, the path of Western Christianity, the above reading should help us to test the validity of that conclusion against the available, if scanty, data on the history of the early Church. And, in the event, the 'fit' seems to be a promising one.

The average Christian is unfamiliar with the early history of his religion. He is aware that many of the early Christians were persecuted and martyred by the Romans, and naturally assumes that this was because they subscribed to the same teachings as himself. He is usually unaware that the early Church was racked by theological bickerings and disputes for nearly two centuries before there was any general agreement on precisely what those teachings were—and notably on the doctrines of the Trinity and the divinity of Jesus.

During the second and third centuries of our era in particular, Nazarene and Pauline protagonists constantly clashed and counter-clashed, while Gnostics and other peripheral groups added their own not inconsiderable voices to an already complicated debate. The Pyramid's general prophecy of a time of indecision, and of attempted, but abortive, reconciliation of the divergent teachings during this period, is amply borne out by the historical records.

The movement founded by Paul the Pharisee was starting to gain ascendancy over the Jewish Christian movement by around the middle of the second century, if the testimony of Justin Martyr is to be relied upon. And while protagonists of Jewish Christian views (such as Papias and Hermas) continued to make their voices heard until even as late as the end of the third century, nothing they could do was henceforth capable of healing the ever-widening split. One of the effects of the wilder second-century Gnostics (such as Basilides and Valentinus) was, ironically, to produce (as it were, by reaction) a movement towards the establishment of some kind of *anti*-Gnostic Christian orthodoxy. A similar reaction was even to set in against the teachings of the later Origen, one of the greatest and most original of the early Church Fathers. And so it was that, by the time of Origen's death in about A.D. 254, the movement towards the establishment of an agreed Christian orthodoxy had acquired a momentum which no human effort was henceforth able to arrest. That the Pyramid's cut-off dates that same development, as we have seen, at A.D. 256 is therefore particularly appropriate.

Even so, since no orthodox creed as such had yet been formulated, independent thinkers still abounded, their views taking up a variety of stances ranging from the near-Nazarene to the out-and-out Pauline. One of the most influential of these thinkers was Arius, who around the end of the third century still had the temerity to propound the quasi-Nazarene view that Jesus was human, not divine—and won over at least one Roman emperor to his view. The reaction, however, was overwhelming—and, in the event, decisive.

For in the year A.D. 313, the emperor Constantine was to issue his edict establishing Christianity as the official state religion. Now, therefore, the teachings of that religion had at last to be definitively codified. At the ensuing ecumenical council in Nicaea in A.D. 325 the various views were put and argued, and the upshot was that the views of Arius were overwhelmingly rejected and the teachings of one Athanasius agreed as the doctrinal basis for the new orthodoxy. Nearly three centuries after Jesus' death, it had finally been decided what he had actually believed and taught. And the resulting Nicene Creed is still recited in many of the more orthodox churches to this day, as a definitive statement of Christian belief.

The Great Pyramid, as we have seen, seems to connect the end of the attempts at reconciliation between Nazarenism and Christianity with the date A.D. 296. Perhaps it is of relevance, then, that Athanasius, the formulator of the Christian creed eventually accepted at Nicaea, was himself born in A.D. 295.

We may conclude, in short, that the datings and symbolisms of the Pyramid's cut-off are quite consistent with our earlier suggestion that the Queen's Chamber Passage represents the 'lower' path of Pauline Christianity.[15]

[15] We should not forget, however, that the original bridging-slab sealing the entrance to that passage was supported by five stone beams, whose wall-sockets still remain (see diagram). Since the spacing of these sockets is distinctly irregular, we could deduce that this—unless entirely arbitrary—may be because they too are symbolic of important events or developments at the time. Indeed, the function of the beams was *to support the bridging-slab* (and thus the floor of the Grand Gallery); and it would be logical to deduce that all five beams could in that case represent specific pro-Nazarene developments. For the benefit of readers who wish to research this particular point, therefore, the apparent 'datings' for the five beams are: No. 1—A.D. 99 to 109; No. 2—A.D. 149 to 168; No. 3—A.D. 180 to 191; No. 4—A.D. 202 to 213; No. 5—A.D. 229 to 243. Since these datings are taken only from measurements of small-scale published diagrams, a margin of error of at least ± 3 years should be allowed for.

The Queen's Chamber Passage (The Path of the Coming Forth of the Regenerated Soul)

The following are the features of the Queen's Chamber Passage:

(1) Direction	due southwards	*Through time*
(2) Slope	horizontal	*Level of attainment*
(3) Width	41·21" (2 RC)	*Rebirth/mortality*
(4) Cross-section	rectangular	*Physical/terrestrial*
(5) Level of floor of first part	1 RC above Plane of Life	*Life/mortality based on the Messianic ideal (?)*
(6) Overall length of passage	approx. 7/6 times length of first part of passage	*Spiritual perfection through preparation*
(7) Level of floor of second part	Plane of Life (notional base of 25th masonry-course)	*The Messianic ideal(?)*
(8) Direction of step at beginning of passage	downwards	*Loss of spirituality*
(9) Height of downward step in body of passage (and also of downward step into Well-Shaft entrance)	20·60659 (1 RC)	*Death or birth: return to mortality by souls taking this path*
(10) Thus, basis for time-scale of whole passage	Royal Cubit instead of Sacred Cubit i.e. 1 *n* (1 RC/100) per year instead of 1" (1 SC/25) per year	

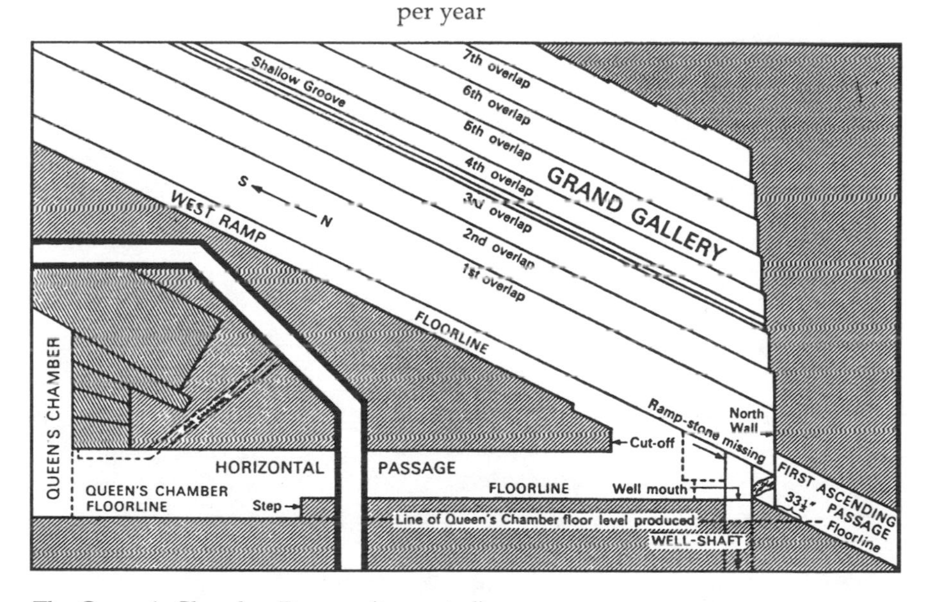

The Queen's Chamber Passage (truncated)

(11) Conversion-formula from inches to *n*	100/20·60659	
(12) But time-scale of first part of passage is further governed by initial, *downward* step		
(13) Hence, time-scale of *first* part of passage is *less* than 1″ per year		
(14) Now height of initial step is	5·32068″	
(15) And nearest base-unit of previous scale is	1″ (= 1 SC/25)	
(16) ∴ In first part of passage, 1″	= (100 × 1)/(20·60659 × 5·32068) years = ·9120669 years	
(17) Termination of time-scale of first part of passage is marked by next step—i.e. the 1 RC step in body of passage		*(Scale-determiner for passage)*
(18) ∴ Time-scale for second part of passage	1 *n* per year	
(19) Thus, in second part of passage, 1″	= 4·8528 years	
(20) Starting-date for floor of first part of passage	A.D. 58 (April)	
(21) Length of floor of first part of passage	1282·81285″	
(22) Time represented	1170·0152 years	
(23) Thus, dating for step in mid-passage	A.D. 1228 (12th April)	
(24) N.-S. width of Well-Shaft	26·7021″	
(25) N.-S. semi-width of Well-Shaft	13·35105″	
(26) Hence, dating for Well-Shaft axis on scale of first part of passage	A.D. 70 (5th June)	
(27) Dating for south edge of Well-Shaft	A.D. 82 (9th August)	
(28) Dating for point of separation from Grand Gallery	A.D. 125	
(29) Length of second part of passage	216·5668″	
(30) Number of years represented	1050·9589	
(31) Thus, dating for entry into Queen's Chamber	A.D. 2279 (23rd March)	
(32) Nature of floor of second part of passage	roughly-finished limestone	*Terrestrial/earthly*

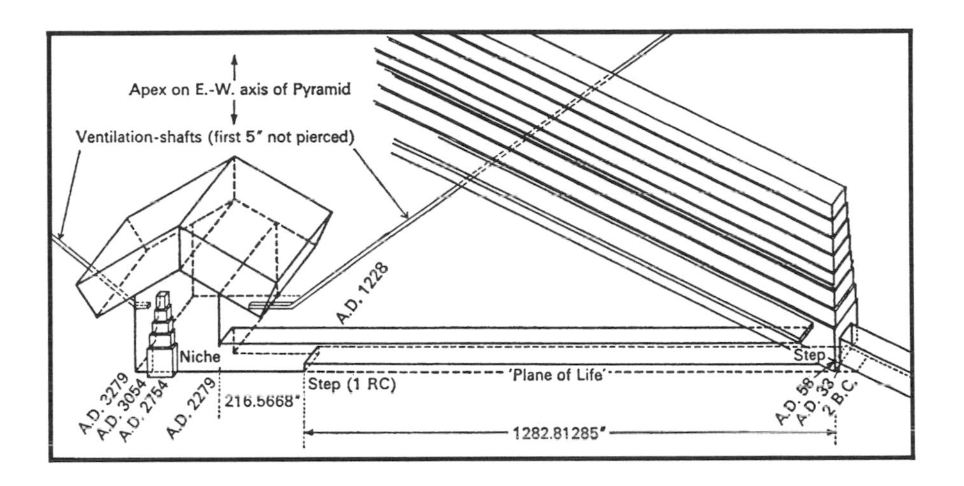

Projection of Queen's Chamber and Passage.

(33) Height of roof of second part of passage	67·5946" (direct code-feature, and equivalent of 2 × 33·5)	*Man-come-of-age, productive of the Messianic presence*

The downward, Royal Cubit step in the Queen's Chamber Passage clearly suggests either a death or a birth—or possibly both. Thus, it could refer to the return to mortality of the souls taking this path; and/or to a specific death or birth; and/or to the 'death' of the Pauline teachings which gave rise to the path in the first place—in which case the way is symbolically cleared for a rebirth of the Nazarene ones.

READING: *The souls of those following the spiritually-static inferior path* (2) (8) *of mortality based on the Messianic teachings* (5) *will continue to reincarnate* (3) *in the physical world* (4) *as time goes on* (1). *But between* A.D. *58* (20) *and* A.D. *82* (27) *those of them who strive towards perfection will undergo a time of death revolving around events in* A.D. *70* (24) (25) (26), *and from* A.D. *125 they will make their final break with the path of the enlightened* (28).

Yet out of the imperfection of this path perfection will come (6?).

A death or birth in the year A.D. *1228* (9) (23) *will mark a return to the true Messianic teachings* (7), *and steps will be taken which will spell the death of the lesser teachings which will have produced this path* (9). *By 'coming down to earth once again'* (9) (32), *the souls who, passing this way, return to mortality at this time* (9), *will create a path which will allow man finally to 'come of age'* (33) *and which will eventually produce the Messianic presence once again* (33).

An era of final decision will be entered in the summer of A.D. *2279* (31).

It is interesting to note that A.D. 1228 falls within two years of the death of Francis of Assisi, and was in fact the year of his official canonisation. Meanwhile, in describing this extraordinary man, at least one historian actually offers the unsolicited comment: 'Men had not lived as St. Francis lived since Christ and His disciples preached in Galilee.'[16]

The Franciscan movement, even more than its contemporary Dominican counterpart, was originally devoted specifically to the rediscovery of the practical meaning of the gospel for ordinary people, in reaction against the Church's tendency to become part of the social establishment, to acquire worldly riches, and to overemphasise priestly ritualism. Moreover, by 1230, a Franciscan school of learning was set up in Oxford by the celebrated Robert Grosseteste, a school whose scholars were subsequently to have great influence on European thought. The most renowned of these was the remarkable Friar Roger Bacon (1214–94)—explosives expert, inventor of both spectacles and telescope, and prophet of the horseless carriage, the steamship and the aeroplane. His rejection of all established authority and his insistence on a return to scientific conclusions based on practical experiment echoed exactly in the secular sphere the Franciscan approach to religion. Moreover, his methodology was later to bear important fruit at the time of the Renaissance and has earned him in some quarters the title 'father of modern science'.

Clearly these thirteenth-century developments in the history of human thought—which may conceivably have been mirrored elsewhere—correspond satisfactorily with the 'coming back down to earth' and the rediscovery of the Messianic teachings, long foretold for this period in the Pyramid's Queen's Chamber Passage. And to the extent that they contain within them the seeds of today's whole scientific culture we may indeed regard them as the first steps in man's 'coming of age'.

The Queen's Chamber (The Chamber of Regeneration/of Rebirth/of the Moon)

The ancient Egyptian *Book of the Dead* seems to identify the Queen's Chamber specifically as a Chamber of Rebirth. Even the name Chamber of the Moon may be seen as having this connotation, since the moon constantly 'dies' and is 'reborn'. In fact the moon's con-

[16] Carter and Mears, *History of Britain*, Oxford.

nection with the tides and the sea—the source of all terrestrial life—as well as with the female menstrual cycle, may likewise be reflected in the ancient name for the chamber, which may thus represent nothing less than the 'womb of the Pyramid'. Examination of the chamber's detailed symbolic features should either confirm or deny the notion:

(1) Shape of chamber	10-cornered, 7-sided (i.e. 10 × 7)	*Millennium of/for spiritual perfection*
(2) Nature of roof	12-stone limestone gable at 30° to horizontal	*For mankind: see* (3)
(3) Symbolism of gable	possibly an upward-pointing arrow-head[17]	*Gateway to a higher destiny*
(4) Position of ridge of gable (and E.-W. axis of chamber)	on east-west axis of Pyramid, and directly below riser of Grand Gallery's Great Step	*A 'turning-point' (?) leading to an enlightened rebirth*
(5) N.-S. distance across chamber	10 RC (or 1,000 *n*)	*See* (7)
(6) Time-scale applicable to distance across chamber	time-scale of last part of Queen's Chamber Passage (q.v.)	
(7) Thus, time signified by chamber	1,000 years (i.e. a 'Millennium')	*(Time-measurement)*
(8) E.-W. length of chamber	11 RC (or 11 × 100 *n*)	*Achievement of the reward*
(9) Distance of west-end of chamber to west (right) of passage-axis	10 RC (1,000 *n*)	*Eternity*
(10) Height of N. and S. walls *and top of Niche* (see (17) to (27))	184·264", or [(4 × 35·76") + 2 RC]	*Terrestrial rebirths of appropriate enlightenment, leading to death*
(11) Height of gable above N. and S. walls and top of Niche (q.v.)	59·49" (code-equivalent: 2 × 29·84")	*Productive of death*

[17] The three 'gables' in the Great Pyramid (the other two being over the Scored Lines in the entrance-passage and over the top Construction Chamber—q.v.) are perhaps too easily seen as such by the average European. Gables and sloping roofs are to carry away rain or snow: they have no place in a largely rainless country, and their inclusion here is thus evidence that their function goes beyond mere roofing. The upward-pointing arrowhead seems a more likely interpretation. And the siting of the Queen's Chamber arrowhead directly below the riser of the Great Step and in the very plane of the summit of the missing capstone seems to lead to the conclusion that this chamber leads in some way to the higher path and the Messianic Millennium (see (4)).

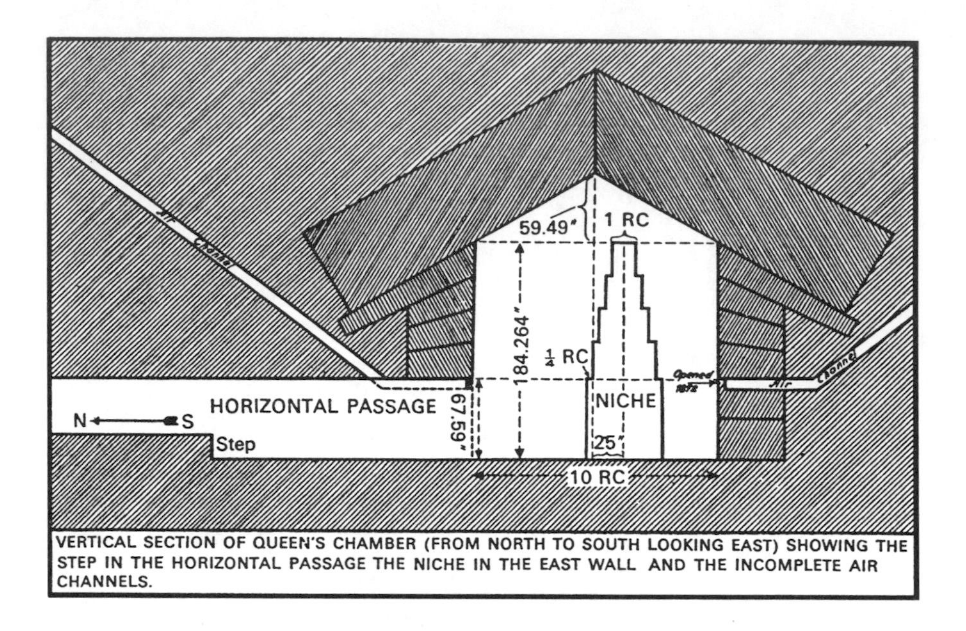

VERTICAL SECTION OF QUEEN'S CHAMBER (FROM NORTH TO SOUTH LOOKING EAST) SHOWING THE STEP IN THE HORIZONTAL PASSAGE THE NICHE IN THE EAST WALL AND THE INCOMPLETE AIR CHANNELS.

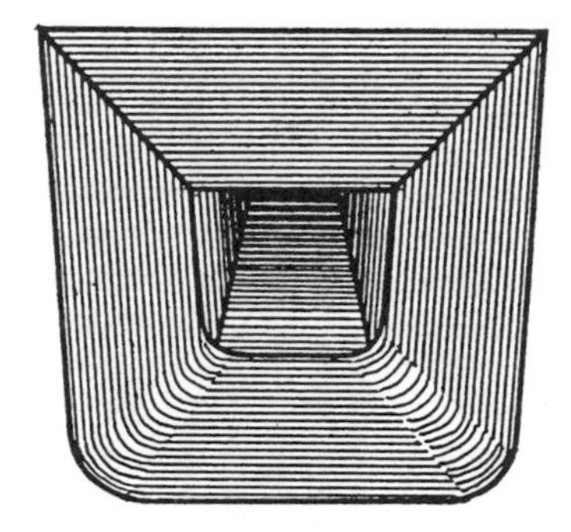

Queen's Chamber Air-Channels—view into cleared lower openings

(12) Nature of floor	roughly-dressed limestone[18]	*Earthly/physical*
(13) Composition of chamber	limestone, heavily encrusted with salt	*Terrestrial: see* (14)

[18] Various researchers have pointed out that the Chamber's rough-cut floor lacks the superimposed slab-floor found in many other 'burial chambers'. And indeed, its level lies some 6·6 P" below that of the bottom of the Pyramid's 25th course of masonry, with which it seems to beg identification (compare chapter 2, pp. 43–4). It seems, therefore, that the addition of a 6·6 P"-thick slab-floor is meant to be assumed, once the Chamber is symbolically 'finished' with the final opening of its air-shafts (see points (28) and (29)). Its level (course 25) would then symbolise the attainment of full initiateship (5^2).

(14) Significance of choice of salt-impregnated limestone	probably symbolic of the sea, the source of physical life, the womb[19]	*Physical rebirth: 'back to the beginning'*
(15) Contents of chamber	originally a lidless, lime-tone coffer—position of N.-S. axis possibly 7 RC (144·2") to west of passage-axis	*Break-out from earthly mortality associated with the 'men-of-men' or 'elect'*
(16) Other features	Niche in east wall. Two ventilation-shafts, in N. and S. walls	
(17) Nature of Niche	a corbelled, 'telescopic' cavity comprising 5 sections, each smaller than the one beneath it.	*Fivefold upward progress achieved through force from within*
(18) E.-W. position of Niche	the most easterly feature of the entire passage-system	*Lack of spirituality/ rebirth*
(19) Position of Niche's N.-S. axis	2 RC to E. (left) of passage-axis	*Death/rebirth*
(20) Position of Niche's E.-W. axis	25" to S. of chamber's E.-W. axis (and thus also of Pyramid's axis)	*A 'turning-point' leading to the achievement of the Messianic ideal*
(21) Position of N. side of Niche	28·2*n* (some 6") before E.-W. axis of chamber	*(Time-measurement)*
(22) Position of S. side of Niche	228·2*n* before south side of chamber	*(Time-measurement)*
(23) Depth of Niche throughout	2 RC	*Mortality*
(24) Distance across Niche at base	3 RC	
(25) Distance across Niche at top	1 RC	*Death: loss of mortality (?)*
(26) Total number and width of corbellings	$8 \times \frac{1}{4}$ RC [or $8 \times (2\ \text{RC}/8)$]	*Subjection to mortality through rebirth*
(27) Possible significance of Niche		*5 rebirths, each containing 'the seed of its successor'*
(28) Nature of ventilation-shafts	upward-sloping air-channels to north and south, originally left uncut at lower end by builders.[20] Probably	

[19] This uterine symbolism is apparently shared by the well-shafts of later tombs, which (according to C. Desroches Noblecourt) signify 'the aqueous region in which the becoming being dwells' between rebirths.

[20] There are signs that there may originally have been a similar obstruction—or at least some kind of constriction—at the entrance of the Chamber itself (all that

	signifying 'paths of escape' from the chamber	
(29) Length of both blockages	5″ (cleared in 1872)	*Obstruction needing five lives to clear*
(30) Cross-section of shafts	part rectangular, part rounded (see diagram p.80)	*Part-physical, part-spiritual*
(31) Average bore of shafts	8″ × 8″	*(Leading to) the 'ultimate rebirth'*
(32) Height of tops of openings, and top of bottom section of Niche, above floor	67·59″ (equivalent: 2 × 33·5?)[21]	*Man-come-of-age; productive of the Messianic presence*
(33) Approximate position of shaft-axes	96″ to west of passage-axis (8 × 12″)	*(Leading to) the rebirth of (all) mankind/of (true) man?*
(34) Length of initial, horizontal portion of shafts	84″ (7 × 12″) approximately	*(Leading to) the spiritual perfecting of man*
(35) Position of west end of chamber	still short by 80·037″ of the Pyramid's N.-S. axis (code equivalent: 8 × 10)	*'Rebirth of the Millennium' still needed for the gaining of full enlightenment*
(36) Nature of datings within chamber	probably symbolic rather than chronological (there being no longer a N.-S. passageway as such)	*(Cross-references to other parts of chronograph (?))*
(37) Symbolic dating of beginning of chamber	A.D. 2279 (23rd March)	*(Conceivably references to the spring equinox)*
(38) Symbolic dating of axis of chamber	A.D. 2779 (23rd March)	
(39) Symbolic dating of south wall of chamber	A.D. 3279 (23rd March)	
(40) Symbolic dating of Niche	apparently based on incre-	

[20] (*continued*) now remains of it being a rough 'pilaster' at the end of the west wall of the entrance-passage). In all three cases the symbolism is strongly suggestive of the uterine membrane that needs to be ruptured before the child (in this case man himself) can emerge from the womb.

[21] The fact that the Chamber's ventilator-tops, like the top of the lowest section of the Niche, are *on exactly the same level* as the roof of the entrance-passage suggests that all four features are closely linked in symbol. The inference is that entry into the Chamber leads directly *either* to some sort of semi-spiritual 'escape' via the air-shafts (which lie to the west, or right, of the passage) *or* to repeated rebirths and a return to mortality (23) (25) (27), as symbolised by the Niche (which lies to the east, or left, of the passage). We are in a 'chamber of alternatives'; upward and downward possibilities are both present.

	ments of $\frac{1}{4}$ Royal Cubit (see (26)), but adjusted slightly so that axis comes 25″ after chamber-axis
(41) Thus symbolic dating of N. side of Niche	25 or (100/4) years[22] before dating for chamber-axis = A.D. 2254
(42) Symbolic dating of S. side of Niche	225 or ($2\frac{1}{4} \times 100$) years[22] before dating for S. wall of chamber = A.D. 3054

Notes on symbolic datings

(43) A.D. 2279 (beginning of chamber)	365 years after S. wall of Grand Gallery (q.v.)	*Culmination of age following the era of the enlightened (?)*
(44) A.D. 2279 (beginning of chamber)	365 years after entry into Subterranean Chamber (q.v.)	*Culmination of the age of hell on earth (?)*
(45) A.D. 2754 (N. side of Niche)	475 (19×5^2) years after beginning of chamber	*Death of the Messianic ideal*
(46) A.D. 3054 (S. side of Niche)	65 (13×5) years after N. wall of King's Chamber	*(See chapter 4)*
(47) A.D. 3279 (S. wall of chamber)	225 ($3^2 \times 5^2$) years after south side of Niche	*The utter perfection of the Messianic ideal*

Levels.

(48) Level of Queen's Chamber floor	Plane of Life i.e., notional base of 25th (5^2) course of masonry, and top of 24th (6×4)	*Basis of possible enlightenment and renewal/ Preparation of the physical giving rise to the Messianic ideal*
(49) Level of exterior ventilator outlets	90th course (9×10) of core-masonry (thickness of course = approx. 38″, or 2×19)[23]	*The utter perfection of the Millennium, but leading to mortality*

[22] Strictly measured in *n*, these figures would be 28·2 and 228·2 respectively.

[23] The average height above the Pyramid's base of the 90th course is notionally 2727·7 P″. But the half-height of the Pyramid's summit-platform is itself exactly 5448·736 P″/2, or 2724·368 P″. The probability is therefore strong that there is an intended link between the two levels—and thus that the Queen's Chamber ventilator outlets have some part to play in defining the height of the summit-platform. (Certainly it is clear from the King's Chamber data, for example, that the Pyramid's *other* two ventilators have such a rôle to play, since the sum of their course-numbers equals the number of courses to the summit-platform). Compare chapter 5 and Appendix H.

(50) Level of exterior ventilator outlets	91st course (7 × 13) of casing-stones (thickness of course = approx. 35″, or 7 × 5″)	*The spiritual perfection of (13?); the spiritual perfection of the initiate*

The twin occurrences of the number 5 in the Queen's Chamber (27) (29) immediately suggest the presence of an initiate. Yet neither occurrence can be dated to any one point in time. Both, in fact, develop gradually. The Niche speaks of five successive 'lives'—and thus of four opportunities to move up the ladder of enlightenment at rebirth (4 × 35·76″). The air-shaft blockages seem to symbolise the fact that one inch of the blockage has to be 'cut away' per life to allow eventual escape at the end of the fifth incarnation (the 'one inch per year' convention is concerned purely with floorline chronology). And Buddhism recognises a similar path of release dependent upon a quadruple experience of enlightenment.

Those who fail to 'turn to the right' at this juncture, however, are subject to the process symbolised by the rest of the east wall's height above the Niche—namely utter death (11). But success in this five-fold process may be seen as resulting in a full initiate, who is then qualified, at death (10) (25), to join the elect of the King's Chamber (15), via the Queen's Chamber air-shafts (30) (31) (32) (33) (34). (Compare section on the King's Chamber.)

At the present time these air-shafts emerge in the 38″-thick 90th course of masonry, and thus symbolise continued subjection to mortality (2 × 19). In fact, it seems probable that they were originally *blocked* at their outer ends—as well as being concealed by the original casing—just as they are known to have been at their inner ends (see chapter 5). But with the culmination of the Messianic Age—and the symbolic completion of the Pyramid to its full design—the air-channels are (I surmise) to be seen as carried through the new casing, when they will emerge in the 91st course, which is 35″ thick. With the final advent of the Millennium they will thus have been transformed at last into paths leading to the spiritual perfection of the initiate (7 × 5).

But that event also has its counterpart in the Queen's Chamber itself. For the builders obviously intended the chamber to be seen as unfinished. They left its air-shaft openings uncut, some of its stones undressed, its floor rough and devoid of the slab-covering usually found in other 'burial chambers'. That floor—as the note to

[23] (*continued*) Meanwhile the information set out on page 182 seems to confirm that the two Queen's Chamber ventilators symbolise paths of non-escape—albeit containing within their design the hope of better things.

point (12) suggests—is probably meant to be seen as some 6″ thick. And in this case it is entirely appropriate that the Chamber's symbolic completion will see its floor brought up to what is in fact the level of the top of the 25th course of masonry—the level characteristic, above all, of the Great Initiate (5^2).

READING: *From the year* A.D. *2279, the souls taking the spiritually static, inferior path founded between* A.D. *46 and 58, will enter an era of final decision, having 'come of age'* (4) (20) (32).

This era will be an earthly (12) (48) *age of a thousand years* (5) (9) *during which the souls in question will experience five crucial incarnations* (10) (14) (17) (18) (19) (23) (25) (26) (27) (29) *connected with their neglect of the Messianic ideal* (45). *During the course of these they will have a chance to leave behind physical imperfection* (48), *to follow the true Messianic path* (20) (48) *and, by 'turning to the right'* (9), *to gain entry to a higher level of existence* (2) (3) (28).

But success in this will necessitate profiting in each successive incarnation from the enlightment gained in the previous one (10) (27) (29). *Failure to continue this process even once will simply result in a failure to 'break out' on dying* (10) (25) *at the end of the fifth life* (29)*—and a return to mortality will ensue* (11). *But success in turning to the right will allow the souls concerned to begin breaking out of the cycle of physical rebirth* (15) (29) *by entering a semi-spiritual* (30) *higher path* (3) (28)*—none other than the long-lost 'path of the enlightened' itself* (4) (29).

Thus, turning to the right during this earthly age (12) *for spiritual perfection* (1) (9) *will bring its reward* (8). *Indeed, it will allow 'mankind-come-of age' to attain the spiritual perfection of the initiate* (32) (34) (50) *by experiencing the ultimate rebirth* (31) (33) *of the Messianic Age* (1?) (5?) (7?) (49). *For that, further experience of rebirth will still be necessary for the final regaining of man's full spiritual nature* (31) (35) (49).

In this reading we have clear confirmation that the Queen's Chamber represents a chamber of life. And indeed, this is no more than one would expect. The chambers of death of the other pyramids all lie at or below ground level and are approached by downward-sloping passages—and for this reason we may well regard the Great Pyramid's own Subterranean Chamber as one such. But the two upper chambers of the Great Pyramid both lie—uniquely—well *above* ground, and are approached by *upward*-sloping passages. It is no more than logical, then, that they, by contrast, should be seen as representing chambers of life.

Again, both upper chambers are equipped with ventilation-shafts. What, one might ask, would the dead want with ventilation-shafts? It is the living who need air to breathe. But then we should remember that the 'living' of the Great Pyramid need to be understood in a special sense: they are the 'spiritually alive' as opposed to the physically moribund. What, then, is the significance of the ventilation-shafts for them?

The answer is surprisingly straightforward. A ventilation-shaft is a channel for air, for breath: but from time immemorial the notion of 'breath' has been intimately bound up with those of 'wind' and 'spirit'. For example, the Greek word *pneuma* was used for both, the German word *atmen* (to breathe) is cognate with the Hindu *Atman* (the name of the 'god within'), and the original Latin meaning of the word 'spirit' itself was actually 'breath'. We can thus deduce without much difficulty that the ventilation-channels are symbolically intended not so much for incoming air as for 'out-going spirit'—they are channels of escape, no less. Meanwhile the fact that the Queen's Chamber and King's Chamber ventilators both mark the half-height of the Pyramid's summit-platform—in the one case geometrically and in the other numerically (see chapter 5 and Appendix H)—would seem to confirm that they are to be regarded as, in a sense, continuous. Escape from the Queen's Chamber may thus lead directly, as we have already deduced, to entry into the King's Chamber.

Yet the roughly-finished Queen's Chamber floor, contiguous with the base both of the Messianic Triangle and notionally of the 25th course of masonry, as well as with the top of the 24th course, makes it clear that that escape depends upon a return to earth and a re-learning and re-application of the Messianic teachings. If there is to be a Messianic Millennium, then it will be based on the efforts of imperfect physical men; and it is in the sphere of man's very mortality (as the Buddha himself taught) that the key to his final escape must be sought.

Note on code interpretation

At this point we can perhaps begin to attach a more precise meaning to the pyramidal level which we have hitherto referred to as the Plane of Life. Clearly, for a start, it represents no finality. Indeed it is not even half-way numerically to the level of the King's Chamber floor, which tops the 50th course of masonry and therefore presumably signifies the Messianic Millennium (10×5). But it is the basis for the Messianic Triangle's upward path towards enlightenment, for the Niche's five upward rebirths, and for the path towards final escape represented by the Queen's Chamber and its air-channels. Meanwhile it is also the level of the top of the 24th course, symbolic of the imperfect physical world (6 × 4). In other words, it represents the lowest level of human evolution at which enlightenment can be attained, and thus marks the *very beginning* of the Messianic process whereby mortals are transformed into 'living beings'. It is in this sense, then, that we should understand the terms Plane of Life and 'plane of potential enlightenment' as listed in our reconstructed code.

The Lower Part of the Descending Passage (The Descent)

We can now continue our exploration of the Descending Passage from the point where we left it—the junction with the Ascending Passage.

(1) Direction	due southwards	*Through time*
(2) Slope	downwards	*Spiritual degeneration*
(3) Angle of descent	26° 18′ 9·7″	*Human evolution*
(4) Cross-section	rectangular	*Physical/terrestrial*
(5) Width	41·21″ (2 RC)	*Mortality/reincarnation*
(6) Vertical component of pi-angle inserted in the passage	37·995″	*Death*
(7) Horizontal component of pi-angle inserted in passage	29·841″	*Death*
(8) Original star-alignment	Alpha Draconis (lower culmination)	*Death/hell (?)*
(9) Distance to floor-intersection with plane of Subterranean Chamber roof (Plane of Death)	2675·006″	*(Time-measurement)*
(10) ∴ Dating of floor intersection	A.D. 1223 (1st April)	*(Dating)*
(11) Distance from above intersection to intersection with Subterranean Passage floorline	283·378″	*(Time-measurement)*
(12) Distance down to bottom end of floor (east side)	286·1″	*Loss of enlightenment*
(13) Distance to bottom end of floor (centreline)	286·835″	*(Time-measurement)*
(14) Thus, notional dating for (11)	A.D. 1506 (17th August)	*(Dating)*
(15) And notional dating for (12)	A.D. 1510 (8th May)	*(Dating)*

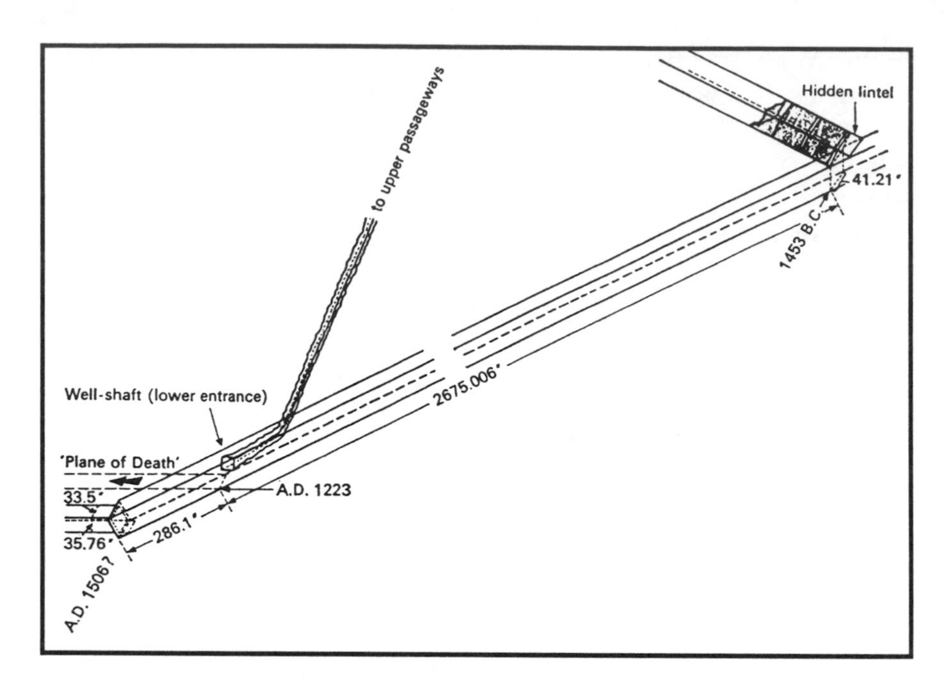

Projection of the lower part of the Descending Passage

(16) Other features	Lower entrance of Well-Shaft (Well of Life), leading to upper passageways, opens to *right* (west) immediately above Plane of Death.	*Escape 'to right' in search of enlightenment and spirituality not possible below Plane of Death*
(17) Nature and dimensions of Well-Shaft opening	rough and ill-defined	*No precise dating; applicable throughout Subterranean Inset (?)*
(18) Cross-section	non-rectangular	*Non-physical/ non-temporal*
(19) Nature of continuation of Descending Passage	a smaller, horizontal passage leading to the Subterranean Chamber (Chamber of Ordeal/ Central Fire)	*Level of restricted (?) attainment*
(20) Nature of 'join' between passageways	virtually non-existent: neither floor, walls nor roof are contiguous, and the plane of the 'join' is not at right-angles to any feature in either passage	*(See* (21) *and* (22)*)*
(21) Possible significance of irregular join	a time of transition, not precisely-dated by the chronograph[24]	

(22) Significance suggested by corresponding feature of Trial Passages (see page 199)	a telescopic join to be adjusted in the light of actual events.

Here the indefinite nature of the Well-Shaft's lower opening, and the apparent link with the Recess in the Subterranean Chamber (see pages 141–2), suggest that the lower end of the Well-Shaft has no chronological dating. Instead it indicates a symbolic level (the Plane of Death) below which it cannot be entered—a level which can however be regained in the western part of the Great Subterranean Chamber (q.v.).

READING: *After the splitting away in 1453* B.C. *of the souls seeking a higher path, the remainder of evolving humanity* (3), *reincarnating* (5) *in the physical world* (4), *will continue to lose more and more of their spirituality* (2) *as time goes on* (1). *Death will continue to be the result of the imprisonment of spirit in mortal flesh* (6) (7) (8).

In A.D. *1223* (10) *these souls will begin to enter a death-laden path* (9) *devoid of all enlightenment* (12)—*a path which will eventually lead to hell on earth* (19). *The decisive events along this path will occur from around* A.D. *1506 onwards* (14) (15) (21) (22).

None the less, a path of self-redemption will be available during this period (16) (17) *to any souls who manage to regain sufficient enlightenment* (12) *to rise above the deathly level to which they have sunk* (16). *Non-physical in nature* (18), *this path will eventually have the effect of transforming them into 'living beings' pursuing, even in the midst of hell, one of the upper paths to immortality* (16).

The date A.D. 1223 seems to refer to the same period as the downward step in the body of the Queen's Chamber Passage (q.v.)—the period which was to see an attempted return to the original Messianic teachings at the time of Francis of Assisi. As we have seen, however, this age also saw the beginning of what might be described as man's coming of age, and in particular the dawning of mental processes which were to lead eventually to our own scientific era.

In other words, the developments represented by the Subterranean Inset may be seen as symbolic of a hell made even hotter by man's scientific progress—and yet at the same time they represent a redemptive possibility, since it is only through a massive increase in scientific knowledge that—it seems—man can hope to under-

[24] No point on the Subterranean Passage floor is vertically aligned with any point on the Descending Passage floor, with the result that no dating on the latter can be used to establish the dating of the former. Compare the similarly 'anomalous' datings apparently applicable to the 'parting of the ways' of the Queen's Chamber Passage and the Grand Gallery (p. 58).

stand and accept intellectually the deep concepts which seem to underlie the Messianic 'rescue-plan'. In the words attributed to Jesus of Nazareth at John 16:12, 'There is still much that I could say to you, but the burden would be too great for you now.'

The Well-Shaft (The Well of Life)

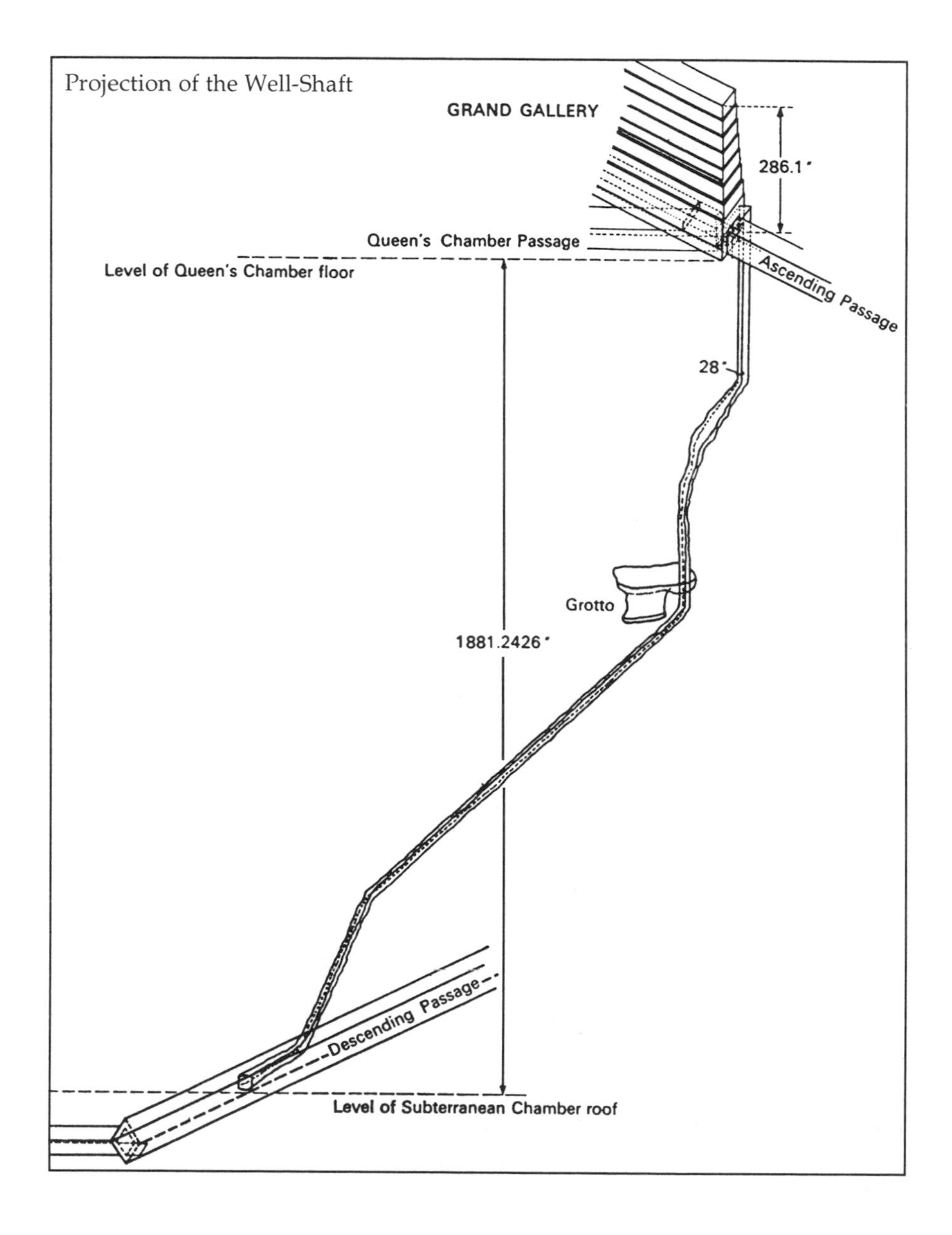

(1) Level of roughly-shaped lower entrance	just above the Plane of Death or level of Subterranean Chamber roof	*Escape towards spirituality possible only* above *Plane of Death*
(2) Level of top	Plane of Life (notional base of 25th or 5^2 course of masonry)	*Well-Shaft leads up to plane of potential enlightenment/Messianic ideal or initiation*
(3) Distance from Plane of Death up to Plane of Life	1881·2426″	*Distance from death to Life; achievement of potential enlightenment*
(4) Direction of lower entrance-tunnel	westwards[25]	*Towards enlightenment*
(5) Direction of upper exit-tunnel	eastwards	*Towards physical rebirth*
(6) Number of intermediate sections	4	*A fourfold path*
(7) Direction of first (bottom) section	steeply upwards and northwards	*Difficult progress towards rebirth*
(8) Direction of second section	less steeply upwards and northwards	*Easier progress towards rebirth*
(9) Cross-section of first two sections	irregular	*Non-physical*
(10) Direction of third section	steeply upwards and northwards	*Difficult progress towards rebirth*
(11) Features of third section	passes vertically through an irregularly-shaped natural 'grotto' before bending northwards; then extremely irregular in shape	*A non-physical/ discarnate period of rest (?) before rebirth*
(12) Approximate level of top of Grotto	level of Granite Plug = level of 7th course of exterior masonry = some 286″ above Pyramid's base-platform	*Spiritual perfection (?)* *Regaining of enlightenment*
(13) Direction of top section	vertically upwards	*'Explosive' spiritual progress*
(14) Cross-section of top section	square	*(Return to) physicality*
(15) Mean bore of top section [But note that N.-S. distance across upper exit	28″ (7 × 4″ or 286·1″/10) = 26·7021″ (2 × 13?)]	*Spiritual perfecting of the physical/Enlightenment through the Millennium (Productive of* (13?)*)*
(16) Length of top section	approximately 300″ (100 × 3)	*The reward of perfection (?)*

[25] The tunnel also tends slightly northwards and downwards.

(17) Nature of walls of top section	limestone blocks	*Physical/earthly*
(18) Position of axis of top section	35·76″ south of bottom section of North Wall of Grand Gallery	*Incarnation based on gaining of Messianic enlightenment*
(19) Height of upper exit-tunnel	1 RC	*Mortality/birth/death*
(20) Height of upward step on to first part of Queen's Chamber Passage floor	1 RC	*Birth/loss of mortality (?)*
(21) Plane of north wall of shaft	plane of top section of Grand Gallery's north wall and beginning of roof proper	*Beginning of the way of the enlightened*
(22) Nature of upper exit	a roughly-shaped hole, seemingly blasted out by force from below, such as to 'raise the roof' of the Grand Gallery by 286·1″	*Ascent of the Well-Shaft involves powerful enlightenment*
(23) Overall directional trend of Well-Shaft	northwards	*Return to mortality/the physical*

READING: *Throughout the period of 'hell on earth' a path of escape will remain open* (1) *for any soul prepared to turn to the right* (4) *and lift itself above the level of blind mortality* (1).

This upward spiritual path (9) (11) *will be a difficult, fourfold one* (7) (10) (6). *None the less, once a start has been made, progress will become easier* (8) *and eventually a discarnate period of recuperation and consolidation* (11?) *will provide the basis for the enlightenment and spiritual perfection needed for the final part of the self-redemptive task* (12).

Followed to its conclusion, this upward path eventually will bring the souls who follow it from blind mortality up to the level of potential enlightenment (1) (2) (3). *Yet this is also a path of rebirth* (19) (20) (23). *For by dint of their achievement of spiritual perfection in the physical world* (15) (16), *the souls in question will return to the earth-planes* (5) (14) (17) (19) (20) (23) *with all the power of the Messianic enlightenment* (18) (22) *and may even succeed in joining the original elect* (21) (22) *in experiencing the ultimate enlightenment of the eventual Millennium* (15).

The fourfold upward path which the Well-Shaft seems to represent interestingly recalls the Theravada Buddhist belief that the *vipassana*-meditator needs to experience the Path (the Zen *satori*) four times before total liberation is finally attained, with various of the physical 'fetters' being released on each occasion. Indeed, there seems to be an echo of this same notion in the Queen's Chamber's Niche (q.v.).

Meanwhile the precise route to be followed by the souls 'emerging' from the top of the Well-Shaft is not entirely clear. There seem, certainly, to be links (18) (21) (22) between the top end of the Well-Shaft and the 'expanded' north wall of the Grand Gallery (q.v.)—in which case the souls concerned may be seen as joining the original Nazarene elect (see page 68).[26]

On the other hand the floor-level reached appears to be only that of the Queen's Chamber Passage—which would suggest that we may here be dealing only with those souls who, while experiencing 'hell on earth', none the less manage to achieve a degree of spirituality based on the Messianic teachings—and whose state could thus, with care, become the basis for eventual entry into the Messianic Millennium (see page 84).

The fairest assessment would therefore seem to be that the Well-Shaft represents a *potential* opportunity to join the path of the enlightened ones—an opportunity which, even if not fully profited from, will still lead at very least to the path of the semi-enlightened. In either event, then, the Well-Shaft represents a Path to Life for those ascending it.

We have already seen, however, that the Well-Shaft just as clearly represents a pit of physical death for those descending it. Not that it is at all illogical that a passage should have a diametrically opposite meaning when traversed in a diametrically opposite direction. What *is* perhaps surprising is that a single feature should apparently exercise two different meanings simultaneously, and that a passage as narrow as the Well-Shaft should symbolically act as a kind of two-way highway. Indeed, the very unusualness of this dual symbolism would seem to suggest the possibility of some direct link between the physical death and spiritual life here referred to. *Perhaps, in short, the two apparently contradictory symbolisms are in some way historically interdependent.*

Now it is known that the lower entrance to the Well-Shaft was left sealed—possibly completely uncut—by the Pyramid's builders. That this was so is evidenced by the fact that the Pyramid's subterranean passages were well-known to men of classical times—men

[26] Indeed, the northward trend of the Well-Shaft could even be taken to signify a process of 'retrospective reincarnation' resulting in actual rebirth *at the time of Jesus of Nazareth* in order to undergo his particular 'spiritual baptism'. There would seem to be little *a priori* objection in logic to such a process, even though the possibility is seldom advanced by writers on the subject. However, the upper Well-Shaft exit comes *after* the end of the Ascending Passage floor at A.D. 58, and it therefore seems more likely that the link is intended to be a purely symbolic one. Point No. (22) itself suggests, like No. (1), a lack of chronological significance.

who were at the same time totally ignorant of the existence of the upper passageways. There are, for example, Roman smoke-inscriptions in the Subterranean Chamber; yet nobody was to set foot in the upper passageways until the year A.D. 820 (see page 18). The Descending Passage's west wall, in other words, gave no clue in those days to the existence of the lower Well-Shaft which lay directly behind it, and this fact would suggest that the designer intended that the Well-Shaft should be seen as having been sunk *from above*. Indeed, it is clear from the diagrams on pages 90 and 159 that the bottom, inclined section of the Shaft descends too far, with the result that the connecting tunnel to the Descending Passage has to ascend again slightly; looked at in the reverse direction, in other words, the lower entrance-tunnel to the upward shaft is, paradoxically, a *descending* one—which is hardly what one would expect if the shaft were meant to be seen as having been tunnelled upwards from below. It is certain, meanwhile, that the designer did intend the shaft to be seen as having been sunk, not 'constructed'—for its upper section was in fact cut through the Pyramid's already-laid lower courses of masonry.

If, then, we are right in assuming that the Pyramid's designer intended us to see the Well-Shaft as having been sunk from above, then it must be admitted that the symbolism of that process is extraordinarily apt in terms of our general read-out. For it reveals that the Well-Shaft's availability as an upward Path of Life is totally dependent upon its having *first* served as a downward pit of physical death.

If the enlightened ones of the first century A.D. had not symbolically 'sunk the well' by willingly accepting physical death as the price of their enlightenment (so the crude symbolism seems to suggest), then the Well-Shaft would never have been opened for later generations to ascend it on their way towards the Light. Indeed, if this reading is a valid one, then there seems to be an interesting link here with conventional Christian doctrine on the subject of Messianic martyrdom; for later men may, in a sense, be said to have been potentially 'saved by their blood'.

As for the precise mechanics of this redemptive process, these are less easy to discern, but later sections of this chapter suggest that the hidden common denominator may well be found in the need for the martyred initiates to return *en masse* to the earth-planes at a later date in order to act as 'midwives' to the redemptive 'new age of the Spirit'. And, meanwhile, the apparent link between the level of the Well-Shaft's lower entrance and that of the roof of the

Subterranean Chamber may suggest a link between the latter chamber and the physical setting for that age.

It thus becomes clear that the Well-Shaft does not, after all, *simultaneously* symbolise the downward Path of Death and the upward Path to Life. Instead, it has to serve first as one and then as the other. It operates on a system of tidal flow, not of two-way traffic. First the well must be sunk; then the 'artesian water' of human souls can enter the shaft and shoot up into the Light.

None the less it will be remembered that the lower entrance-tunnel, which leads the re-vivified soul westwards at last into the redemptive Well-Shaft, initially takes it both backwards and downwards before it can start its eventual steep ascent towards the 'Regions of Light'. This arrangement corresponds closely, in symbol, to the experience of many a seeker for the Truth who has found his initial steps along the Path of Enlightenment to be difficult and disheartening ones, apparently taking him even deeper 'into the mire'—a veritable 'dark night of the soul', with no apparent light at the end of the tunnel. But persistence, it seems, tends to be rewarded; the apparent darkest hour comes just before the dawn. And it is via the Well-Shaft that the returning initiates of the Grand Gallery—the 'disciples' of John 21—will lower their mighty net 'to the right of the boat' and haul up into the morning sunlight the '153 fishes' who are the enlightened of the New Age.

The Grand Gallery (The Hall of Truth in Light)

(1) Direction	due southwards	*Through time*
(2) Slope	upwards	*Spiritual progress*
(3) Angle of ascent	26° 18′9·7″	*Human evolution*
(4) Mean roof-height	286·1″ above Ascending Passage roof-line	*Gaining of enlightenment*
(5) Width of floor between ramps	41·21″ (2 RC)	*Mortality/reincarnation*
(6) Width of top section and roof	41·21″ (2 RC)	*Mortality/reincarnation*
(7) Height of top section and portion between ramps if 'fitted together'	52·7″ (4 × 13?)	*The terrestrial* (13?)

(8) But (5), (6) and (7) are the dimensions of the Ascending Passage

The lower end of the Grand Gallery from above the entrance of the Queen's Chamber Passage. Note corbelling, groove for 'sliding floor', and upper Well-Shaft entrance.

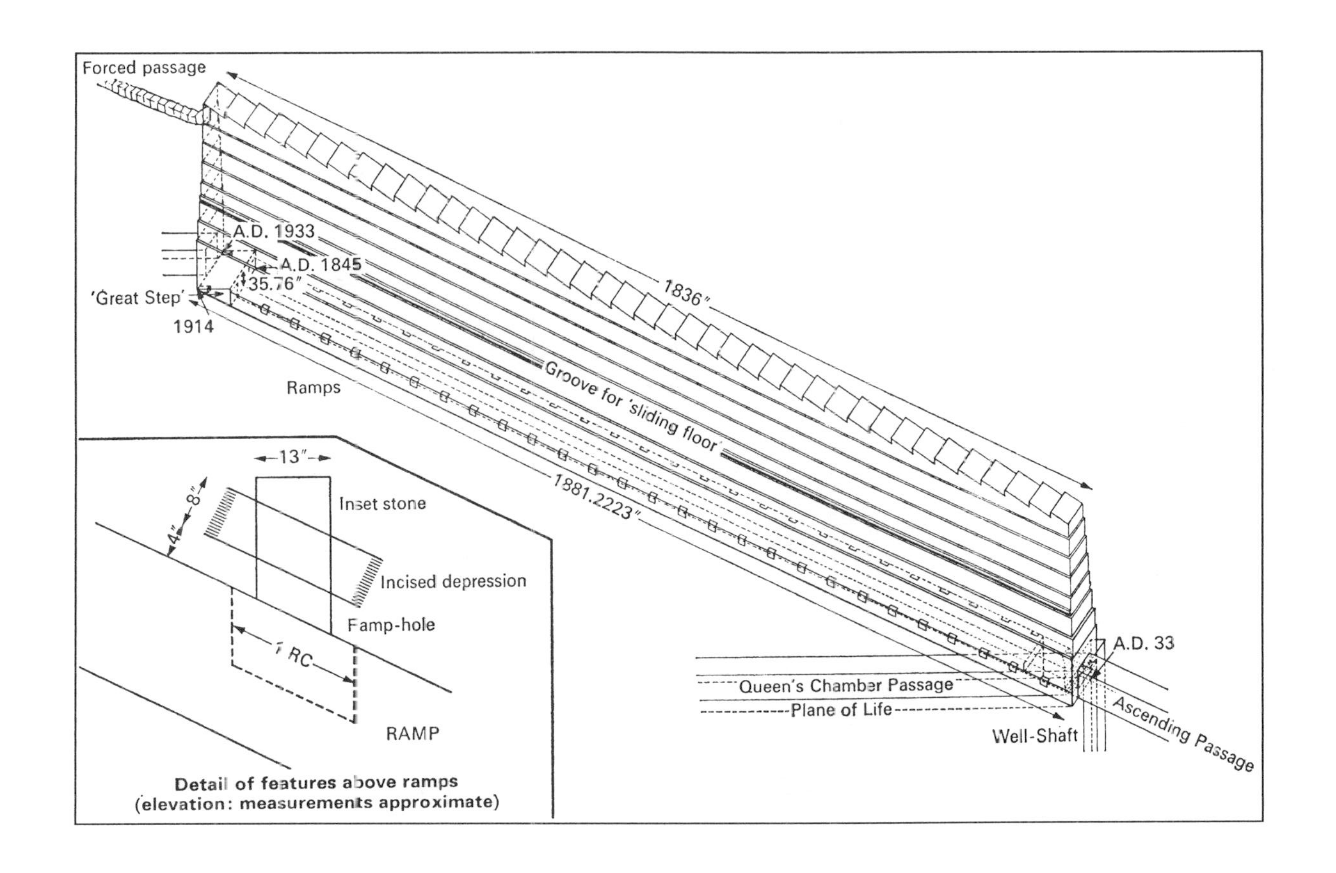

Projection of the Grand Gallery, incorporating roof detail.

(9) Thus, the significance of the Gallery is that of the Ascending Passage 'expanded telescopically'[27]

(10) Cross-section of Gallery	a 'telescopically-expanded rectangle'	*'Expanded' physical consciousness? / a 'supra-physical' path*[28]
(11) Architectural form of 'telescopic expansion'	corbelled vault	
(12) Number of 'expansion-sections' between ramps and top section	7	*Spiritual perfection/the spiritually perfect*
(13) Number of corbelled 'overlaps' in S., E. and W. walls	7	*Spiritual perfection/the spiritually perfect*
(14) Number of corbelled sections above base-walls	7	*Spiritual perfection . . .*
(15) Height of base-walls	89·8″	
(16) Total height of 7 'expansion-sections' between ramps and top section	286·1″	*Gaining of enlightenment*
(17) Average width of overlaps	3″	*Perfection*
(18) Number of 'expansion-sections' in *north* wall	6	*Preparation*
(19) Mean height of ramps	21″ perpendicular to slope (code-equivalent: 3 × 7)	*Utter spiritual perfection*
(20) Width of each ramp	20·61″ (1 RC)	*See pp. 101, 103*
(21) Features of ramps	each is pierced by a succession of rectangular holes cut vertically down against the base-walls	
(22) Dimensions of each hole	6″ wide by 20·61″ long (1 RC)—alternately horizontally and on slope: 10″ deep	*Preparatory rebirths of ascending mortals on way to Millennium (?)*
(23) Wall-feature above each hole, except the two in the Great Step (q.v.)	an upright, inset 'milestone' crossed by a rectangular depression parallel to slope of Gallery (see diagram p. 97)	*See note 29*
(24) Depth in wall and width of each 'milestone'	13″ × 10″	*The (souls?) of the Millennium*
(25) Approximate dimensions	8″ wide by 25″ long	*The rebirth of the Great*

[27] In other words, the Gallery, if 'collapsed', would simply form a continuation of the Ascending Passage which leads into it.

of each incised depression	(ends left deliberately indeterminate)	*Initiate/Messianic ideal*
(26) Height of top and bottom of each depression above ramp	12" and 4" approximately[29]	*Physical mankind*
(27) Number of ramp-holes (designed) including Great Steps	56 (8 × 7), or 28 (7 × 4) on each side of Gallery	*Rebirth of spiritual perfection/Spiritual perfecting of the physical/*
(28) Number of ramp-holes (actual)	55 (11 × 5), through removal of bottom one on west ('good') side by irruption of Well-Shaft entrance tunnel	*Achievement of the initiate(s)*
(29) Feature half-way up E. and W. walls	a slot running the length of both sides of the Gallery, apparently for a 'sliding-floor'	*No need to 'redescend' during course of Gallery; suspension of reincarnation*
(30) Width of slots (top to bottom)	6"	*Preparation*
(31) Nature of roof	40 (8 × 5) clearly-defined limestone slabs, set ratchet-wise (see diagram p. 100)	*(Suspended) rebirth of the initiates/Great Initiate until top end of Gallery*
(32) Length of roof	1836" (153 × 12)	*The enlightened of mankind*
(33) Notional length of floor, N. wall to S. wall	1881·2223" (code-equivalent: 99 × 19)	*Path from enlightenment to escape/Culmination of mortality (end of death)*
(34) Feature at top (S.) end of Gallery	the Great Step	*A 'great step' in human evolution; scale-change*
(35) Height of riser of Great Step	35·76"	*Enlightened incarnation . . .*
(36) Height of riser above ramps	approx. 12·3"	*. . . as man (?)*
(37) Distance from base of Great Step riser to notional top (S.) end of Gallery floor	68·744"	

[28] Comparison with point (20) and page 103 shows that the Gallery in fact loses 2 Royal Cubits of width as it ascends—which suggests that it actually involves 'loss of mortality' in some way.

[29] (21) to (26) suggest the ramp-holes symbolise the destruction of physical death during the Millennium. The 'millennial souls' those 'fit for the kingdom', symbolised by the 'milestones') are 'borne upwards', without descending repeatedly into mortality (the ramp-holes), by the evolutionary dynamic of the Great Initiate on his way to physical rebirth (the incised depressions) sometime during that Final Age. The 'depressions' may be seen as a kind of 'railway-track' carrying the 'milestones' across the ramp-holes to the top end of the Gallery. Compare page 103.

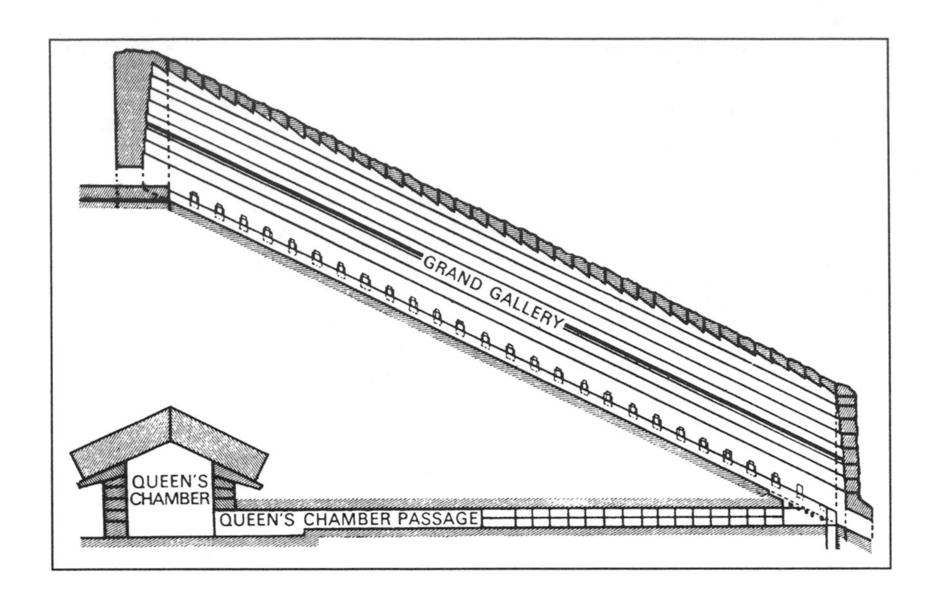

Grand Gallery and Queen's Chamber Passage

(38) Justification for assuming continuation of Gallery floorline to S. wall beneath Great Step	two rectangular holes, 1 RC long, cut in Step to meet floorline (see p. 101)[30]	
(39) Thus, chronological significance of Great Step	scale-change, *superimposed upon Grand Gallery floorscale*, preparatory to King's Chamber Complex	
(40) Position of Great Step riser	exactly on Pyramid's E.-W. axis, and thus directly above ridge of Queen's Chamber gable	*A turning-point (?)*
(41) Dating of Grand Gallery floor beneath N. wall	1st April A.D. 33	
(42) Dating of lower floor break-off, north edge of Well-Shaft, and beginning of Grand Gallery roof proper (see diagram above)	1st April A.D. 58 (see pp. 58, 65)	
(43) Notional dating for floor recommencement	1st April A.D. 152	

[30] These holes, though 1 RC long like the ramp-holes and of similar width, have no inset-stones in the wall above them. If, then, the inset-stones represent a series of 'skipped incarnations' (see note 29 page 99), the holes in the Great Step could well refer to an incarnation that *cannot* be skipped, i.e. a return to mortality (compare points (29), (50) and points (20) and (21) on page 110).

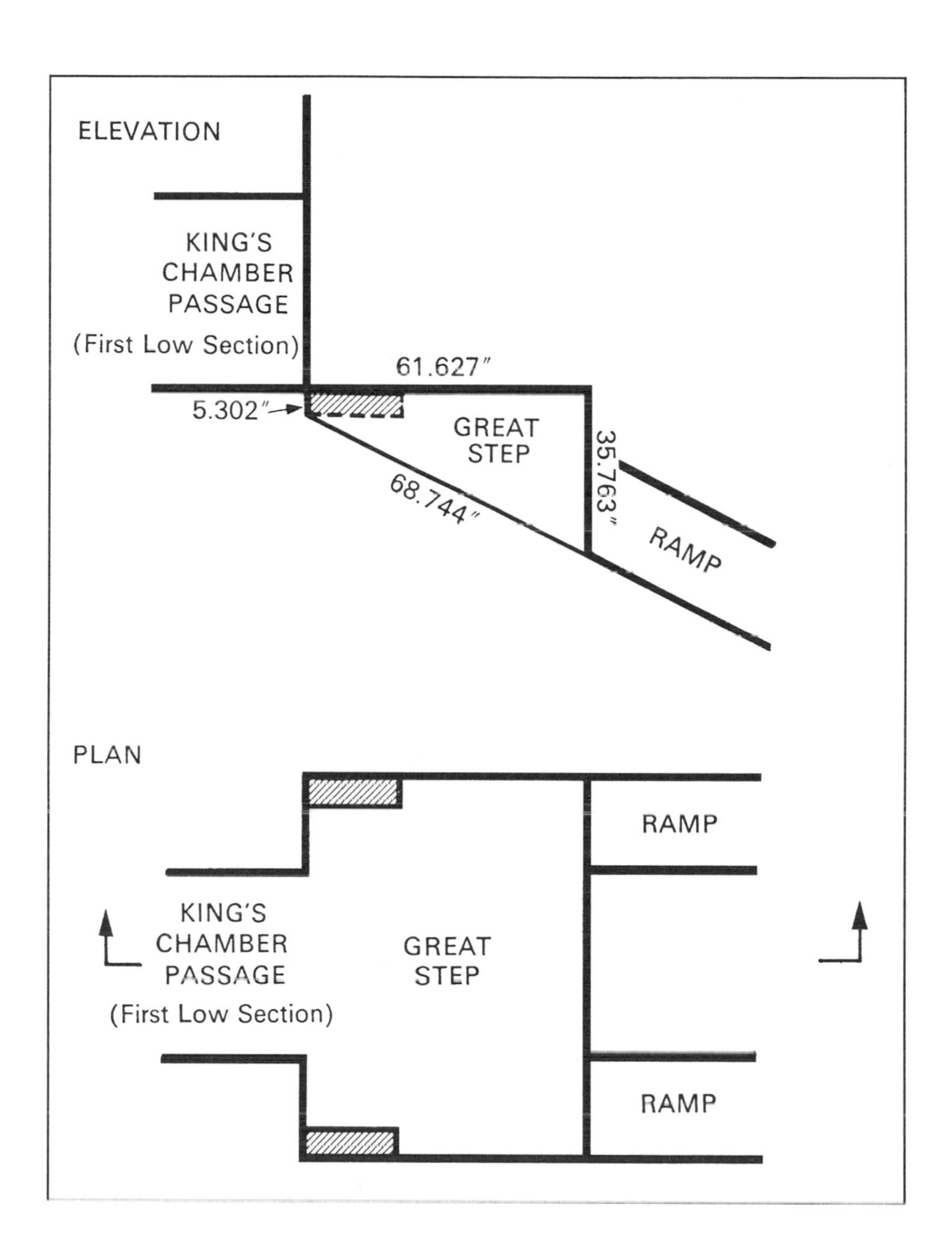

The Great Step (Elevation and Plan)

(44) Notional dating for top (S.) end of Gallery floor beneath Great Step at entrance to King's Chamber Passage	1881·2223 years after 1st April A.D. 33 = 22nd June 1914 (summer solstice)	
(45) Corroborative nomenclature for entrance to King's Chamber Passage in Egyptian *Book of the Dead*	the Royal Arch of the Solstice	*Summit of year/age (?)*
(46) Meanwhile, dating for Great Step riser	22nd June 1914 minus 68·744 years = 23rd September 1845 (autumn equinox)	
(47) Greatest width of Grand Gallery	82·42637" (4 RC)	*See* (48)
(48) Significance of (47)	apparently the 'reincarnating-room' of two souls simultaneously (2 RC × 2)	
(49) Width of top and bottom sections	41·2" (2 RC)	*Reincarnating-room of* one *soul*
(50) Thus, further symbolic significance of Grand Gallery	possibly a 'giant soul' taking another soul 'on its back' between A.D. 33 and 58, bearing it through time on a higher level (see (29)) and setting it down again some time after A.D. 1845 (see p. 103)	
(51) Further feature of Gallery	A small passageway cut from the top (S.) end of the top section (east side) through to the bottom Construction Chamber (see p. 129 and diagrams pp. 97, 107 and 125)	*A direct path to the spirit-planes for the fully enlightened*[31]
(52) Nature of passageway	a rough, irregular, but basically rectangular limestone passage	*Non-temporal path via the physical*

The apparent symbolism of the forced passage between the Grand Gallery and the lowest of the Construction Chambers seems to fit in well with our overall exegesis thus far. From the topmost (i.e. most spiritual) level of the Grand Gallery, it opens to the east—and thus in the direction of rebirth—but instead of the usual *physical* rebirth it leads directly to the lowest of the spirit-planes. This is, perhaps,

[31] Perhaps specifically for those who have come up the similarly rough Well-Shaft, or Well of Life.

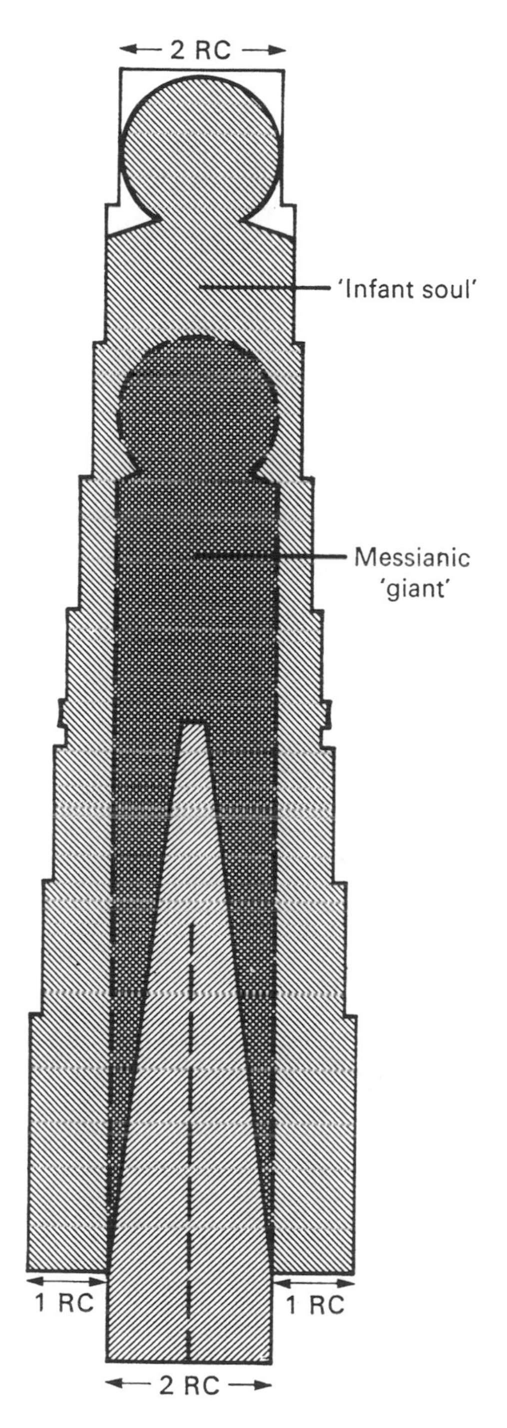

Cross-section of Grand Gallery with St. Christopher-like Messianic symbolism superimposed. The ancient story of the friendly giant who carried the infant Jesus across the flood could itself represent an adaptation of the notion of the infinitely tall heavenly Messiah bearing the soul (traditionally represented in early Christian art in the form of an infant) across the waters of rebirth and mortality. (See features 47–50.) This in turn could help to explain why, in the Greek Church, St. Christopher was formerly depicted with a dog's head reminiscent of Anubis—for Anubis likewise, in Egyptian legend, was son of the divine Osiris and *guardian of the dead*. This symbolism is also paralleled in the much later tomb of Tutankhamun, where the god Menkheret was shown bearing the shrouded soul of the young king through the 'limbo' of the spirit-world towards its coming rebirth (*Tutankhamen*, by C. Desroches Noblecourt, p. 248).

no more than one would logically expect of what appears to be the path of the foremost spiritual initiates.

Doubts have been expressed in the past as to whether the forced passage was part of the original design. Its aptness to the Pyramid's general message suggests that it probably was, and corroboration for this view is supplied by the fact that the lowest of the Construction Chambers is the *only* one containing no quarry hieroglyphics—which suggests that, unlike the upper chambers, it was cleaned for the benefit of intended future visitors. And this in turn would argue the existence of some means of access for them.

READING: *The souls of those who learn fully to accept the Messianic enlightenment between* A.D. *33 and* A.D. *58* (4) (41) (42) *will enter an upward evolutionary path* (2) (3) *reserved for the spiritually perfect* (12) (13) (14) (19) *enlightened ones of mankind* (16) (32).

First, however, they will have to turn to the right and accept physical death (42) *as the price of their initiateship* (28). *But, as a result of this, they will succeed in partially breaking out of the reincarnation cycle in the physical world* (9) (10) (31) (32). *Indeed, through the power of the Messianic initiative* (25) (47) (48) (49) (50), *the truly enlightened ones will not again taste death and mortality* (22) (29) (31) (note 28) *until they rejoin millennial mankind* (24) (36) *in incarnations of appropriate enlightenment* (35) *from the year* A.D. *1845 onwards* (46).

In fact there are a few (51) *who will attain such a high degree of spiritual perfection* (12) (13) (14) *that they will enter a direct, non-temporal route from the physical to the spirit-planes without the need for any further physical rebirths* (51) (52).

For the rest of the enlightened, the upward path will lead directly towards the eventual culmination of mortality (33)*—i.e. the perfection of mortal man, the consummation of the evolutionary process—through the perfection of the coming Millennium* (24) (39) *and the return of the initiates* (25) (31). *The great step leading to these events will have its beginnings in the year* A.D. *1845* (39) (46), *and will be complete by the summer of 1914* (44).

The message of the Grand Gallery seems to bear several clear links with concepts said to have been propounded by Jesus of Nazareth. First of all there is the statement at Matthew 16:21 and 24–8: 'From that time Jesus began to make it clear to his disciples that he had to go to Jerusalem . . . to be put to death and to be raised again on the third day'[32]. . . Jesus then said to his disciples, "If anyone wishes to be a follower of mine, he must leave self behind; he must take up his cross *and come with me*. Whoever cares for his own safety is lost; but if a man will let himself be lost for my sake, he will find his true self. What will a man gain by winning the whole world, at

[32] For the probable significance of the expression 'on the third day' see chapter 7.

The now-dilapidated Great Step at the upper end of the Grand Gallery, showing the entrance of the King's Chamber Passage. Note ramp-holes, corbelling and groove for 'sliding floor'.

the cost of his true self? Or what can he give that will buy that self back? For the Son of Man is to come in the glory of his Father with his angels and then he will give each man the due reward for what he has done. I tell you this: there are some of those standing here who will not [again[33]] taste death before they have seen the Son of Man coming in his kingdom." '

This passage suggests that Jesus was already well aware of his own impending death—and of the fact that, in fulfilment of some pre-existing 'plan', he himself had to bring that death about. Moreover, he appears to have foreseen that violent death would likewise overtake all his faithful followers—*indeed, that this too was something that needed to be consciously sought*. And the apparently contradictory concluding sentence of the passage makes perfect sense if the word 'death' is understood in the sense of mortality and thus of rebirth.

Meanwhile, all three ideas reflect faithfully the Pyramid's apparent message as already outlined above. And, in particular, the notion of a long discarnate period prior to the coming Millennium, for the souls of the fully-enlightened, finds a pyramidal echo not only in the symbolism of the Grand Gallery, but also in that of the Well-Shaft (see page 92).

Finally, in the same chapter of Matthew's gospel, Jesus is reported as claiming that 'this gospel of the kingdom will be proclaimed throughout the earth as a testimony to all nations; and *then* the end (of the present age) will come' (24:14). Similarly it is interesting to note that the Grand Gallery's roof-measurement of 1836" (153 × 12) could be interpreted as signifying not only the enlightened ones of mankind but also the enlightening of mankind—a notion almost identical to that propounded by Jesus.

These clear parallels between the Pyramid's apparent message and Jesus' own foreknowledge of some kind of world-plan which had to be fulfilled, once again raise the question of whether the Pyramid itself might have been the source of that foreknowledge. Could this, indeed, be the true significance of Jesus' statement reported at Matt. 24:35, 'Heaven and earth [i.e. the present world-order] will pass away: *my words will never pass away*'? Were Jesus' 'words', in fact, already enshrined, even before he uttered them, in the enduring stone of the Great Pyramid of Giza?

[33] My interpolation—P.L.

The King's Chamber Passage (The Passage of the Veil and Chamber of the Triple Veil)

(1) Level of Passage's floorline	top of 50th (10 × 5) course of masonry, 153 courses below summit-platform	*The Messianic or initiatory Millennium The enlightened*
(2) Thickness of 50th course	28″ (7 × 4)	*The spiritual perfecting of the physical*
(3) Initial feature of passage	upward Great Step (see p. 90)	*'Great Step' in human evolution*
(4) Dating for Great Step riser	23rd September A.D. 1845	*(Dating)*
(5) Height of Great Step riser	35·76277″ (286·1″/8)	*Rebirth of the enlightened*

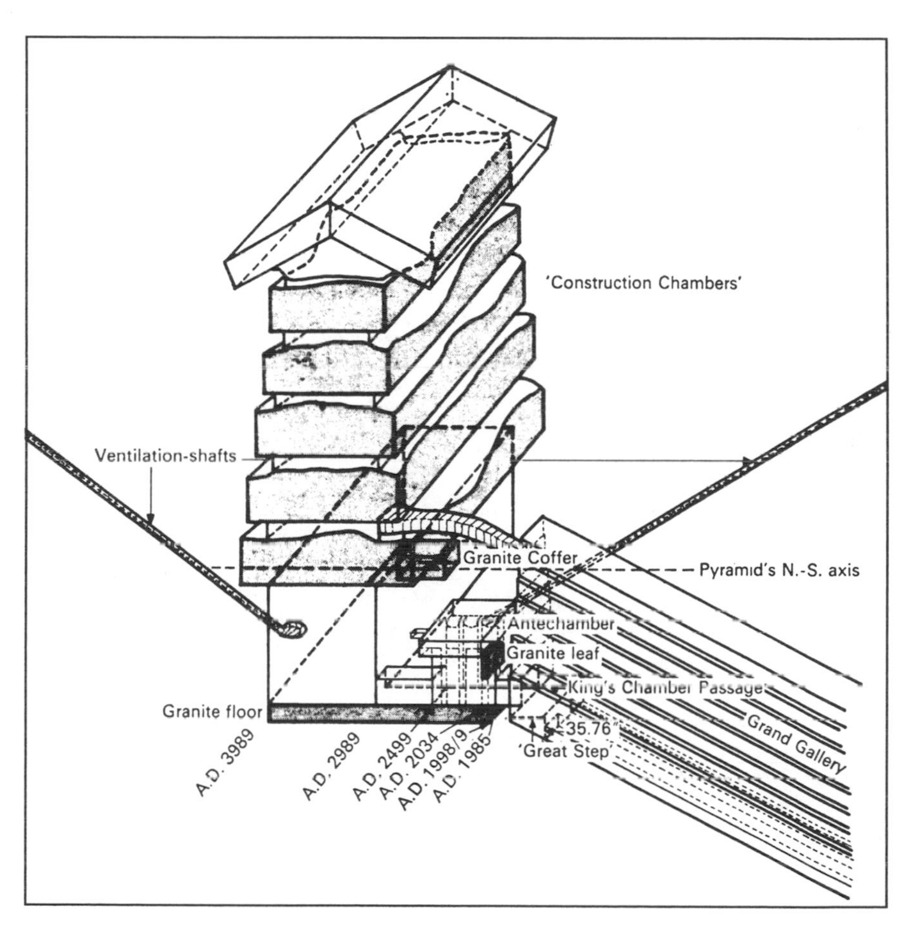

Projection of the King's Chamber Complex

(6) Chronological significance of Great Step	'upward' scale-change at 23rd September 1845, giving *more* than one year per inch, and based on the quantities 35·76″ and 25″ (the Sacred Cubit)	
(7) Thus, on new scale, time for each inch	35·76277/25 years = 1·4305 years	
(8) N.-S. distance across top of Great Step	61·6266″	
(9) Thus chronological time across Great Step	61·6266 × 1·4305 years = 88·157516 years	
(10) Therefore, dating for entrance of King's Chamber Passage	20th November A.D. 1933[34] (superimposed upon Grand Gallery floor-dating of 22nd June 1914)	
(11) Feature at entrance	Notional 'hidden step' (see pp. 101 and 109) linking Gallery floorline to floorline of King's Chamber Passage	*(Further scale-change)*
(12) Feature at entrance	Grand Gallery south wall, lowering roof-height to 41·21″ (2 RC)	*Return of the enlightened to mortality*
(13) Height of 'hidden step'	5·3015″ (code-equivalent: 5)[35]	*Messianic leader?*[36]
(14) Function of 'hidden step'	cancellation of previous scale-change (see code)[37]	

[34] This date comes some 19 years after the corresponding Grand Gallery floor-dating of 22nd June 1914, and 1900 (100 × 19) years after A.D. 33.

[35] Compare the height of the small *downward* step at the beginning of the Queen's Chamber Passage.

[36] This apparent isolated reference to a 'Messianic leader' seems oddly out of place at this juncture. However, projection of the Grand Gallery floorline past this point to where it actually cuts the King's Chamber Passage floor gives an 'inverted' 'Messianic triangle' commencing (on the King's Chamber Passage floor-scale) on 20th November 1933 and terminating on 10th August 1944. If this be taken to refer to some kind of anti-Messiah, then the temptation is considerable to associate the triangle with Adolf Hitler, whose whole *Weltanschauung* bristled with typical Jewish Messianic imagery—complete with chosen people, the return to the homeland, racial purity, the notion of a Messianic saviour and the dedication to a new race of man and a thousand-year kingdom. Indeed, it could well have been a subconscious jealousy of Jewish Messianism that lay at the basis of Hitler's pathological anti-Semitism. None the less it must be pointed out that the dates indicated are somewhat inaccurate by Pyramidal standards, both being some ten months out. The nearest significant date to 20th November 1933 seems to be Germany's final withdrawal from the ill-fated League of Nations in October of that year.

[37] That the 'hidden step' does not produce its *own* floor-scale is evident from the following facts:

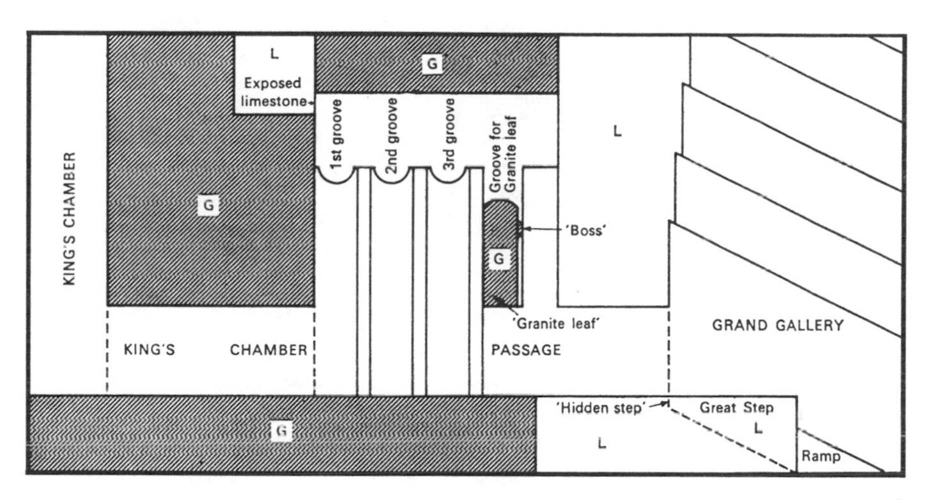

The King's Chamber Passage (G = granite: L = limestone)

(15) Therefore, scale from 'hidden step'	as previous scale = 1″ per year

First Low Section

(16) Direction	southwards	*Through time*
(17) Slope	horizontal	*Level of attainment*
(18) Cross-section	square	*Physical/terrestrial*
(19) Nature of masonry	limestone	*Earthly*

[37] *(continued)*

(a) it is an upward step whose height has already been taken into account in calculating the upward Great Step's time-scale, and (b) the King's Chamber Passage floorline—the one currently under consideration—does not itself rise or fall at this point. Thus, the 'hidden step' can have only one step-function left to it—the cancellation of the previous floor-scale at the point indicated by the designer via the rectangular holes in the Great Step (i.e. at the entrance of the King's Chamber Passage).

As for the designer's reasons for incorporating the Great Step's brief, 88-year 'speeded-up section' in the first place, these are presumably to be sought in the need to 'gain' some 19 years of chronological time without encroaching either on the symbolic length of the Grand Gallery (see points (32) and (33), p. 99) or on that of the King's Chamber Passage proper (see point (97), p. 118). Indeed, the very design of this extraordinary yard-high obstacle suggests not so much a step as *an extrusion of the King's Chamber Passage into the Grand Gallery*, such as would accord entirely with the theory just advanced (compare the two similarly 'telescopic' joins in the Subterranean Inset). Whether the figures 19 or 88 (11 × 8, or 8 × 11?) are here deliberately significant is open to question.

(20) Width of Passage	41·21″ (2 RC)	*Death/reincarnation*
(21) Height of Passage	41·21″ (2 RC)	*Death/reincarnation*
(22) Length of first low section	52·02874″ (4 × 13″?)	*(Time-measurement)*
(23) Therefore, dating for beginning of Antechamber	52·02874 years after 20th November 1933 = 30th November 1985	*(Dating)*

Antechamber

(24) First feature	roof-height increases by 108·23382″ (code-equivalent: 9 × 12)	*Perfecting of mankind*
(25) Nature of roof	granite	*Spiritual*
(26) Number of slabs in roof	3	*Perfect*
(27) Direction through Chamber	southwards	*Through time*
(28) Slope	horizontal	*Level of attainment*
(29) Cross-section	basically a rectangular passage with its 'lid taken off' and an irregularly-shaped upper portion added, with certain features extending downwards as far as 3″ below the floor	*Incursion 'from above' into the physical world, bringing perfection to its very basis*
(30) Height of east wainscot	103·03296″ (5 RC)	*Initiation / the Great Initiate?*
(31) Height of west wainscot	111·8034″	
(32) Width of 'shelf' above each wainscot	12·0214″ (code-equivalent: 12)	*Influences 'from above' descending among mankind(?)*
(33) Thus, width of ceiling	65·25603″	
(34) Width of passageway through lower part of Chamber	41·21″ (2 RC)	*Mortality/reincarnation*
(35) Height of Chamber	149·44701″ = [Pyramid's vertical ht + base-side]/100	
(36) Length of Chamber	116·26025″ = 365·242″/π	*An age influenced by the eternal*
(37) First floor-feature	beginning of granite floor	*New spiritual basis*
(38) Distance from beginning of Chamber to start of granite floor	13·22729″	*(Time-measurement)*

(39) Thus dating of start of granite floor	13·22729 years after 30th November A.D. 1985 = 21st February A.D. 1999[38]	*(Dating)*
(40) Chronological significance of granite floor	commencement of new time-scale at 1*n* per year (see code)	
(41) Thus on new scale, each inch	= 100/20·60659 years = 4·8528 years[39]	
(42) Next major feature	the 'Granite Leaf'	*Spiritual . . .*
(43) Nature of Granite Leaf	2 blocks (upper and lower) lowered and cemented into rectangular grooves in the sidewalls, the upper block bearing a 'boss' or 'seal' on its north face and having an irregularly-shaped top as though 'broken off from above'. The boss resembles a bas-relief of the sunrise or of the rainbow, and has exactly the same thickness as the Leaf's two rebates on its north side (see pp. 109 and 116)	*. . . double (?) . . . . . . incursion into the physical world . . . . . . 'sent down from above' . . . . . . the sunrise of a new age (?)*
(44) Dimensions of surface of boss	3″ high × 5″ wide	*The perfect One-who-is to-come (see* (51)*)*
(45) Approximate dimensions of boss's base	8″ wide × 5″ high	*Rebirth of the Great Initiate (see* (51)*)*
(46) Thickness of boss and *side-rebates*	1″ exactly[40]	*The divine*

[38] Thus, the granite floor begins 153·413354 years along the limestone King's Chamber Passage floor from the beginning of the Great Step. This clear reference to the presence in the earth-planes of the enlightened (153) reflects the fact that the same floor is 153 courses of masonry below the Pyramid's summit-platform (see point (1)), and that the two low sections of the passage are together 153·07503″ long. Meanwhile it confirms, gratifyingly, the correctness of our earlier calculation of the Great Step time-scale, as well as our reading for the top end of the Grand Gallery on pp. 104–6. Note also how the beginning of the granite floor represents a 'delayed reflection' of the beginning of the granite roof—a specific instance of the Hermetic law of spiritual antecedents summed up by the words 'As Above, so Below'.

[39] Owing to the relative imprecision of the data presently available for the remaining floor-measurements of the Antechamber, to the fineness of the time-scale, and to cumulative calculation error, all datings given from this point on should be regarded as correct *only to the nearest month.*

[40] In other words, the Granite Leaf itself was originally 1″ thicker, but was planed down uniformly to leave only the boss and the side-rebates at the original thickness.

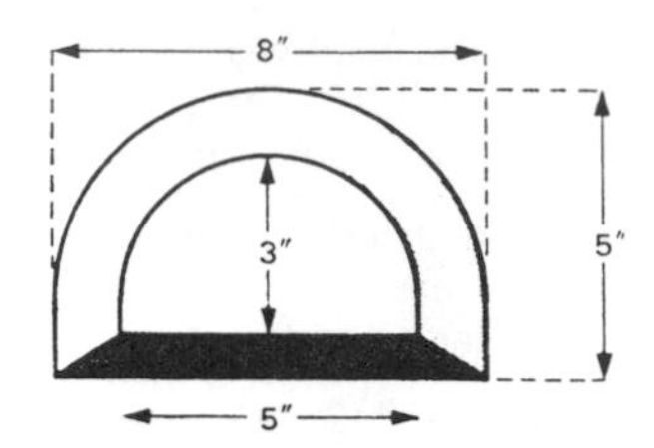

The 'Boss' or 'Seal' (measurements approximate)

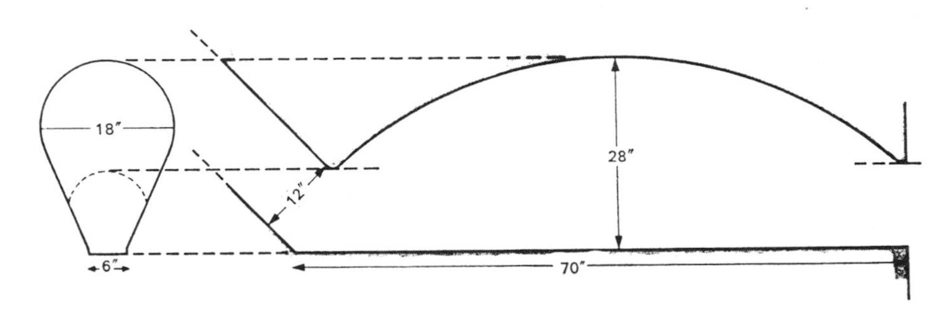

King's Chamber—detail of south ventilator-opening (measurements approximate)

(47) Position of centre of boss	1″ to right (west) of centre of Granite Leaf	*Leading towards the divine*
(48) Position of centre of Leaf	1 RC from each side-wall	*Mortality*
(49) Position of centre of boss	1 Sacred Cubit from east end of Leaf (embedded in wall)	*Messianic ideal leading away from mortality*
(50) Level of bottom of boss at base	5″ above horizontal joint between upper and lower slabs	*The Great Initiate*
(51) Level of bottom of boss at surface	$33\frac{1}{2}$″ above bottom of Leaf	*The Messianic presence*
(52) N.-S. position of boss's surface	1 RC south of N. wall of Chamber = 1 RC minus 13·22729″ beyond beginning of granite floor (= 7·3793″)	
(53) Vertical distance from floor to bottom of Granite Leaf	41·21″ (2 RC)	*Mortality/a mortal*
(54) Notional dating for floor beneath beginning of boss	7·3793 × 4·8528 years after 21st February 1999 = 35·810388 years = (14th December) A.D. 2034	

(55) Thickness of boss and rebates	1″ exactly = 4·8258 *n*	*(Time-measurement)*
(56) Thus, notional dating for N. side of Granite Leaf proper	(21st October) 2039[41]	*(Dating)*
(57) Dressed-down thickness of Leaf	15·75″ (code equivalent: 3 × 5) = 76·431859 *n*	*The perfect One-who-is-to-come*
(58) Thus, notional dating for S. side of Granite Leaf	(28th March) A.D. 2116[42]	
(59) Next feature	a retaining granite pilaster on each side of passage	*New spiritual environment*
(60) E.-W. depth of pilasters	3·25″ (code-equivalent: 3)	*(Leading towards) perfection*
(61) Width of passage between pilasters	41·21″ (2 RC)	*Mortality/reincarnation*
(62) N.-S. thickness of pilasters	3·75″ (code-equivalent: 3)	*Perfection*
(63) Next feature	a rectangular groove in each granite wall of passage, from top of wainscot to 3″ below floor level, apparently to receive a 'portcullis' similar to the Granite Leaf, but this time descending below floor-level[43]	*Descent of a further 'Messianic figure' into mortality, resulting in the sealing off of the passage*
(64) Notional dating for N. side of 'portcullis-grooves'	3·75 × 4·8528 years after (28th March) A.D. 2116 = (8th June) A.D. 2134	

[41] This date is forty years after 1999. Rutherford points out that forty years is frequently treated as a 'time for judgement' in the Hebrew scriptures. In terms of the Pyramid's code. however, the reference appears to be to the rebirth of the Great Initiate (8 × 5).

[42] Available measurements for the remaining chronological features of the Antechamber are correct only to the nearest hundredth of an inch—in some cases only to the nearest tenth. A margin of error of ± 6 months ought therefore to be assumed until the south wall of the Chamber is reached.

[43] The fact that the three 'portcullises' are designed to descend to a point 3″ *below* the granite floor—i.e., that the granite floor succeeds in passing through the Antechamber *despite* the three 'portcullises' (see p. 114)—suggests that the passage is triply barred, in symbol, to all but those who base their lives on the 'things of the spirit'. The chamber's three 'veils', in short, may be seen as filters—corresponding, perhaps, to the three 'lenses' of the girdle-stones in the Ascending Passage. Only those capable of passing through each filter in turn are permitted entry to the King's Chamber. The symbolism is thus of an intense screening-process, as indeed was often hinted at by Jesus of Nazareth.

The south end of the Antechamber, showing the 'portcullis-grooves' (much dilapidated) extending below the granite floor. The King's Chamber opens to the right beyond. Note semicircular hollows at tops of right-hand grooves, and limestone section at top of south wall.

(65) Depth of grooves	3·25″ (code-equivalent: 3)	*Perfection*
(66) Features at groove-tops	grooves *on western side* end at top of granite wainscot in 8.75″-long semi-circular hollows terminating 12″ to west of passage-wall	*Rebirth of the spiritual in human form*
(67) Diameter of each semi-circular hollow	17·25″	
(68) Thus, circumference of each hollow	17·25″ × $\pi/2$ = 27·463″ (code-equivalent: 9 × 3)	*Utterly perfect spirit*
(69) N.-S. distance across grooves	21·5″ (code-equivalent: 3 × 7)	*Utter spiritual perfection*
(70) Thus, notional dating for south side of portcullis-grooves	21·5 × 4·8528 years after (8th June) A.D. 2134 = (9th October) A.D. 2238	
(71) Next feature	a second pair of granite pilasters similar to the first	*Continuing spiritual environment*
(72) But N.-S. width of pilasters	5·3″ (code-equivalent: 5)	*Progress of initiates*
(73) Next feature	a second pair of granite portcullis-grooves extending to 3″ below floor-level, complete with semi-circular hollow at top of W. groove	*Descent of a third Messianic figure into mortality, further sealing the passage; spiritual descent into human form*
(74) Notional dating for N. side of grooves	5·3 × 4·8528 years after (9th October) A.D. 2238 = (28th June) A.D. 2264	
(75) N.-S. distance across grooves	21·5″ (code-equivalent: 3 × 7)	*Utter spiritual perfection*
(76) Thus, notional dating for S. side of grooves	21·5 × 4·8528 years after (28th June) A.D. 2264 = (29th October) A.D. 2368	
(77) Next feature	a third pair of granite pilasters identical to the last pair	*Continuing spiritual environment*
(78) Thus, N.-S. width of pilasters	5·3″ (code-equivalent: 5)	*Messianic progress/ progress of the initiates*
(79) Next feature	a third pair of granite portcullis-grooves, identical to the last	*A fourth Messianic figure, as above*
(80) Notional dating for N. side of portcullis-grooves	5·3 × 4·8528 years after (29th October) A.D. 2368 = (18th July) A.D. 2394	

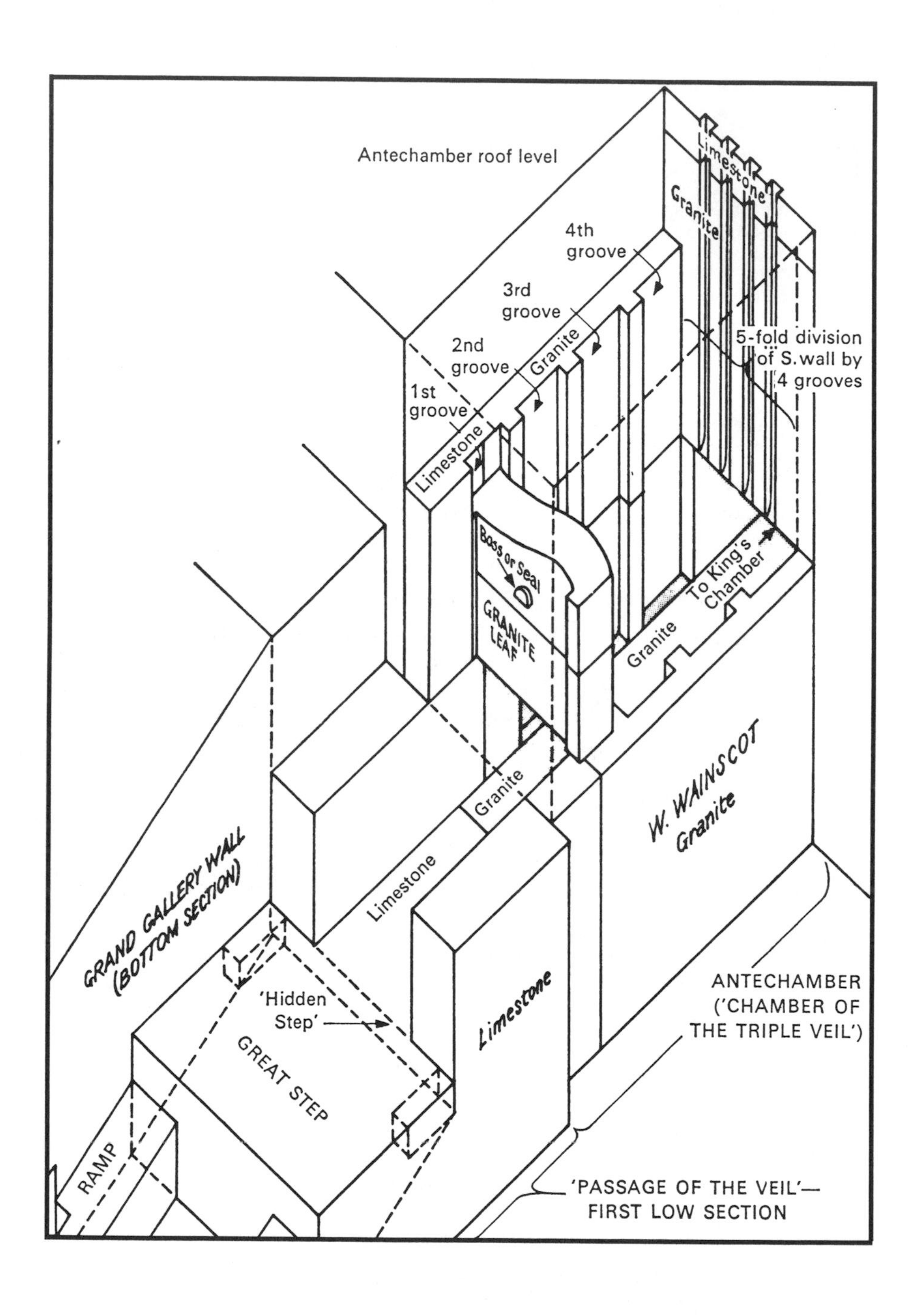

Exploded isometric projection of Passage of Veil and Antechamber

(81) And notional dating for S. side of grooves	21·5 × 4·8528 years after (18th July) A.D. 2394 = (18th November) A.D. 2498	
(82) Next feature	south wall of Antechamber	*End of preparatory Messianic era*
(83) Length of Antechamber's granite floor	5 RC (see diagrams p. 125)	*(Time-measurement)*
(84) Beginning of granite floor	21st February 1999 (see (39))	*(Dating)*
(85) Thus, *exact* dating for south wall of Antechamber	500 years exactly after 21st February 1999 = 21st February 2499[44]	
(86) Features of S. wall	4 vertical rounded grooves running from roof of chamber to roof of second low portion of passage, and dividing the S. wall equally into 5 vertical strips	*The spiritual way of the physical initiates*
(87) Width of grooves	4″	*Physical/terrestrial*
(88) Depth of grooves	2·8″ (7 × 4/10?), but tapering, scoop-like, to nothing through the bottom 8″, as though to 'skim off' the contents of the Passage	*Spiritual perfecting of the physical through the Millennium; 'raising of man' at rebirth*
(89) Length of grooves	108·23382″ (code-equivalent: 9 × 12)	*The utter perfection of man*
(90) Thus, length of full-depth portion of grooves	approximately 100″	*Reward*
(91) Further feature of S. wall	*top* 12″ *only* are of lime-stone, otherwise granite	*Physical man 'raised' by the spirit*
(92) Thus, height of granite portion	96″ approximately (8 × 12)	*(Re)birth of spiritual man*

Second Low Section

(93) Next feature	second low portion of Passage	
(94) Width, height, cross-section, slope, and masonry	as for first low section, but in granite	*Further rebirth/mortality but through a spiritual environment*

[44] Comparison with point (81) validates our earlier assumption of a tolerance of ± 6 months.

(95) Length of second low section	101·04629″ (code-equivalent: 100 + 1?)	*More than a reward (?)*
(96) Thus, dating for entry into King's Chamber	101·04629 × 4·8528 years after 21st February 2499 = 2nd July, A.D. 2989	
(97) Total length of both low sections combined	52·02874 + 101·04629″ = 153·07503″ (code-equivalent: 153)	*(For) the enlightened*
(98) Inference from points (53), (61) and (94)	King's Chamber Passage should be regarded as measuring 2 RC high by 2 RC wide throughout, with an 'irruption from above' during the period indicated by the Antechamber	*A path for mortals, influenced 'from above' during the age depicted*
(99) Distance from beginning of granite floor to S. end of coffer in King's Chamber[45]	= distance from centre of Antechamber to S. wall of King's Chamber = 365·242″ (see diagrams p. 125)	*An age leading to immortality*

READING: *The great step* (3) *leading to the foundation of the Messianic Age* (1) *will start to make its influence felt from the autumn of* A.D. *1845* (4) *and from this time the enlightened* (1) (97) *will start to be reborn once again in incarnations inspired by the Messianic enlightment* (3) (5) (39, note).

After the closing of the age of discarnate enlightenment in the summer of A.D. *1914* (10) (12) *a time of preparation for the new Messianic Age* (1) (36) *will start in earnest from late* A.D. *1933* (10). *From this time all those who have achieved enlightenment*[46] *must undergo a period* (16) *of continual reincarnation* (20) (21) (27) (94) (97) *in the physical world* (18) (19) (39, note).

In late A.D. *1985* (23) *spiritual or cosmic events will occur* (25), *calculated to bring perfection to the very basis of reincarnating human life* (24) (26) (29) (34) (60) (62) (63) (65) *through spiritual influences from above* (29) (32) (98) *and an irruption of the eternal into the temporal sphere* (36).

From early A.D. *1999* (39) *a Messianic era* (83) *reflecting the new spiritual initiative will start to force man to base his life on spiritual instead of physical foundations* (37). *Then, in late* A.D. *2034* (54), *the bow-like sign of the Great Initiate, the One-who-is-to-come, will appear in the cosmos* (44) (45) (46) (47) (49) (50) (51), *and by the autumn of* A.D. *2039* (56) *this messenger from eternity* (43), *full of Messianic perfection* (44) (57), *will have taken up his physical rôle in the earth-planes* (43) (48) (53).

In the spring of A.D. *2116 this Messianic figure will depart* (58). *But in the summer of* A.D. *2134* (64) *a new Messianic emissary of supreme spiritual perfection*

[45] This is assuming that the coffer is placed centrally in the Chamber along the Pyramid's north-south centreline, as was almost certainly the architect's intention.

[46] Indeed, the souls of *all mankind*—see p. 146.

(68) (69) *will arrive* (63), *his function being to re-lay the very foundations of human life and finally to shut off the enlightened from any possible relapse into blind mortality* (63).[47] *The human* (66) *or physical* (63) *phase of this cycle of his spiritual existence being completed* (66), *this figure too will depart in the autumn of* A.D. *2238* (70), *only to begin a new cycle of physical existence just 25 years later, in the summer of* A.D. *2264* (73) (74).

Departing again in the autumn of A.D. *2368* (76), *this Messianic leader will make a third appearance in the summer of* A.D. *2394* (80), *and with the end of his mission in early* A.D. *2499* (85) *the era of the preparatory Messianic initiative will draw to a close* (82).

From that date, with the passage leading back to blind mortality now triply blocked behind them,[48] *the reincarnating enlightened ones* (92) (94) (97) *will set out on the final spiritual path* (94) *of ultimate reward* (90) (95), *which will eventually result in the raising of the physical initiates to utter spiritual perfection* (86) (87) (89) (91) *through the catalysing experience of the earthly Millennium* (2) (88). *The age of the fourfold Messianic initiative will lead directly to that Millennium* (1) (99), *which will commence in the year* A.D. *2989* (96).

The Antechamber's overall measurements clearly mark it out as containing within it the key to all the Pyramid's dimensions. We have already established, for example, that its length and height (36) (35) are directly related to the Pyramid's own vital dimensions. Indeed, as the lower diagram on page 125 also shows, its east wall and granite floor mark out between them a square (ABCD) of side 5 Royal Cubits;[49] while the chamber's side-elevation also clearly invites the inscribing of a circle of circumference 365·242 P" to touch both its north and south walls. Moreover, square and circle, as now constructed, have exactly the same area—namely 5 square Royal Cubits. The clear code implication (q.v.) is that the Antechamber represents a Messianic Age for the bringing of the physical world to spiritual perfection—as our reading already suggests quite independently. But it also becomes apparent that one purpose at least of the Antechamber is to demonstrate the relationship between Primitive Inch and Royal Cubit.

The Granite Leaf itself takes this process a stage further, for its measurements clearly demonstrate both Primitive Inch, Royal Cubit *and Sacred Cubit* (46) (47) (48) (49). In short, the measurements of the Granite Leaf may be seen as the key to those of the whole Pyramid.

But in exactly the same way as the Antechamber and its Granite

[47] The notional portcullis, once 'lowered', will block the Passage completely.

[48] By the three 'lowered' Messianic portcullises.

[49] Hence the code assumption that on granite floors a Royal-Cubit-based timescale applies.

Leaf seem to supply the *geometric* key to the Great Pyramid, so also they clearly supply its *symbolic* key. The key to the perfecting of mortal man and planet earth, in other words, lies in the events symbolised by the Antechamber, and above all, perhaps, in the figure symbolised by the Granite Leaf itself. And the latter's dimensions, together with the sign inscribed upon it, speak unmistakably of a returning Messianic leader. In particular the boss itself speaks again and again of the Messianic presence (see points 44, 45, 49, 50 and 51) and seems to represent either the sunrise[50] or the rainbow, with their respective symbolisms of a new age or an end to death.[51] Or possibly we can see in it the Messianic aura or 'halo' about the head of a man some 6 ft. 3 in. in height. In short, the boss invites identification as none other than the biblical 'sign of the Messiah'—a notion which in turn necessarily equates the Granite Leaf with the Christian notion of the 'second coming' and the long-awaited advent of the One-who-is-to-come.[52]

But there is a piece of even clearer evidence to support this supposition. For perhaps the boss's most obvious resemblance is to the Egyptian hieroglyph 𓏏 (t)—a sign which originally denoted a loaf of bread. Taken in conjunction with the 'broken off' top of the Leaf and the boss's patently Messianic dimensions, the Antechamber's seal could thus be held to represent some kind of Messianic 'bread sent down from heaven' preparatory to the final Chamber of Resurrection. *But this is precisely the imagery reportedly used by Jesus of Nazareth to describe his own Messianic function.* 'I am that living bread which has come down from heaven,' he says in chapter 6 of John's gospel, '. . . if anyone eats this bread he shall live for ever . . . I will raise him up on the last day.'

Again, the single sign 𓏏, when used hieroglyphically and not as an ideogram, was used in Egyptian to signify a discrete word *in only one known instance*—as the word 'father' in the ecclesiastical title [hieroglyphs] 𓏏 or 𓏏 [hieroglyph], 'god's father'.[53] Yet at verse 27 of the chapter already

[50] Compare the Egyptian ideogram [hieroglyph] in [hieroglyph] 'hill of the sunrise' and [hieroglyph] 'appear in glory' (Gardiner, *Egyptian Grammar*, 3rd Edition, p. 489), which seems to offer strong supporting evidence for the notion.

[51] Compare the biblical story of Noah's flood, and see pp. 259–70.

[52] Petrie, the Egyptologist, always denied the uniqueness of the boss, claiming that many similar excrescences are to be found in the Pyramid. All those he lists, however, are *in the King's Chamber Complex*, and the numbers listed for each chamber are *all multiples of the number 5*. A specifically Messianic significance thus still seems likely, even if Petrie's claim were substantiated and the bosses in question were found to be as finely shaped and of similar dimensions. Nobody else, however, appears to have noticed anything remotely comparable.

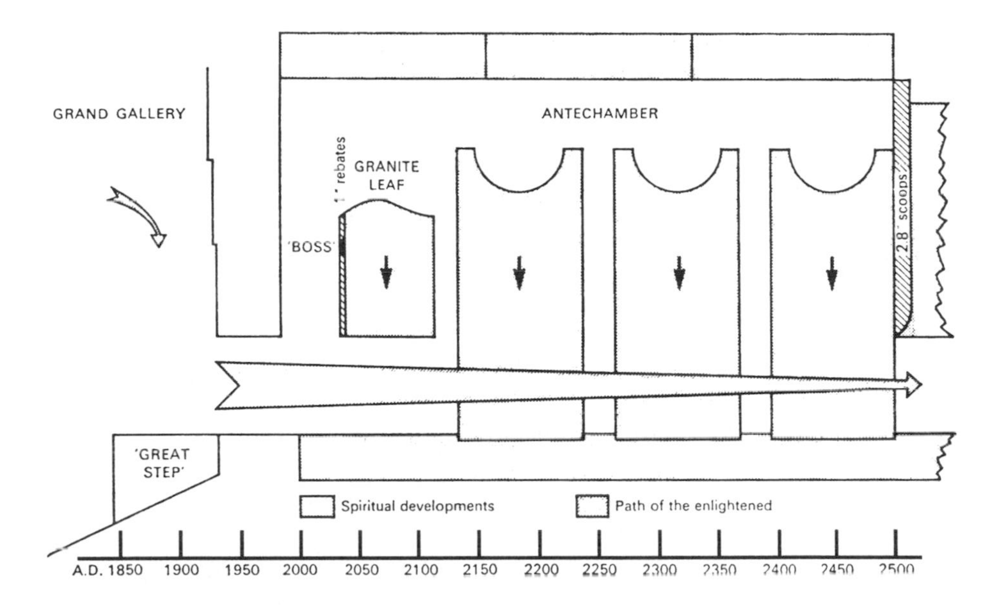

Constant-scale timechart of Antechamber. Not to architectural scale.

already quoted we find Jesus reported as saying, of that same 'bread from heaven': 'This food the Son of Man will give you, for he it is upon whom *God the Father* has set the *seal* of his authority.' It is as though we have here some sort of cryptic, esoteric riddle for the benefit of 'those who have ears to hear'. Certainly it would be difficult to find a more astonishing series of direct parallels. Indeed, as in the case of the celebrated 'I am the way; I am the truth and I am life' (John 14:6)—conceivably a direct reference to the Pyramid's Messianic Triangle[54]—we seem once again to have direct evidence that Jesus himself (or at very least the author of John's gospel) was aware of the Pyramid's message and symbolism, and set out consciously to fulfil it.[55]

[53] Gardiner, *Egyptian Grammar* (3rd Ed.), p. 555.

[54] Compare p. 63.

[55] Nor does the above exhaust the catalogue of extraordinary circumstances surrounding the 'bread' sign. Whether, like both its Hebrew and Greek equivalents, it was ever called *tau*, we can only guess. But in Greek the latter took the form of the sign 'T'—a sign which from time immemorial seems to have had some kind of sacred significance, subsequently perpetuated in the Christian cross (indeed, chapter 9 will even suggest that the links between 'seal' and cross may be specific). Meanwhile the resemblance between the words *tau* and *Tao* is so great as to suggest interchanging experimentally words such as bread, way, truth, life and Tao in the Taoist and Christian scriptures respectively. The results are so interesting as to lead one to wonder whether some single underlying factor might explain this whole series of apparent links.

Meanwhile the Granite Leaf shows the Messianic advent as taking the form of a spiritual incursion into the earth-planes—the spiritual 'taking over', perhaps, of a physical man by a spirit-being of unimagined power, such as apparently happened during the baptism of Jesus of Nazareth. But that incursion will not be a once-for-all affair. For the process is shown as being repeated three more times. The notion of a fourfold Messianic visitation may shock some readers and be unfamiliar to others—yet there are clear hints of such a possibility in the Judæo-Christian scriptures, and notably in the allegorical stories of Noah and Moses. 'All these things that happened to them were symbolic, and were recorded for our benefit as a warning. For upon us the fulfilment of the ages has come.' So, once, wrote Paul the Pharisee of the story of the Mosaic Exodus,[56] and chapter 9 of the present volume is devoted specifically to the investigation of this notion.

As for the datings given above, they are the only datings that are completely compatible with the data set out in our hypothetical code. Other hypotheses, however, would naturally produce other results.

If we assume with Rutherford, for example, that the effect of the Great Step is to leave the time-scale exactly as it was, then all the above datings (after A.D. 1845) need to be brought forward by some nineteen years, fixing the beginning of the granite floor at A.D. 1979 and the 'sign of the Messiah' at 2015.[57] If, on the other hand, the Great Step were merely ignored as a deliberate 'blind' by the designer, then the 'hidden step' would supply its own scale-change, whose effect would be to put back the datings given very considerably. In this case the granite floor would commence in around A.D. 2260, with the first Messianic advent dated at around A.D. 2300.

Even sticking to the Great Step scale-change suggested above, however, it might still be argued that the scale of 1" per year should apply throughout the King's Chamber Passage and right into the King's Chamber, since no actual 'step' has intervened. This would not be in accordance with the code as postulated—indeed it seems to fly in the face both of the obvious symbolism of the 5 RC-long granite floor in the Antechamber and of the 10 RC (1,000-year) width of the King's Chamber. None the less, application of such a thesis would date the first Messianic advent at A.D. 2012, the second at 2030, the third at 2057 and the fourth at 2084. But in this case it

[56] 1 Cor. 10:11.

[57] Rutherford, for his part, postulates that granite floors are totally without time-scale.

seems fairly obvious that the dates are too close together to make sense.

History, of course, draws derisive attention to the various incorrect time-scale assumptions. For example, Davidson and Aldersmith's calculation—albeit arrived at with some misgivings—that the 'world would end' in 1953 has already misfired. We can only wait, and watch the 'signs', and remember the biblical warnings that things may happen later than we think.

The King's Chamber (The Chamber of the Open Tomb/of Resurrection)

By turning right on entering this chamber, the visitor can at last proceed westwards as far as, and even beyond, the north-south axis. (See diagram page 107).

(1) Nature of chamber	entirely of granite	*Spiritual era*
(2) Slope of floor	horizontal	*Level of attainment*
(3) Level of floor	top of 28″-thick (7 × 4) 50th course of masonry (10 × 5)	*Spiritual perfection of the physical: Millennium of the initiates*
(4) Number of sides (including roof and floor)	6	*Preparatory only*
(5) Number of corners	8	*Rebirth*
(6) Length (E.-W.)	$2 \times 365{\cdot}24235''/\sqrt{\pi}$ = 20 RC (2 × 10)	*An age produced by the seed of the eternal; (productive of) the Millennium*
(7) Breadth (N.-S.)	$365{\cdot}24235''/\sqrt{\pi}$ = 10 RC	
(8) Height	$\sqrt{5 \times 365.24235''}/2\sqrt{\pi}$ = floor-diagonal/2 = 230·3871″	
(9) Number of wall-courses	5	*(Chamber of) initiates*
(10) Number of stones in walls	100	*Reward*
(11) Level of base of walls	5″ below floor-level	*Based on Messianic initiation(?)*
(12) Level of top of walls	notional top of 60th (6 × 10) course of masonry	*Even the Millennium is only a preparatory phase*
(13) Number of beams in roof[58]	9	*Utter perfection*

[58] The largest beam weighs an estimated 72 tons.

(14) Distance of S. wall from mid-point of Antechamber	365·24235″	*Termination of preparatory age*
(15) Ratio of N. or S. wall diagonal to length of Chamber to cubic diagonal of Chamber[59]	3 : 4 : 5	*Perfect/physical/initiates*
(16) Number of air-shafts to Chamber	2 (in N. and S. walls)	
(17) Bore of upper sections of shafts	9″ × 9″	*(For the) utterly perfect*
(18) Average distance from axis of entrance passage to axis of shaft-openings	80″ (8 × 10)	*Rebirth of the Millennium*
(19) Height of top of north ventilator opening above floor	41·21″ (2 RC)[60]	*For mortals*
(20) Slope of shafts beyond initial horizontal sections	upwards[61]	*Further spiritual progress*

North shaft

(21) Dimensions of horizontal part of north shaft	8″ wide × 5″ high × 112″ long (code-equivalent: 8 × 2 × 7?)	*(Leading to) rebirth of initiates: rebirth productive of spiritual perfection(?)*
(22) Cross-section of north-shaft	rectangular	*(Leading to) the physical/ earthly*

[59] In other words, this ratio applies to the three sides of the largest right-angled triangle which can be inserted diagonally in the Chamber with its base along the N. or S. edge of the floor, thus forming a slope from floor to roof. See Rutherford pp. 1010–12 for complete details.

[60] The top of the north ventilator-opening, like those in the Queen's Chamber, lies on exactly the same level as the roof of the Chamber's access-passage, and it seems likely that this level was originally also that of the tops of the side-walls of the south ventilator opening (which is now much mutilated). As in the Queen's Chamber, then, there would seem to be a close symbolic link between these features. The obvious inference, therefore, is that entry into the Chamber leads directly to 'escape' via one or the other of the air-shafts (provided, of course, that a 'turn to the right', or west, is duly made on entering the Chamber).

[61] The King's Chamber's north air-shaft is reported by the Edgars as taking 'a number of short, sharp bends, each succeeding bend tending upward and toward the north-west, before it finally bends northward to proceed directly to the outside of the Pyramid at a steep angle.' It could be that each of these bends is intended to represent a 'point of transition' in the soul's continuing upward path after leaving the King's Chamber—and thus corresponds to the granite interstices between the six chambers of the Pyramid's vertical 'granite house' (see pp. 129–34).

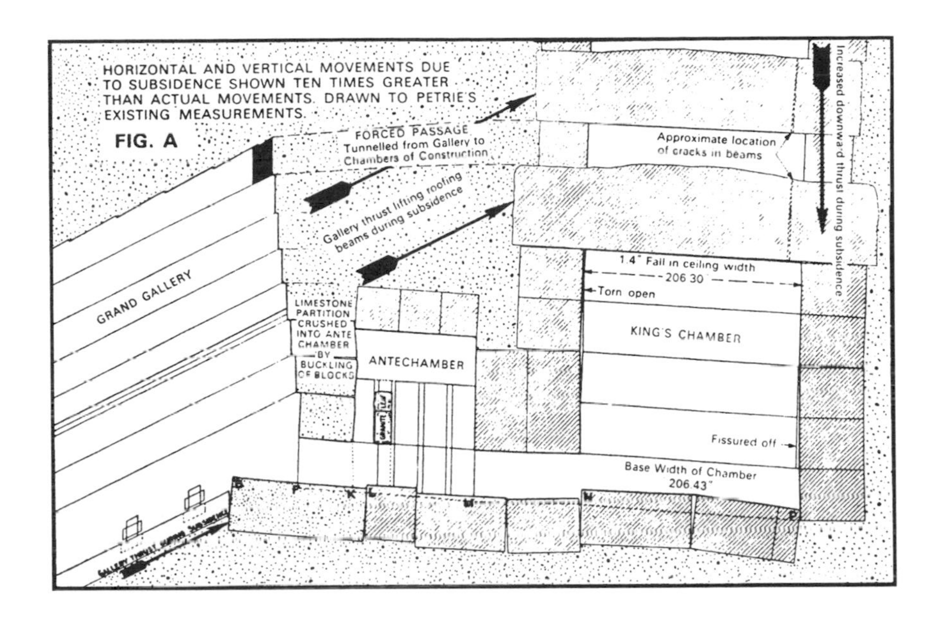

The King's Chamber complex (A) distorted by earth movements and (B) restored

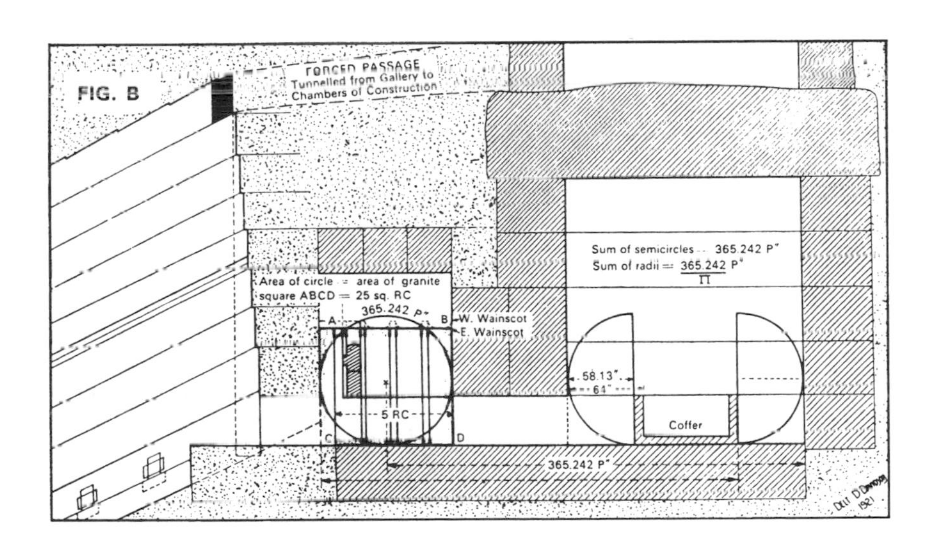

(23) Level of core-masonry outlet	101st (100 + 1) course	*More than a reward(?)*
(24) Notional thickness of 101st course	32″ (8 × 4)	*Physical rebirth*
(25) Level of shaft's casing-outlet	103rd (100 + 3) course	*Failure to break out*[62]
(26) Notional thickness of 103rd course	29″ (code-equivalent: 29·85″)	*Death/mortality*

South shaft

(27) Level of core-masonry outlet of south shaft	102nd (100 + 2) course[63]	
(28) Notional course-thickness	28″ (7 × 4)	*Spiritual perfecting of the physical*
(29) Level of casing outlet	104th course (100 + 4, or 8 × 13?)	*Path of escape (see note 62)*
(30) Notional thickness of 104th course	26″ (2 × 13?)	*Productive of* (13?)
(31) Cross-section of south shaft at lower end	circular—diameter approx. 12″	*Spiritual man*
(32) Nature of lower opening	an irregular, 'domed', horizontal section (see diagram p. 112)	*(See* (33)*)*
(33) Symbolism of lower opening	possibly an oven, or the pregnant womb: see also the boss (pp. 112, 120–1)[64]	*For cooking of 'bread'; gestation of soul before liberation*

[62] The true significance of the casing-outlet in the 103rd course is likely to be not so much its level above the base as *its distance below the summit-platform atop the 203rd course.* For it is clear that the addition of 100 courses ('the reward') to the level of the north air-shaft's outlet indicates a level *within the 203rd course.* It therefore signifies that, even with the reward added to it, the soul taking this path 'fails to break out'. Only via the south ventilator does the addition of the reward to the soul emerging in the 104th course bring it up to the level of the capstone itself—the notional 204th and final 'course' denoting the completion of the evolutionary Plan for Planet Earth.

[63] The core-masonry outlets of the King's Chamber ventilators thus define numerically the topmost course (course 203) supporting the Pyramid's summit-platform—for the sum of courses 101 and 102 is 203. That this fact is demonstrated by the *core-masonry* levels suggests, moreover, that the original Pyramid *had* no casing-outlets, and that all four ventilators were originally left sealed. The designer apparently foresaw, in other words, that his monument would at some future date be stripped of its casing.

[64] The cross-sectional dimensions given on page 112 (showing a minimum width of 6″ expanding to a maximum of 18″) suggest that this ventilator symbolically

(34) Nature of coffer	an uninscribed, lidless, rectangular, granite sarcophagus	*Spiritual escape from physical mortality*
(35) Exterior length of coffer	width of King's Chamber minus length of Antechamber = 89·80568″ (code equivalent: 3 × 29·84″?)	*Utter death (?)*
(36) Exterior width of coffer	38·69843″ (2 × 19)[65]	*Productive of death*
(37) Exterior height of coffer	41·21319″ (2 RC)	*Death*
(38) Sum of all three dimensions of coffer	sum of Chamber's three dimensions/5 = 169·7173″ (13 × 13)	*'Ultimate resurrection' through (gaining of) initiateship*
(39) Thickness of sides	6″ approximately	*Incompleteness/ Preparation*
(40) Thickness of bottom	7″ approximately	*Based on spiritual perfection*
(41) Thus, interior length of coffer	approx. 77·8″ (code-equivalent: 11 × 7)	*Achievement of spiritual perfection*
(42) And interior breadth of coffer	approx. 26·7″ (2 × 13?)	*Productive of* (13?)
(43) Meanwhile total N.-S. thickness of coffer's sides	total E.-W. thickness of sides = 12″ approximately	*Mankind*
(44) Position of coffer	moveable within Chamber[66]	
(45) Probable intended position of coffer	mid-way between N. and S. walls of Chamber, with its axis N. and S. on the Pyramid's centreline	*See* (46)
(46) Thus, distance of coffer's axis to west of King's Chamber Passage axis	286·1″	*Re-attainment of enlightenment*

[64] (*continued*) allows the imperfect (6″) to rise at last, like bread in the oven, to perfection (2 × 9″). The remaining features (shown in the side-elevation) confirm this interpretation in that they symbolically permit (spiritual) man (12″) to emerge from the Chamber as a result of the spiritual perfecting of the physical during the Millennium (7 × 4; 7 × 10). Compare feature (31), which shows that the emerging 'man' is spiritual in nature. Incidentally, the association of the foetus in the womb with bread in the oven is still popularly current today.

[65] In other words, the coffer is slightly wider than the lower end of the Ascending Passage where it is tapered to hold the Granite Plug. Even if the Granite Plug had not been built-in, therefore, the coffer could not have been brought up the passage, and must have been 'built-in'.

[66] This would suggest that the coffer has no specific chronology.

(47) Distance of ends of coffer from Chamber's N. and S. walls	58·1013" = 365·24235"/2 π	*An age influenced by the eternal*
(48) Thus, the significance of N.-S. position of coffer	'splitting apart' of space and time (see lower diagram p. 125)	
(49) Distance from N. wall to interior of coffer	64·1" approximately (code-equivalent: 8^2)	*The 'ultimate rebirth'*
(50) Distance from coffer-axis to west wall (i.e. distance Chamber projects westwards of Pyramid's axis)	105·42314" (code-equivalent: 3 × 7 × 5)	*Still short of the utter spiritual perfection of the full Initiate*
(51) Distance from limestone N. wall of Antechamber to granite south end of coffer	365·24235"	*End of age*
(52) Dating of entry into Chamber	2nd July A.D. 2989 (see p. 118)	
(53) Significance of 10 RC width of Chamber's granite floor	10 × 100 years = 1000 years.	*Millennium: time-measurement*
(54) Thus, notional dating for S. wall	2nd July A.D. 3989[67]	
(55) Next feature of passage-system	five chambers *above King's Chamber*	*Five further spiritual planes*

READING: *Earth's great spiritual* (1) *Millennium* (3) (6) (7) (53) *will commence in the summer of* A.D. *2989* (52).

It will be an age for bringing the physical world to spiritual perfection (3) (13) (17) (21) (28) (51) *through the Messianic influence of the eternal planes* (6) (7) (8) (11) (47).

During this age the souls of the enlightened, gathered from all corners of the globe (43), *will once again be physically reborn* (5)—*but this time it will be to experience the reward (10) of the ultimate, millennial rebirth* (18) (49). *Even at this stage, however, it will still be fatal to look back,*[68] *for failure to complete the course will lead, at death* (19) (35) (36) (37), *to further physical rebirth* (22) (24) *in search of spiritual perfection* (21). *Those who succeed in finally recovering their lost enlightenment and in achieving perfection* (13) (17) (41) (46) *will at last succeed, at death* (19) (35) (36) (37), *in effecting their total escape from mortality*

[67] It is problematical how far N.-S. distances inside the terminal chambers are meant to be taken as chronologically significant. None the less the apparent reference to a thousand-year period fits the notion of the Millennium perfectly. A.D. 3989 is 5441 years after 1453 B.C.—the Pyramid's dating for the beginning of the Ascending Passage and (apparently) the Israelite exodus from Egypt. This period of time is interestingly close to the height of the Pyramid's summit-platform in Primitive Inches (5448·736).

[68] Compare Jesus of Nazareth at Luke 9:62, 'No one who sets his hand to the plough and then keeps looking back is fit for the kingdom of God.'

and the physical world (20) (32) (33) (34) (48) *and a translation to a higher plane of existence* (20) (55). *This transformation will be based on their own spiritual perfection* (41) *and on their achievement of full initiateship* (9) (11) (38).

Yet even with this final escape from the physical world, the development of the souls of men will not stop, for their initiation will still be incomplete (4) (12) (50). *They will merely have been reborn into the lowest of the spirit-planes* (31) (33) (34) (55).[69]

With the above reading, then, we can at last positively identify (in terms of our hypothetical code) the symbolic occupant of the coffer: it is the souls of the enlightened who have finally achieved spiritual perfection (41) as initiates (38) and have thereby escaped 'upwards' from mortality (37). It is apt that the ancient Saïte recension of the Egyptian *Book of the Dead* should apparently refer to the King's Chamber as that of the 'Open Tomb'. And the 'Resurrection' of which it also speaks, seems to refer to nothing less than man's final birth into a new dimension of existence entirely.

The Construction Chambers (The Secret Places of the Hidden God)

This feature comprises a remarkable series of hidden chambers (not fully re-discovered until 1837) stacked one on top of the other above the King's Chamber. Orthodox theory has it that they were intended in some way to protect the King's Chamber from the enormous weight of masonry above. But the theory seems singularly unconvincing, even though that has tended to be their effect. Why, for example, protect the King's Chamber in this way, and not the Queen's Chamber? Why construct a series of five chambers, when one would presumably have had the same effect? And why cap the resulting 'granite house' with a single limestone gable, when a multiple granite one would have been more durable?

Once again we are driven to the conclusion that all these features—including the choice of stone for the crowning gable—are there primarily for symbolic purposes. In terms of our hypothetical code, for example, we have hitherto assumed that limestone represents the physical world and granite the spiritual. Presumably, then, we should continue so to assume. Moreover we now seem to be dealing (see above) in terms of human souls which have forsaken the physical world and gained entry into the spirit-planes.

[69] See next section.

It seems logical at this point to make two basic assumptions. The first is that only spirit—represented by granite walls—can symbolically 'contain' spirit, while the physical—represented by limestone—no longer presents any obstacle to its passage. And following on from this, the fact that all but the top two of the Construction

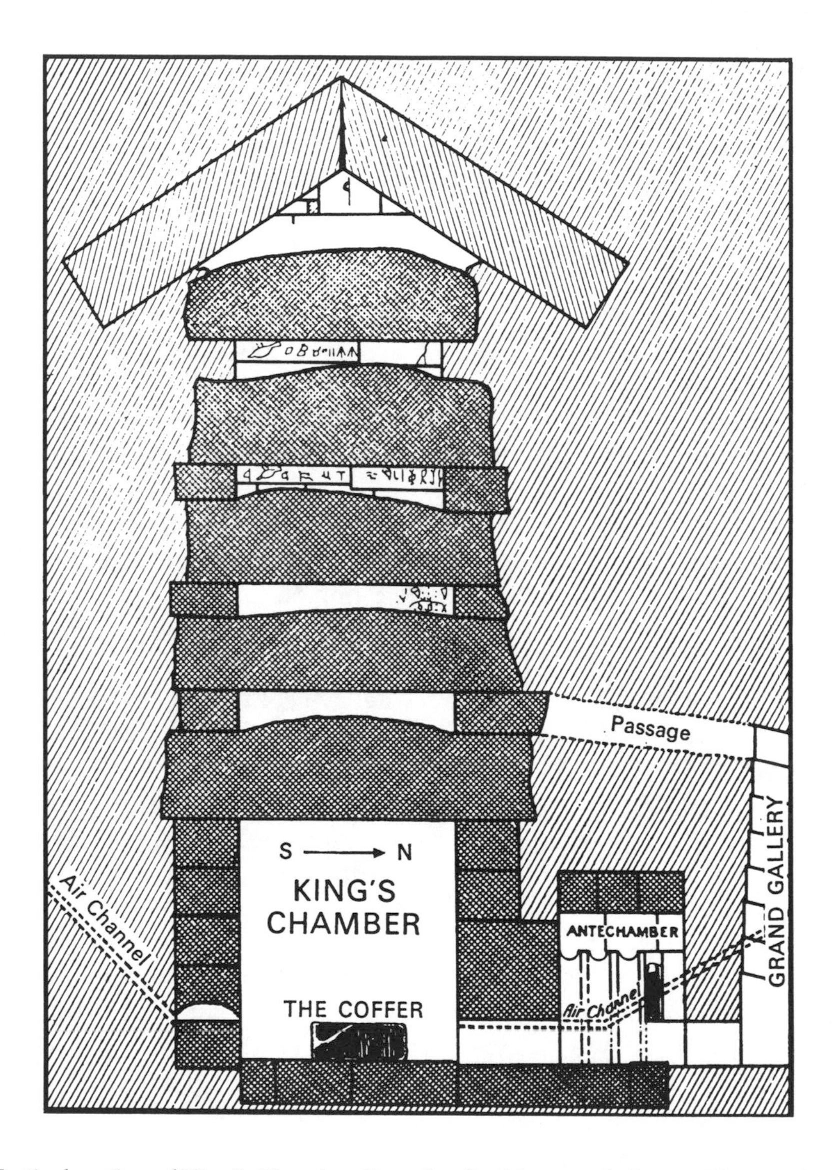

Vertical section of King's Chamber Complex (looking west). Crossed lines indicate granite; single lines limestone.

Chambers have north and south walls of granite leads us to our second basic assumption. It is that the discarnate souls symbolically passing through these chambers will still need, as it were, to 'progress' from north to south through each chamber before ascending further—though no longer, one assumes, through the particular dimension we call 'time', as the extraordinarily irregular floors of all five chambers amply confirm.

On the other hand, the east and west walls of all the Construction Chambers are of limestone. This fact in turn suggests that for the souls in question the east and west walls simply 'do not exist'. We may interpret this notion to mean that the discarnate entities in question no longer need to operate in terms of 'good and bad', of 'positive and negative': they have finally left behind relativity and the 'tree of the knowledge of good and evil', and have freed themselves from what Buddhism calls the 'world of opposites'. Man, in fact, has returned to his spiritual, 'pre-fall' state as a denizen of the world of primal absolutes.

On this basis we can now list the relevant data for these chambers in order to derive from them the usual reading.

Chamber 1 (lowest Construction Chamber)

(1) Position of Chamber	directly above flat granite roof of King's Chamber	*Plane achieved by escape from King's Chamber coffer*
(2) Nature of floors of all Construction Chambers	highly uneven granite beams.	*See* (3)
(3) Significance of (2)		*No time-scale: a 'spiritual' plane*
(4) Nature of N. and S. walls	vertical flat granite	*Spiritual progress needed*
(5) Nature of E. and W. walls of all Construction Chambers	vertical flat limestone	*No longer subject to relativity*
(6) Average height of 4 lower Chambers	120″ (10 × 12)	*'Eternalising' of mankind (?)*
(7) Nature of roof of Chamber 1	horizontal flat granite	*Entry to higher spirit-plane*
(8) Notional number of corners	8	*Rebirth . . .*
(9) Notional number of faces	6	*of incompleteness/ preparation*

(10) Special feature	a rough tunnel ($26\frac{1}{2}''$ × 32″) breaks into N. side of chamber from top section of Grand Gallery[70]	*Non-physical path for the totally enlightened*

Chamber 2

(11) Position	above *and slightly to west of* Chamber 1	*Entry to higher plane depends on further spiritual effort*
(12) Nature of N. and S. walls	vertical flat granite	*Non-relative*
(13) Nature of roof	horizontal flat granite	*Entry to higher spirit-plane*
(14) Notional number of corners	8	*Rebirth . . .*
(15) Notional number of faces	6	*of incompleteness/ preparation*

Chamber 3

(16) Position	above *and slightly to west of* Chamber 2	*Entry to higher plane depends on further spiritual effort*
(17) Other features	as for Chamber 2	*Non-relative plane, still incomplete, leading higher still*

Chamber 4

(18) Position	above *and slightly to east of* Chamber 3	*Higher plane reached without further spiritual effort*
(19) Nature of N. and S. walls	vertical flat limestone	*Nothing further to achieve spiritually, but still incomplete, leading higher still*

[70] All Construction Chambers, except No. 1, contain Egyptian quarry hieroglyphics—among them, it is believed, the cartouche of the historical pharaoh Khufu. This fact, plus the rough entrance tunnel, suggests that only the bottom chamber was ever intended to be seen by visitors. The memory of the remaining 'hidden' chambers was presumably intended to be preserved by oral transmission until such time as techniques for rediscovering them should be developed—such as the cosmic-ray detection apparatus presently installed in the Second Pyramid.

(20) Other features	as for Chamber 3	

Chamber 5

(21) Position	directly above Chamber 4	*Automatic consequence of previous plane*
(22) Nature of N. and S. walls	limestone and irregular—indeed virtually non-existent	*Non-finite*
(23) Nature of roof	salt encrusted limestone gable (24 stones—2 × 12 or 6 × 4)	*Source of all life; preparation of the physical producing (true?) man*
(24) Possible significance of gable	an upward-pointing arrowhead[71]	*'Upward' transformation*
(25) Notional number of corners in Chamber	10	*Millennium . . .*
(26) Notional number of faces	7	*. . . of spiritual perfection*
(27) Position of topmost chamber	the summit of the entire passage-system	*Completion of human evolution*

General

(28) Number of chambers	5	*For the initiates: initiation*

READING:*The discarnate souls of the enlightened, having escaped from mortality* (1) *and entered the non-temporal* (2) (3), *non-relative* (5) *planes of the spirit* (2), *will pass upwards through higher and higher spiritual levels* (1) (7) (13) (20) *of initiation* (28) *to produce the final apotheosis of man* (6).

The attainment of the second and third of these levels will represent the fruit of further spiritual endeavours on the levels preceding them (11) (16). *But attainment of the fourth plane will mark the cessation of spiritual endeavours* (18), *and will lead inevitably to entry into the topmost plane of all* (21). *With nothing more to achieve* (22), *the souls emerging on to this plane will finally ascend* (24) *into the non-finite* (23) *realms of spiritual perfection* (25) (26). *Man will finally have realised his true identity* (23), *and the Messianic Plan for the evolution of True Man will have reached its ultimate fulfilment* (23) (27).

[71] See p. 79. The symbolism of the gable seems to suggest that 'true man' has at last left behind the physical world and returned to the 'source of all life'—as represented by the salt-incrustation.

Man, in short—the 'prodigal son' of the spiritual world—will finally have 'come home' and laid claim to his everlasting inheritance.

The Subterranean Inset

The Subterranean Inset comprises the group of horizontal passages and chambers extending southwards from the bottom of the Descending Passage. The Great Subterranean Chamber itself seems to be cognate with the so-called *Book of the Dead*'s Chamber of Ordeal or Chamber of Central Fire.

(1) Direction	southwards	*Through time*
(2) Slope	basically horizontal	*Level of attainment*
(3) Cross-section of passages and chambers	basically rectangular	*Physical/terrestrial*

Subterranean Chamber Passage

(4) Width	33·5204" (code-equivalent: 33·5")	*The Messianic presence*
(5) Height	35·7628" (286·1"/8)	*Unenlightened incarnations*[72]

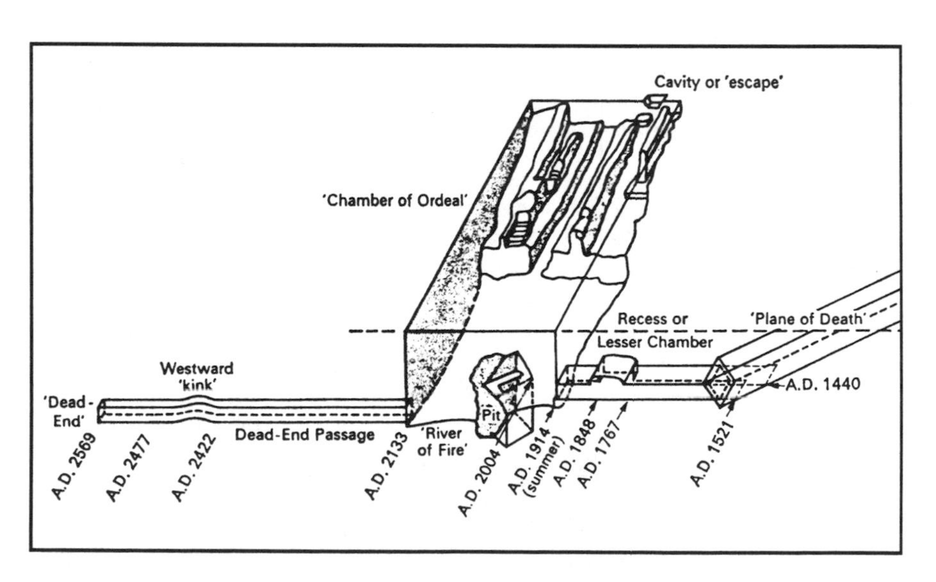

Projection of the Subterranean Inset.

(6) Position of axis	$\frac{5}{8}$″ *west* of Descending Passage axis	*Messianic path through rebirth (?)*
(7) Initial feature	upward, non-vertical step	(*Trigonometrical scale-change*)
(8) Level of lowest part of step-bottom	1162·60251″ below Pyramid's base = 3652·42″/π	*An age (influenced) by the eternal (?)*
(9) Significance of point (7)	trigonometrical scale-change based on projection of horizontal measurements onto continuation of slope	
(10) Thus, on new scale, time for 1″	1·11549 years (*more* years per inch)	
(11) Peculiarity of step and whole join between Subterranean and Descending Passages	no feature is at right-angles to any other	
(12) Significance of (11)	precise dating subject to adjustment (compare Trial Passages' apparent suggestion of a 'telescopic fit' on page 199)	
(13) Method of adjustment	via experimental historical 'fit'[73]	
(14) Centreline distance from *intersection of floorline with Descending Passage floor*[74] to furthest extremity of floor at entrance to Great Subterranean Chamber	352·2933″	
(15) Thus, time represented by (14)	392·9803 years	
(16) Meanwhile distance from same point to north side of Lesser Subterranean Chamber	220·3984″	

[72] See point (12), p. 87. In the Subterranean Inset we are in the downward mode: the distance 286·1″ has already been experienced in a downward direction, and the succeeding measurement of 35·76″ is likewise experienced during the *descent* to the level of the passage-floor and must also be thought of as a downward measurement.

[73] Compare the marine orientation-method involving the taking of successive soundings on a given heading and at given intervals, and then compare the line of readings with charted depths at the same intervals in search of a 'fit'. Heading and distances must of course be known in both cases.

[74] This is, of course, the only point of contact between the two floorlines.

The junction of the Descending and Subterranean Passages. Pyramidologist John Edgar sets the scale.

(17) Thus, time represented by (16)	245·8526 years
(18) Distance across Lesser Subterranean Chamber	72·35187″
(19) Thus, time represented by (18)	80·70792 years (see corroborative dimensions on diagram on p. 138)
(20) Apparent significance of square Lesser Chamber	a physical 'roof-fall' (see diagram)—conceivably a 'time of turbulence'[75]
(21) Apparent significance of Greater Chamber	a 'time of hell on earth'(?)[76]
(22) Experimental historical 'fit' suggests datings as follows:	
(a) Extremity of floor at entrance to Great Subterranean Chamber	(August) 1914[77]
(b) South end of Lesser Chamber	(February/March) 1848
(c) North end of Lesser Chamber	(June) 1767
(d) Intersection of Subterranean and Descending Passage floorlines	(August) 1521
(23) Thus, maximum 'discrepancy' between Subterranean Passage and Descending Passage datings for this intersection (see page 87, points (14) and (15))	15 years[78]

[75] The measurements of 6″ and 18″ (3 × 6, or 2 × 9) shown in the diagram may conceivably refer to a striving to turn 'imperfection' into 'perfection' during the period indicated—a violent upsurge of idealism, in fact.

[76] This significance is suggested both by the chamber's appearance and by the ancient Egyptian nomenclature above.

[77] If we assume that the 'sliding-fit' is not likely to extend more than, say, twenty years or so in either direction, then the entrance to the Subterranean Chamber must come at some point between A.D. 1879 and 1919. The obvious 'change of epoch' would therefore seem to be the beginning of the First World War in 1914—an event also apparently referred to at the beginning of the King's Chamber Passage. It is then found that the Subterranean Passage's other features fit earlier historical events with considerable aptness (see pp. 144–6), as well as fitting in well with later features of the chronograph. On the other hand, no other 'fit' appears to work at all. The datings above thus seem to be confirmed.

[78] The figure certainly seems to be reasonable, being neither so small as to be insignificant, nor so large as to be improbable.

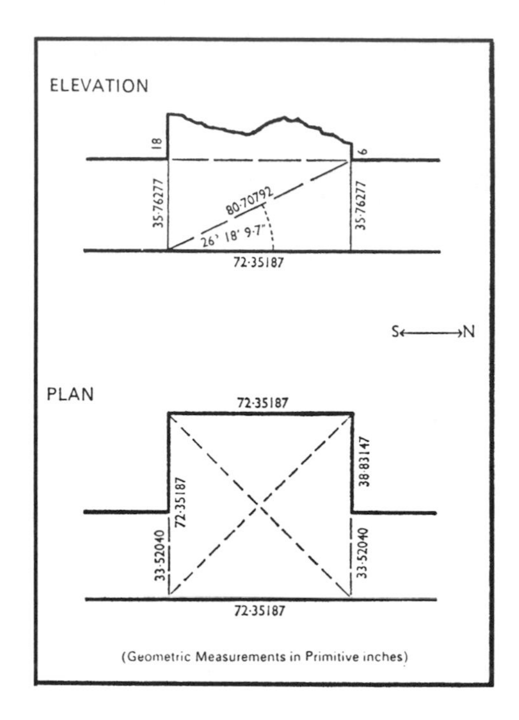

Lesser Subterranean Chamber (elevation and plan).

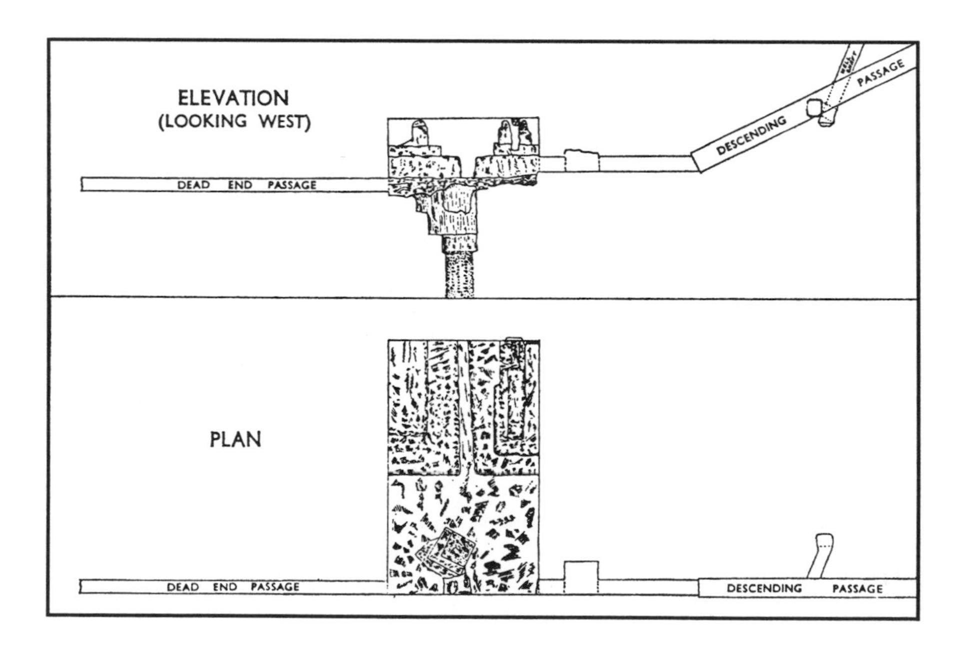

The Subterranean Chambers with adjacent passages.

(24) If intersection of Subterranean Passage roofline with Descending Passage floorline is also taken to be significant, this point now marks	80·70792 years before (August) 1521 = (July/August) A.D. 1440	
(25) Thus, possible significance of Descending Passage floorline denoting period A.D. 1440–1521	the historical period producing or leading to the events of the Subterranean Inset	
(26) Dimensions of square Lesser Chamber:		
(a) side-length	72·35187″ (code-equivalent, 6 × 12 or 8 × 9)	*Imperfection of man: rebirth of utter perfection*
(b) distance of west wall to *west* (right) of passage	38·813147″ (code equivalent: 2 × 19)	*Productive of death*
(c) height of north and south walls above passage roofline	6″ and 18″ (3 × 6, or 2 × 9) respectively	*Imperfection (leading to) utter imperfection or producing perfection*

Great Subterranean Chamber

(27) Feature at entry into Chamber	irregular downward drop or step, while roof-level rises by 89·80568″ (3 × 29·84″, or exterior length of King's Chamber coffer)	*Alternatives of sudden decline or 'break-out' from time and space (see diagrams page 125)*
(28) Justification for assuming that drop is a downward vertical step	Passage floor-level at southern exit from Chamber is exactly 2 RC below passage floor-level at northern entrance—the only point where a clear vertical drop occurs.	*Death or birth: the return to physical existence of the souls taking this path*
(29) Thus, new scale from downward step	25 × 1·11549/41·2131 years per inch[79] = ·6766598 years per inch = *less* years per inch	
(30) Distance to south wall of Chamber from extremity of entrance-passage floor	322·7711″[80]	

[79] It should be remembered that the code states that the formula shall take the *nearest* base-unit of measurement to the height of the step in question as the basis for calculation. This has, therefore, to be the Sacred Cubit of 25″, as modified by the earlier, trigonometrical scale change at (9).

[80] The *north wall* of the Chamber actually comes some 5″ before the extremity of

(31) Time represented by (30)	218·40622 years	
(32) Dating for south wall of Chamber	December A.D. 2132/ January 2133[81]	
(33) Nature of Chamber	The largest of the Pyramid's Chambers (big enough to contain all the others), its western extremity is by far the most *westerly* point in the whole passage-system.	*An era of possible redemption*
(34) Nature of walls and roof	flat, largely rectangular	*Earthly/physical*
(35) Nature of floor	extraordinarily irregular	*Timing approximate*
(36) Nature of eastern part of floor	relatively flat, but containing a deep, semi-rectangular pit with a notional 'river-bed' flowing into it from the east wall	*Path leading to physical 'ordeal'*
(37) Length of side of pit	'nearly seven feet' (Rutherford)	
(38) Length of diagonal of pit	100″	*Reward*
(39) Depth of upper part of pit	67·59″ (code-equivalent of 2 × 33·5)	*Man-come-of-age: productive of Messianic presence*
(40) Depth of lower part of pit (below ledge in southwest corner)	41·2″ (2 RC)	*Death/mortality*
(41) Total depth of pit	108·8″ (code-equivalent: 9 × 12)	*Utter perfection of man*
(42) Mean width of S.W. ledge	20·6″ (1 RC)	*Mortality/rebirth*
(43) Total distance from roof of Chamber down to bottom of pit	306″ (2 × 153)	*(Down) Produces loss of enlightenment/(Up) Produces enlightened ones*
(44) Alignment of pit	roughly NNW-SSE, i.e. not that of Chamber	
(45) Significance of (44)	timing approximate?[82]	

[80] (*continued*) the floor of the entrance-passage, and may thus represent a date some three years earlier. Alternatively the projection of the passage floor into the Chamber may suggest a further 'telescopic' (and thus marginally adjustable) join, as at the beginning of the passage.

[81] If, as Rutherford suggests, the floor-entry into the Chamber in fact marks 1st August 1914, then the south wall marks the night of 29th/30th December 2132.

[82] Measuring in a straight line across the Chamber, the north side of the pit appears to be dated at about A.D. 2004 and its mid-point at 2032.

(46) Nature of western part of floor	a sudden high plateau of rock lying mainly to the *west* of the Pyramid's axis, and bearing a number of high ridges running east to west and separated by deep gullies—the whole platform being divided by an even deeper gully running off to the west (right) abreast of the pit	*Human civilisation, the result of(intellectual) enlightenment; its achievements and its temporary collapse*
(47) Feature in west wall	a small recess or 'escape' set into the wall against the roof—and thus the most *westerly* feature of the whole passage-system	*See* (52)
(48) Approximate width of escape	36″ (3 × 12)	*(Leading to) human perfection*
(49) Extreme depth of escape	18″ (6 × 3, or 2 × 9)	*Preparation of the perfect/productive of utter perfection*
(50) Height of escape at entrance	28″ (7 × 4)	*Spiritual perfecting of the physical*
(51) Height of escape within entrance	12·5″ (1 Sacred Cubit/2 ?)	*(Capable of producing) the Messianic ideal (?)*
(52) Possible significance of escape	symbolic of the Well-Shaft, whose bottom breaks off to the west immediately *above* the same roof-level, and whose bore is also 28″	
(53) Apparent access to escape	by turning *right* opposite the pit and up the gully leading to the Chamber's west wall	*Spiritual reform at time of 'ordeal'*
(54) South exit from Chamber	entrance of the Dead End Passage	

Dead End Passage

(55) Width	29·8412″	*Death/mortality*
(56) Height	29·8412″	*Death/mortality*
(57) Cross-section	square	*Physical*
(58) Finish of walls	rough	*Earthly*

(59) Position of axis	1·2114″ to *east* of Subterranean Chamber Passage axis—the most *easterly* passage-axis in the Pyramid (other than the Niche, q.v.)	*Path of utter mortality* *Rebirth*
(60) Slope	horizontal	*Level of attainment*
(61) Direction	southwards	*Through time*
(62) Extreme length	645·5422″	*(Time-measurement)*
(63) Length of time represented	436·81245 years	
(64) Thus dating for S. end of passage	436·81285 years after December A.D. 2132/ January A.D. 2133 = autumn A.D. 2569	*(Dating)*
(65) Feature of passage	a westward 'kink' just over 35 ft. from the entrance, subsequently returning to the line of the passage	*Temporary reform*
(66) Extent of kink	6″ *to west*	*Incomplete reform*
(67) Length of kink	84″ (?) (code-equivalent: 7 × 12)	*(influenced by) spiritual perfecting of mankind (?)*
(68) Nature of end of passage	roughly squared-off	*(See* (69)*)*
(69) Apparent significance of 'blind end'	passage 'stops' rather than finishes	
(70) Total floorline-distance to extremity of passage from original entrance to the Pyramid	5448·736″	*Earthly imperfection/ story unfinished*

Point (43) above suggests that the Subterranean Chamber's Pit is a direct function of the enlightenment of those who experience it. But one can read its overall depth from the Chamber's roof in a downward *or* an upward direction. Thus, falling (downwards) into the Pit from the level of the Plane of Death must clearly be connected with loss of enlightenment; while climbing (upwards) out of it again to roof-level (surmounting the Plane of Death) must, by the same token, be reflected in a regaining of enlightenment. Indeed, the inexorable southward progression of time through the Chamber makes it clear that the Pit *must* be climbed out of sooner or later. Everybody, in short, must either cross or circumvent this 'river of fire', and the very fact of doing so cannot help but produce a measure of enlightenment or purification. During the Final Age the 'gold' must, in the biblical metaphor, be 'tried in the fire'. Compare Paul's words at 1 Cor. 3:13, 15, '. . . that day dawns in fire, and the fire will test the worth of each man's work . . . If it burns, he will have to bear the loss; and yet he will escape with his life, as one might from a fire'. Paul seems to have foreseen what the Pyramid likewise reveals, namely that the Pit—the coming 'river of fire'—is not final, but has an essentially redemptive function.

READING: *Between* A.D. *1440 and 1521* (22) (24) *events will occur which will produce* (25) *a period of uniformly unenlightened physical reincarnations* (2) (3) (5)[83] *for those who have taken the downward path—a path through time* (1) *which will lead directly to the age of 'hell on earth'* (21).

Yet even at this lowly level the Messianic presence will still make itself felt (4) (6?),[84] *for this age too is a function of the Divine Plan* (8).

Between A.D. *1767 and 1848* (22) *the souls in question will undergo a period of turbulence* (20) *during which imperfect man will endeavour to regain some of his perfection* (26) *by physical means resulting in death* (26).

The onward path will then be resumed until, in the summer of 1914 (22), *all imperfect human souls start to reincarnate* (28)[85] *to tumble into an age of physical* (34) *ordeal* (21) (27) *which will bring with it man's chance of breaking out of the cycles of time and space* (33) (27).[86]

During this age, those who fail to 'turn to the right' in time (46) *will have to pass through a 'river of fire'* (36) *which will sweep 'man come of age'* (39) *into a deathly* (40) *'bottomless pit'. This will be the inevitable reward* (38) *for the unenlightened* (43). *Yet man's reaction to this very experience may lead him back towards enlightenment* (43) *through its perfecting and purifying effect* (41). *It will also produce the Messianic presence* (39). *Indeed, any souls who 'turn to the right' at the time of the great ordeal* (46) (53) *will still be able to regain more than the measure of enlightenment they originally lost* (33) (47), *for the Well of Life (q.v.) will still be waiting for them* (52) *with its Messianic promise* (51) *of the spiritual perfection of physical man* (48) (49) (52).

But for those who still fail to make the effort to 'turn to the right' (54) *a new path of utter death* (55) (57) *and physical degeneracy* (57) (58) (59) *will commence with the new year of* A.D. *2133* (32). *Despite a half-hearted effort at self-improvement* (67) *between* A.D. *2422 and 2477* (65) (66) (67), *they will relentlessly persist with this lowly path through time* (61) *up to and beyond the year* A.D. *2569* (64) *when, for these souls, the scope of the present prophecy comes to an end* (68) (69).

The dates A.D. 2422 and 2477, quoted in the last part of the reading, are approximate datings for the 'kink' in the Dead End Passage, and it is worth noting that they are fairly close to the terminal dating given for the Antechamber (q.v.). This raises the possibility that the

[83] See page 87, point (12).
[84] Compare Matthew 28:20, 'I am with you always, to the end of time.'
[85] See page 145.
[86] Compare the significance of the coffer on page 125.

kink may foreshadow a *general* effort towards reform encouraged specifically by the entry of the enlightened into their final path of spiritual escape—or at least the prospect of that imminent event. The passage then continues just far enough to show that there is no longer any possible escape from it to that higher path. For the latter is finally sealed in 2499—just seventy years (the 'standard' human life-span) before the terminal date of the Dead End Passage. 'Life goes on' at this lowly level—such is the apparent symbolism of this feature.

Meanwhile various interesting links, both with known historical events and with other parts of the Pyramid's chronograph, are to be observed in the data and reading for the Subterranean Inset.

To start with, the years 1440 to 1521—here depicted as the 'seedbed' of our present age—do indeed appear to qualify for that description. The year 1453 saw the final fall to the Turks of Constantinople, the last bastion of the eastern Roman Empire. That event was to have thunderous consequences. For a start, it was largely responsible for the flight of the scholars to Italy, bearing the knowledge and skills which had been largely lost to Europe ever since classical times. And from this new invasion of long-lost ideas sprang the great age of the Renaissance, with its overthrowing of accepted ideas, its spirit of open-mindedness and its return to the old written sources—tendencies long ago foreshadowed by the great Friar Bacon, and which, in turn, helped to produce the revolution in religious doctrine known to us as the Reformation. Among the most decisive moments in the founding of this latter movement were Luther's publication of his celebrated 'Ninety-Five Theses' in 1517, his excommunication in 1520 and the outlawing of him by the Diet of Worms *in 1521*. And it was in that same year that the reformers, now finding their feet as a movement, first called themselves 'Protestants' at the Diet of Spires.

But it should be remembered that what made possible the return to written sources underlying both Renaissance and Reformation was the invention of the printing-press, which took place in Germany *in the year 1440*.

Meanwhile the fall of Constantinople had had other effects. Because it had effectively closed the ancient spice and silk routes to the orient, men in Europe started to turn their minds to reaching the east by other routes. And it was not long before—again under the influence of the ancient knowledge, though this time in the form of geography and astronomy—the idea was born of reaching the east via a westward circumnavigation. To this idea we owe the American voyages of Columbus in 1492 and 1498, and of Cabot in 1497. In 1498 Vasco da Gama finally reached the orient by sea—albeit eastwards—while, in 1519, Magellan at last set out on the first global circumnavigation, and from these events centuries of European domination of the globe were eventually to spring.

The period 1440 to 1521, then, did see the laying of the foundations of the present world-order; and to the events of that period we can trace most of the fundamental developments and tendencies of our own age.

The 'turbulent idealistic period' marked out by the Lesser Subterranean Chamber between 1767 and 1848 is equally well corroborated by history. From the British imposition of the American tea-tax in 1767 we can trace the War and Declaration of Independence of 1776. The earth-shaking French Revolution of 1789 was followed by two more revolutions in 1830 and 1848—a year which was to see no fewer than six revolutions in Europe—while 1847 saw the appearance of Marx's *Communist Manifesto*. In the meantime, the Napoleonic wars had wrought havoc in Europe, culminating in the Battle of Waterloo in 1815. And yet this same period was to see the foundation of virtually all the world's existing Bible Societies—evidence of a search for a more spiritual kind of perfection which may indeed be directly associated with the 33·5″ width of the Subterranean Passage, symbolic of the continuing Messianic presence throughout the period under consideration.

Finally, the tumbling into the Chamber of Ordeal, in the summer of 1914, aptly corresponds to the beginning of the end of the old world-order which so clearly dates from about this time—whether in Europe or the Far East. The psychological trauma of the First World War, the crumbling of established values, the toppling of European dynasties, the Russian and Chinese revolutions, the theories of Einstein, the sudden explosion of scientific, military and commercial technologies, the advance of medicine, the consequent surge in world population—shaken by such upheavals, the old world could never be the same again.

Nor can it reasonably be objected that the prophecy of an end to the old order is something which we ourselves have 'fed into' the chronograph. The Chamber of the Ordeal is indisputably *there*, as it has been for the last four thousand years or more: all we have done is to fit prophecy to apparent fulfilment in order the more finely to date the process. And from this dating the pinpointing of the periods 1440 to 1521 and 1767 to 1848 follows quite automatically, simply confirming the justness of the proposed 'fit'.

Again, the inevitable conclusion to be drawn from the Pyramid's predictions is that there will be a 'general' rebirth from 1914 onwards: for rebirth is symbolised at this time not only by point (28) above, but also by points (18), (19), (20) and (21) on pages 109–10, and by the 'continuing reincarnation' symbolised by the Queen's Chamber Passage, especially from the year 1228 onwards. For the souls taking all three passageways rebirth is foreshadowed for the beginning of the Messianic era—a notion which accords completely with John 3:3ff. One is forced to deduce that a population explosion of unparalleled proportions during the age of 'hell on earth' is absolutely necessary for the fulfilment of the Messianic Plan. Indeed, the final destruction of the old order will no doubt spring directly from the strains which it will impose. Moreover if, as some suggest, civilisations have already existed on earth whose scientific achievements were equal or superior to our own, then it is no more than one might expect that more scientific advance has occurred in the last century—presumably through the awakening of ancient knowledge in souls now re-entering incarnation—than in the rest of known human history put together.

Again, an almost inevitable result of a general reincarnation of all souls, however long discarnate, would be an exaggerated 'generation-gap', since many children born at this time would presumably show mental and spiritual tendencies totally incomprehensible to their elders and more in tune with developments at the date of their last incarnation. Additionally, disruptive and destructive tendencies are likely to be exacerbated by the vastly increased leisure-time which results inevitably from the advance of scientific materialism. In most of the 'trade-marks' of our own age, in short, it is not difficult to read the 'signs of the times' and to identify it as the critical period represented in all three levels of the Pyramid's passageways as a time of universal reincarnation.

The Great Subterranean Chamber in particular must therefore represent none other than our present age, with all its ups and downs and the warning of a 'bottomless pit' to come—perhaps the very test from which Christians have long prayed to be spared in their Lord's Prayer. Yet out of it all, it seems, some redeemed souls will come, while others will simply soldier on into the future in their old accustomed way. And by A.D. 2133—the year before the Antechamber's second Messianic visitation—the 'path of escape' will at last close its doors on unregenerate humanity.

After A.D. 2569, it seems, the Pyramid has no more to say of these unredeemed souls. There is simply a silence. Perhaps their way, too, will be written in a Pyramid. But it will be some other Pyramid. For the Great Pyramid of Giza will have finished its task, and can at last be left in silence to moulder amid the sands.

Note: As point (70) of the Subterranean Inset data makes clear, the distance of the extremity of the Dead End Passage from the Pyramid's entrance is exactly 5448·736″. But this distance, representing in code the 'incomplete or imperfect world', is also the height of the incomplete Pyramid's summit-platform above the base. To this height, there has to be added a further 364·2765″ when the full-design Pyramid's crowning capstone is added (whether physically or merely in symbol) with the completion of the evolutionary plan for planet earth.[87]

But in this case, the passage-length itself may also be regarded as 'incomplete'. To it, likewise, a further measurement must presumably be added on the completion of the Plan. Which immediately poses the question—is there some feature of the Pyramid's plan which has yet to be added, something of crowning importance in the rock beneath the Pyramid *some 364¼ P″ beyond the end of the Dead End Passage?* If so, then it may lie either on the same level as that passage or—more likely, perhaps—directly in line with the slope of the Descending Passage (as the Trial Passages suggest—see page 199). In which case, whatever it is—whether chamber or time-capsule—has been waiting now for some four thousand years at the bottom of a deep shaft in the rock, 791 Primitive Inches below the floor level of the Dead End Passage and 1993½ Primitive Inches (some 166 ft. 3 inches) beneath the Pyramid's base. This level corresponds to a point, now permanently inundated, some 19 feet below the ancient level of High Nile[88]—a situation which reflects with intriguing accuracy the description of the interment of Khufu which Herodotus received from the contemporary Egyptian priesthood. The royal burial-chamber, they affirmed, was to be found *on a subterranean island surrounded by the waters of the Nile*. What that chamber really contains—if it exists at all—is a matter for conjecture. But how much longer, one wonders, will the long sleep of Khufu—whoever or whatever he is—remain undisturbed?

[87] Interestingly enough, this distance factorises as 7 × 4 × 13, and thus apparently shows the capstone as symbolising, among other things, the 'spiritual perfecting of the earthbound soul'. It is this event alone (so the Pyramid's overall height clearly suggests) that can bring to final fruition the 'incomplete or imperfect world' signified by the height of the summit-platform. The concept fits our general exegesis with great aptness.

[88] Or some 15 ft. above Low Nile (Nile-levels for the third millennium B.C. based on Pochan, who cites a published sedimentation-rate of 13 cm. per century).

The Great Subterranean Chamber (detailed reading)

The detail of the gigantic Great Subterranean Chamber (the chamber allegedly described by the *Book of the Dead* as the Chamber of Ordeal or Chamber of Central Fire) is considerable, and given that it had to be carved out of solid rock one must naturally assume that this detail is deliberate. The popular theory that the irregularities in the floor indicate that the Chamber was left unfinished simply does not accord logically with the flatness of its walls and ceiling, nor indeed with the fact that a further passageway (the Dead End Passage) was bored southwards *from the far side of the chamber*. Again, if the builders' methods involved finishing the roofs of subterranean chambers before excavating downwards to the floor, then one would have great difficulty in explaining why, in the Lesser Subterranean Chamber, it is the roof that has been left unfinished while the floor and walls are all flat. Unless, of course, *all* these features are symbolic.

Consequently we are forced to conclude that the contours of the Great Subterranean Chamber (see diagram on p. 150) were deliberately shaped to correspond to the 'ups and downs' of the apparently crucial age in question. And that age, as we have now established, is none other than our own.

How, then, are we to read those contours? We might assume from the code that the sheer irregularity of the floor indicates a lack of definite time-scale—but then the flat roof and walls tend to belie this idea, and to suggest that the various contours can, on the contrary, be read off against some sort of time-scale. After all, we have a 'starting-date' and a 'finishing-date'. Distance across the Chamber is therefore calculable against the time-scale which we have already established for the Chamber. In this way it should be possible to date the various features.

Thus it seems that, by taking a series of north-south 'slices' through the Chamber, we might hope to identify and date some of the events and tendencies apparently predicted for the period. We may accept that the floor's roughness and unevenness indicates a measure of approximation—and our results themselves are likely to be even more approximate, in view of the high premium which any attempt at dating must necessarily place on our interpretation of the measurements and diagrams of those who have investigated the Chamber.[89] Moreover there are indications that the datings for

[89] The conclusions which follow are based entirely on the data provided by Rutherford and the Edgars.

the whole Chamber may be adjustable to the extent of some three years.[90] Consequently the minimum tolerance for any datings arrived at will be of the order ± 3 years, and discrepancies of up to ± 5 years would not be too surprising. Meanwhile, we have to decide which north-south slices to take. The first obvious 'slice' must be that represented by the southward projection through the Chamber of the Subterranean Passage axis. This line will presumably correspond to the path of those who continue along the path marked out by that passage—a path, above all, of unenlightened materialism, such as is perhaps characterised by the leading Western civilisations in our own day.

The next fairly obvious slice to take is one lying sufficiently far to the right (west) of the first path mentioned (though still in the lower, or eastern part of the Chamber) to avoid the pit entirely. This slice presumably also represents a possible path through the Chamber, but in this case those who take it will, in symbol, have moved significantly to the right *vis-à-vis* the Subterrannean Chamber Passage. They can therefore be identified, perhaps, as those who reject the materialist philosophy of the rest of mankind and instead attempt to perfect themselves through a search for more spiritual values.

The high, western part of the Chamber, on the other hand, does not represent a 'path' at all—if it did, its precipitous east-west ridges would effectively block it. Instead it seems more appropriate to see it as a high plateau of bedrock with a number of artificial structures raised upon it. In short, it begs identification with the 'bedrock' of contemporary civilisation and the achievements which man has based upon it.

Hence the obvious third slice to take would seem to be one lying about half-way towards the west end of the Chamber, and passing through the 'bedrock' just east of the ridges which it supports. Since this slice would thus be lying almost exactly on the Pyramid's north-south centreline, we may see in it the rediscovery of a measure of enlightenment—albeit the intellectual enlightenment characteristic of civilisation, rather than its spiritual counterpart.

In order that our fourth slice should be representative of the achievements of civilisation, we would seem to need to choose a line passing through all the main 'ridges' of the western part of the Chamber. The selection of any given line is bound to be arbitrary but a point almost half-way between the third slice and the western

[90] See note 80 pages 139–40.

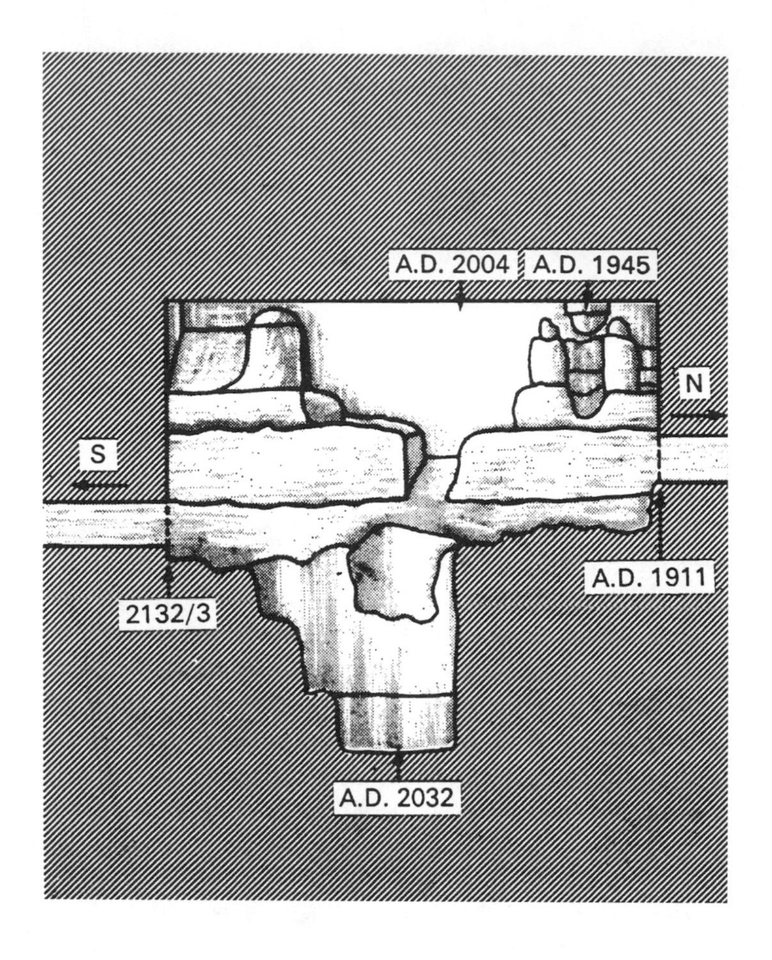

The Great Subterranean Chamber (looking west). Upper features lie at west end of Chamber, lower features at east end.

end of the Chamber would appear—on the basis of the unforthcoming diagrams available—to be fairly representative. It may be that alternative slices slightly to the east or west of this may be significant in their own right: or they may represent genuine alternatives to the one we have chosen—alternative 'histories of twentieth-century civilisation', so to speak, depending on the extent to which that civilisation tends towards the material or the spiritual in its dealings. At all events, it should be remembered that this fourth slice is arbitrary in a way which those we have hitherto chosen are not. Man's achievements in the arts and sciences may here be represented—with especial reference, perhaps, to the state of his basic life-support technologies.

Our final slice is entirely free of such arbitrariness and needs little identification, for it is clearly delineated by the line between the floor and the west wall of the Chamber. In view of its extreme westerly nature by comparison with the rest of the passage-system, this slice must inevitably have some connection with the level of spirituality or enlightenment during the period under consideration—and more particularly, perhaps, with the effect of that spirituality or enlightenment upon practical conduct.

The even more westerly recess or 'escape' in the west wall may symbolise the Well-Shaft, as has already been suggested. On the other hand it may also symbolise an event or period of special spiritual significance involving—like the Well-Shaft—a return to basic spiritual principles.

Meanwhile the soul—or possibly human society in general—seems to have open to it a number of alternative paths through the Chamber. To start with, admittedly, the choices are limited. After tumbling into the Chamber over a step whose notional height helps to indicate that all souls must return to mortality at this time (see page 145), man can either go right ahead and stumble into the 'river of fire' which will sweep him—albeit westward—into the pit; or he can edge to the right—or west—of his own accord, in which case the deepest part of the pit may yet be avoided. But further westward he cannot go without great difficulty: man's civilisation and its achievements (so the Chamber's layout seems to suggest) actually stand in the way of his search for spirituality. And it is only 'through' the temporary breakdown of that civilisation and its achievements—i.e., via the deep gully dividing the Chamber's western plateau in two—that the souls of men can finally reach the western end of the Chamber and so make good their spiritual 'escape'.

The chart on page 153 sets out these various tendencies against what appears to be the underlying time-scale, and on this basis it seems worth attempting a rather more detailed analysis of the developments foretold by the Chamber. It should be borne in mind, however, that the Chamber seems to have what we have described as an 'adjustable join' with the passage leading into it (see note 80 pages 139–40) and that its features themselves are extremely rough and irregular. Consequently a tolerance of at least ± 3 years should be applied to all datings given. It should also be remembered that general predictions are likely to be valid (like Asimov's fictional science of 'psychohistory') only for the mass of humanity, and not for specific individuals or small groups within it.

Subject to this general caveat, we can proceed to a read-out of the

Great Subterranean Chamber's 'outline of history' for our own age.

READING: *The years 1914 to 1918 will see a steep material decline. Between 1921 and 1932, however, man's civilisations and technologies will have the chance of achieving a great deal, and after a temporary setback between 1932 and 1939, followed by a deep twelve-year 'trough', there will be another period of rapid potential progress between 1951 and 1965.*

But 1968 will mark the start of a decline in the achievements of man's advanced societies, and the downward slope will steepen dramatically around 1971. By 1978 there will be a further collapse to what might be termed 'subsistence level'. Yet civilisation itself will survive this collapse, showing only a slight downward tendency for a further twenty-six years.

Then, in around 2004, 'the bottom will fall out of the world'. Both world-civilisation and its technologies will quickly collapse to rock-bottom by 2010,[91] *and will remain at that level for at least fifteen years.*

Thus far, it must be acknowledged, the detailed predictions of the Great Subterrannean Chamber seem to fit both historical fact and the reasonable forecasts of contemporary experts to a remarkable degree. The periods of both World Wars seem to be clearly identified, as do the two periods of postwar recovery—though not, of course, in specific terms. The series of world economic setbacks in the nineteen-thirties are well predicted. A general sapping of the morale and self-confidence of the leading world-societies—with all that that has led to—can certainly be traced back to around 1968, when the 'drop-out' phenomenon among the young first started to have a major impact. And the seventies, when the great world population, resources and pollution problems started to make themselves universally felt, are correctly identified as a time of growing crisis. The prediction of further setbacks before 1980, and of a major breakdown of civilisation shortly after the end of the present century, is no more than a confirmation of what experts in a number of fields have long been saying, and indeed of what most intelligent people have already in large part come intellectually to expect—even if not yet fully to believe.

Meanwhile a number of what appear to be spiritual or moral developments are also marked out by the Chamber for the same period—though whether these are meant to be seen as complementary to the above predictions or as alternatives to them is open to some doubt. They include a remarkable but short-lived spiritual or moral revival between 1930 and 1937, and a period of strong spiritual influences starting in 1936, culminating in around 1945 and termi-

[91] Compare the 'destruction of the world' forecast for 24th December 2011 by the ancient Mayan calendar.

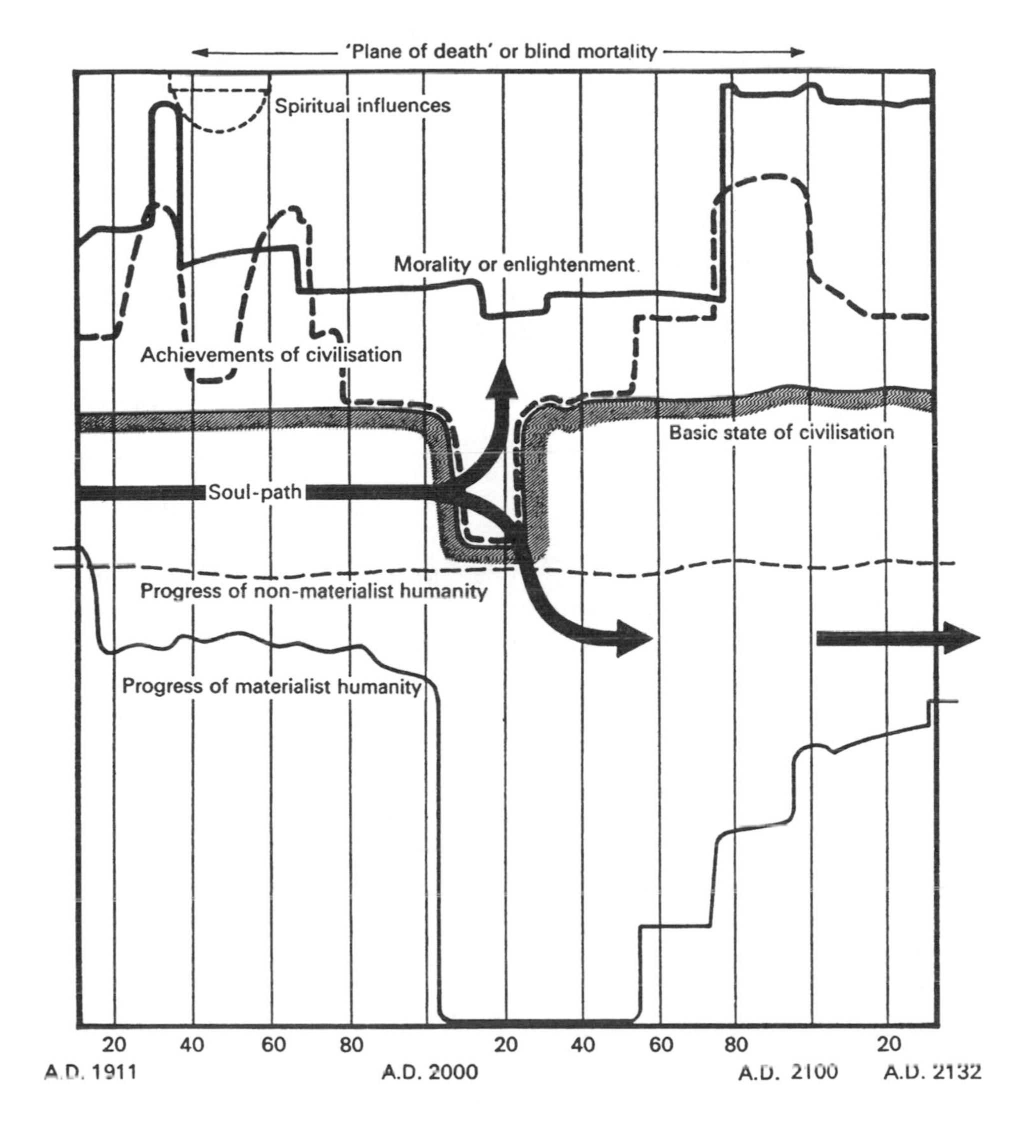

Timechart of the Great Subterranean Chamber.

nating in 1955. At the same time a gradual improvement in general morality or enlightenment seems to be indicated, leading to an abrupt collapse around 1967 and a further collapse with the commencement of the 'river of fire' in around 2014.

These 'spiritual' predictions are naturally less easy to check for accuracy than their more material counterparts, yet the period 1945 to 1955[92] was the period of the discovery of the now-famous Dead

[92] Dates which encompass the re-establishment in 1948 of the state of Israel.

Sea Scrolls, whose influence in the spiritual sphere has already been profound—even though it does not yet appear to have penetrated very deeply into the citadels of the religious Establishment. Throwing new light on the origins and teachings of the Essenic Movement, they have already done much to put infant Christianity into a truer social and historical perspective and to illuminate the very background and assumptions of Jesus of Nazareth himself. Consequently they have permitted something of the 'return to basic spiritual principles' already suggested by the symbolism of the Subterranean Chamber's 'escape'.

Whether the partial collapse predicted for 1967 has any connection with the inception of the so-called 'permissive society', and the increasing rejection by the young of established standards of conduct, is something which the reader must decide for himself. Certainly this date is a reasonable one for the developments in question, which are perhaps not unconnected with the simultaneous loss of morale and self-confidence already noted as having started to occur among established Western societies at this time.

The predicted entry into the 'river of fire' in 2014 would suggest that the coming 'time of ordeal' will start later for the spiritual than for the more materially-minded. On the other hand this staggering of events actually seems a fairly improbable development, and may suggest that the spiritual path to which we have been referring should in fact be seen as an *alternative* path to the more material one outlined above for our society, and not as complementary to it. The implication in this case would be that, if civilisation had taken a more enlightened or spiritual path than it actually has, then (as the chart above suggests) technology might have made less rapid progress in the nineteen-fifties, the major world-crises of population, raw-materials and pollution would have been tackled earlier (around 1962) and with much more success, and the inevitable world-wide collapse predicted for the early twenty-first century would have been both later in date and less severe in its effects. And it would seem, a *priori*, that these are by no means unreasonable as a set of deductions.

Accepting, therefore, that man in general appears to have chosen a path through the Chamber lying slightly to the right—or the 'good' side—of centre, we are now in a better position to complete our detailed read-out.

READING *(continued): By around the year* A.D. *2025 world-civilisation will be re-established, and by 2055, with material conditions at last rapidly improving, human technology will revive to at least its former level. Then, in around 2075, there will be a sudden explosion of progress on all fronts, and a new civilisation of extraordinary vigour will arise whose physical aspects will last until the year* A.D. *2100, and whose spiritual achievements—which may be of unprecedented magnitude—may even endure until the very end of the era predicted. This is dated at* A.D. *2132–3, just prior to the second Messianic visitation.*

If the above read-out is at all valid, then we have to expect a fifty-year period of crisis or ordeal marking the end of the present age (astrologically that of Pisces) and the beginning of the new (that of Aquarius). And it is in the very midst of this period of 'hell on earth', shortly after the re-establishment of some sort of civilisation, that—if our earlier exegesis on pages 118–19 is to be relied on—we should expect the first of the predicted Messianic visitations. As Jesus of Nazareth himself put it at Matthew 24:29–30, 'As soon as the distress of those days has passed, the sun will be darkened, the stars will fall from the sky, the celestial powers will be shaken. *Then* will appear in heaven the sign that heralds the Son of Man. . . .' The degree of accord between the two predictions is in fact little short of remarkable, and we have little choice but to see the Great Subterranean Chamber as representing the era, *par excellence*, of the biblical 'last times' and of the initial Messianic return.

The reader must of course judge for himself the likely validity of the read-out suggested. Regarding timing, the author's own impression is that the datings given are so far running, on average, some one to three years ahead of actual events. For purposes of extrapolation into the future, therefore, it would seem advisable to add some three years to the datings suggested above.

Meanwhile two associated points are perhaps worth noting. First, the astronomical date given for the actual beginning of the Age of Aquarius by the French Institut Géographique National is A.D. 2010. Second, if we compare the above predictions with the conclusions of the Club of Rome—a well-known 'think-tank' of seventy eminent international scientists and businessmen—we find that their published, computerised prognoses of world population, pollution and resources figures point specifically to a time of dire crisis *around* A.D. *2020, and culminating in the year 2050*—both of which dates fall comfortably within the period already delineated by the Great Pyramid's chronograph as the 'time of ordeal'.

Modern Features

The hypothetical would-be initiate—whose footsteps we have been tracing through the Pyramid—has now completed his visit, for the builders left no further passages or chambers to explore, so far as is known.

Nevertheless the modern visitor will notice that there are a number of rough tunnels and excavations in the Pyramid which, with a handful of other features, have not been mentioned above. These are all quite modern in origin. For the reader's information, however, the various 'alterations' which the Pyramid has suffered since its construction are listed below in roughly chronological order:

Feature	*Originator*	*Approximate date*
Al Mamoun's Hole	Caliph Al Mamoun	A.D. 820

This forced passage, at the level of the seventh course of masonry, is the entrance at present used by tourists. It is situated in the centre of the Pyramid's north face, and joins the original passageway at the Granite Plug.

Rough excavation beneath King's Chamber coffer	Caliph Al Mamoun	A.D. 820
Almost total removal of original casing	Arab Mosque-builders	13th century onwards

Following Al Mamoun's attack on the Great Pyramid's casing, it seems that the builders of Cairo's mosques preferred to rob the Giza pyramids for their fine limestone, rather than select their own materials from the original quarries in the Moqattam hills. The much harder granite blocks of the second and third pyramids did not interest them, however, and most of these fine stones from Aswan still litter the ground in the vicinity of their parent-monuments.

Rough tunnel eastwards from Queen's Chamber niche	Unknown 'treasure-seekers'	?
Widening of outer end of north airshaft to King's Chamber	Unknown 'treasure-seekers'	?
Hieroglyphs on upper right-hand gable-stone above entrance-tunnel	Friedrich-Wilhelm IV of Prussia (Lepsius expeditions)	19th century
Rough tunnel westwards from first low section of King's Chamber Passage	Captain Caviglià (Italian)	1817 and 1837

Caviglia (who was responsible for the first modern clearance of the lower Well-Shaft and of the lower passage leading to the Great

Subterranean Chamber, both formerly filled with rubble) excavated this limestone tunnel in order to follow the course of the north air-shaft of the King's Chamber, in case a further chamber should lie at the end of it. In so doing he discovered the extraordinary multiple bends in it described on page 124, note 61.

Enormous gash in masonry of south face	Colonel Howard Vyse	1836 onwards
Rough vertical tunnel to four upper Construction Chambers	Colonel Howard Vyse	1836 onwards
Excavation under N.W. corner of Queen's Chamber gable	Colonel Howard Vyse	1836 onwards
Shaft descending vertically from bottom of subterranean pit	J. S. Perring	1838

Vyse and his colleague Perring were perhaps the most significant of the Pyramid's early nineteenth-century investigators. Vyse in particular was quite ruthless in his excavation of what he saw as a 'pagan' monument, and unashamedly blasted his way through the ancient masonry with liberal quantities of gunpowder.

Steel mast on summit-platform	Astronomers Gill and Watson	1874

The top of this mast was designed to show the theoretical position of the apex of the completed Pyramid.

Apart from the above, the passageways and chambers in general have suffered much at the hands of the original excavators, torch-bearing visitors and tourists anxious to leave their mark in other ways—not to mention the effects of at least two earthquakes (in A.D. 908 and 1301 respectively) and geological subsidence amounting to about a foot under the Pyramid's centre. The installation of hand-rails, wooden gangways and rudimentary electric lighting during the present century by the Egyptian Department of Antiquities has further mutilated the Pyramid's interior, albeit with the best of intentions. That it should still be possible, despite all the above, to arrive at a highly accurate idea of the Pyramid's original features and dimensions, is perhaps a not inconsiderable testimonial to the skill and thoroughness of the ancient builders and those who have laboured for so long to measure and survey what still remains.

Noon-Reflections of the Original Casing

If our hypothetical would-be initiate had left the Pyramid, after his exploration of its interior, at noon on the day of the summer solstice, he would no doubt have been interested to observe that the sun's rays, reflected by the polished white-limestone casing-stones, cast no shadows, but instead created *a star-shaped reflection* on the desert around the Pyramid—a reflection whose seasonal variations were distinct enough for the ancient Egyptians to use them as a kind of calendar. The phenomenon is illustrated on page 337, and may have more than a little to do with the fact that the ancient Egyptians are known to have called the Pyramid *Ta Khut* (The Light).

The Bethlehem-Angle

If our ancient visitor had then chosen to head away from the Pyramid in a north-easterly direction, regardless of swamps, sand or sea, at an angle of 26° 18′ 9·7″ to the Pyramid's east-west equator, he would have passed directly through the town of Bethlehem—led, as it were, by the 'Light' or 'star' described above. The apparent link with the traditional nativity story of the Three Kings is taken up in chapter 9.

The Quadrantal Angle

Alternatively, if the visitor had chosen to head north-east along the Pyramid's quadrantal angle (that marked off by the intersection of the Pyramid's square base by a circle of equal area and having the same centre) he would have found himself passing through the western outskirts of Jerusalem. This might possibly suggest, in terms of the code, that the city in question may have some connection with a 'sunrise' designed to bring the physical world into line with the spiritual, conceivably as the future Messianic capital.

The 'Heavenly Messiah'

If, on the other hand, the ancient visitor, on leaving the Pyramid, had headed southwards towards Sakkara and King Zoser's archetypal step-pyramid, and had turned round one final time to visualise to himself the complex of chambers and passageways within King Khufu's monument, then he might have become aware that the passages and chambers of the Pyramid etch out between them in elevation a figure remarkably similar to that of a man wearing a

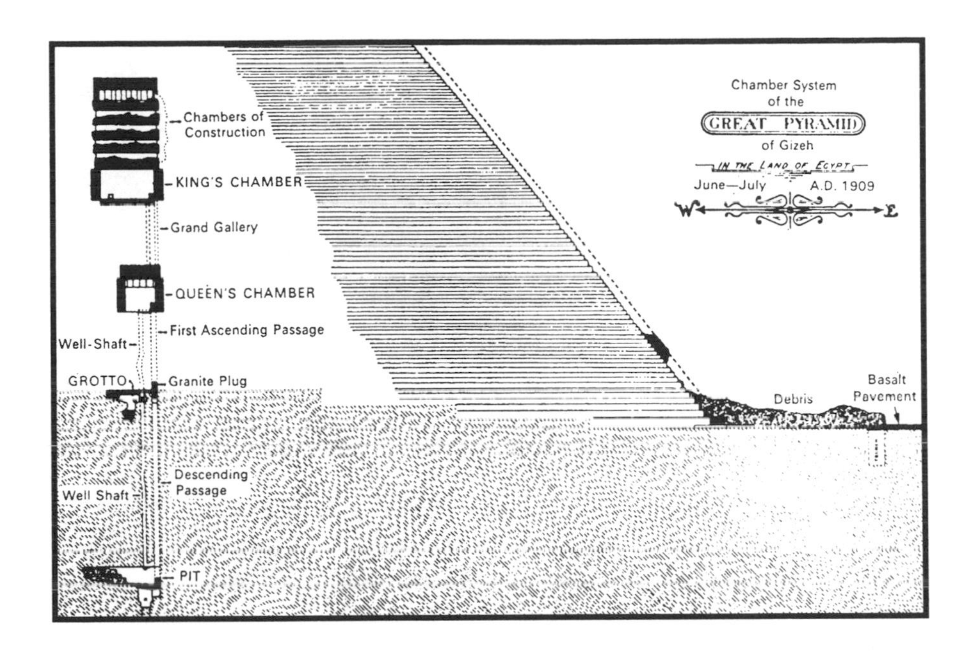

Elevation of the Passage-system and Chambers (looking north)

fivefold crown, looking towards the west, and crushing beneath his heel the contents of the Subterranean Pit. As the diagram above makes clear, head, stomach, genitals and feet are all clearly identifiable, with a clarity considerably greater than that of the primeval star-pictures with which the astronomically-minded ancients were familiar. One wonders whether some such realisation might not at some point have been the basis for the ancient esoteric Pharisaic notion of the infinitely-tall Heavenly Messiah—the 'first Adam' or Archetypal Man, cognate with the *Logos* of John's gospel and there referred to specifically as 'the light'. This notion is also found in the christology of Paul the Pharisee, and to it we possibly owe in large measure the Christian doctrine of the divinity of Jesus of Nazareth.

It is by no means unlikely, of course, that such avid enthusiasts of the hidden lore as Paul and the Pharisees were already familiar, via the Egyptian priesthood and their own sacred writings, with the contents of the Great Pyramid and its message for man. Indeed, we have already discovered strong hints that one of the authors of John's gospel may have been an initiate of the mysteries, and the evidence adduced in chapter 9 suggests that Moses too was a hierophant. As for Jesus himself, we have clear evidence of his familiarity with the Pyramid's message, *and also of his having visited Egypt.*

Recent Revelations

Since this book was originally published, research on the Great Pyramid has continued unabated. Cosmic-ray detection experiments have failed to find further chambers, but investigations involving magnetometers and gravimeters claim to have detected sand-filled voids behind the walls of the Queen's Chamber Passage. Both the pyramid and its surroundings have also been probed by researchers from Waseda University in Japan with sensitive ultrasonic scanners. As a result, the south-western boat-pit has been found to contain, like its south-eastern neighbour, a great pharaonic 'solar boat', while further underground tunnels and chambers have been detected near both Pyramid and Sphinx.

More recently, efforts to improve the Great Pyramid's ventilation have at last resulted in a proper exploration of the long-neglected Queen's Chamber air-shafts. With the aid of a small tracked robot, the engineer Rudolf Gantenbrink was able to confirm my original conviction that the shafts do indeed extend most of the way to the exterior of the structure. The southern one, however, proved to stop short after some 65 metres at what appears to be a stone door or portcullis only some 25 metres short of the Pyramid's surface.

This unexpected find was in due course to be taken up by Robert Bauval and Adrian Gilbert in their immensely stimulating book *The Orion Mystery* (Heinemann, 1994). In the course of it they proposed a stunning new theory that would see the pyramids of the west bank's 'land of the dead' as direct reflections on the ground of the stars of the constellation Orion (= Osiris), with the main Giza pyramids corresponding to the imperfectly aligned stars of Orion's belt and the Nile itself reflecting the celestial Milky Way.

On this basis, the southern airshafts in particular would demand to be seen not as ventilators but (as originally proposed in this book) as *escape-shafts* – in this case for the Osirian soul as it wings its way via the Queen's Chamber's southern airshaft directly towards the star Sirius (= Isis), or via the King's Chamber's southern airshaft directly towards the star Alnitak in Orion's belt. The traditional designations 'Queen's Chamber' and 'King's Chamber' would then acquire an unexpected new validity.

At the same time, the astronomical alignment of the King's Chamber's south airshaft is claimed to suggest a date for the Pyramid's construction of 2450 B.C. Yet my own astronomical application of the building's dimensions in my book *The Great Pyramid:*

Your Personal Guide (Element, 1987) had already suggested a date for this of as early as 6250 B.C. Clearly, then, the lessons that the monument has to teach us are far from exhausted yet.

4

Feedback: The Question of Validity

IN THE COURSE of the previous three chapters we first of all suggested that the Great Pyramid might contain a message based on some kind of architectural and numerical code. We then listed a number of items which appeared to be of possible relevance to this concept, and proceeded to undertake a 'total' read-out of the Pyramid's features and dimensions on the basis of that list.

The time has now come to look back at the read-out in an attempt to assess the likely validity of the suggested code. What are the characteristics that we would expect to find in our read-out, if our initial code-hypotheses are valid?

First of all we would expect the message to 'make sense'. Its ideas or storyline, in other words, should not be totally disjointed or illogical, nor should its underlying values and assumptions appear to change arbitrarily from moment to moment. And by this initial criterion at least, our read-out in chapter 3 would seem to be acceptable.

By the same token, we should expect the message to be clear and relatively straightforward. There should be little or no room for doubt or ambiguity—for, whatever else the designer of the Great Pyramid may have been, he certainly does not seem to have been woolly-minded. And here again the message appears to fulfil our expectations—which is not to say, of course, that the clarity of the proposed read-out cannot be further improved upon as a result of subsequent research.

Next, we would expect the message to accord with known facts and to be consistent with at least a proportion of man's reasonable assumptions. In the case of what appears to be a basically chronographical message, therefore, the message should square with the events of recorded history, and should reach conclusions for the future consistent with at least some sources of prediction. And indeed we have already seen, again and again, the almost uncanny concordance between the Pyramid's chronograph and known historical events; while the information set out in chapter 9—combined with the expectations of a constantly growing number of 'conventional' experts in various fields—makes it clear that the second, predictive part of this particular criterion is likewise amply fulfilled by our read-out.

A further potential hallmark of the message's validity would tend, oddly enough, to be a measure of unorthodoxy. If the religio-philosophical content of the message as derived turned out to be a total vindication of, say, nineteenth-century Anglicanism or eleventh-century Theravada Buddhism, then this would suggest that the reading had been 'fiddled' to fit a preconceived schema in the translator's mind. Nineteenth-century Anglicanism, after all, was not the same as ninth-century Christianity; still less did it resemble twentieth-century Catholicism or the Christianity of the early Church Fathers. It would be nothing less than astonishing, in other words, if it subsequently turned out that one particular nineteenth-century Christian sect had stumbled—quite by chance, as it were—on the 'truth', whereas no other religious sect in the whole of history had ever done so.

And in terms of this criterion, too, the read-out arrived at in the last chapter seems to be basically satisfactory. It suggests, after all, that no existing religious 'system' has a monopoly of the truth. The conclusion seems inescapable that Jesus of Nazareth, far from opposing the teachings of figures such as Krishna and the Buddha, actually 'stood on their shoulders'—a universalist standpoint apparently reinforced by the Pyramid's whole external design. And yet, at the same time, the message which emerges is not in any conventional sense the Christian one: the basic reincarnationist doctrine, the multiple Messiahs, and above all the depiction of Christianity as an *inferior* path—all these mark out the Pyramid's message (as here reconstructed) as a distinctly unorthodox one. And this, on the whole, must *add* to the likelihood that the message itself may be valid.

Perhaps the major criterion of validity, however, is that the message should be important and purposeful. It is quite unthinkable that the message of a structure so immense as the Great Pyramid should be anything else—assuming, that is, that its function was to leave a message in the first place. How far, then, does our read-out as reconstructed meet this vital criterion?

What is clear above all from our exegesis in the last chapter is that the Pyramid's message represents nothing less than a blueprint for the final culmination of the present evolutionary phase of planet earth. It depicts the process whereby mortal man becomes a 'living creature' and is finally 'made into the image of God.' And it shows very clearly that that final apotheosis is something which only *man himself* can carry through—and even then only at the cost of many reincarnations and great physical self-sacrifice. Help there may be from other dimensions, but it is basically man who must be the agent. Salvation there may be, but there can be no other saviour but the son of man himself.

Clearly, then, if this is the burden of the message, it is as important as any that has ever been proclaimed. And, moreover, it is vital that it *should* be proclaimed. For if man has a paramount purpose to fulfil and a destiny to help bring about, then he must first learn what that purpose is. In the words of Edgar Cayce, 'Mind is the builder.' First he must know, then he can act—for there is no other to act on his behalf. And as scripture itself proclaims (John 8:32), from the day when man shall learn the truth, that truth will start to set him free.

So far, then, our reconstruction of the Pyramid's message seems to satisfy without difficulty the various criteria which we would expect such a message to meet. At this point, therefore, it seems appropriate to examine the message and its code from a more technical viewpoint—that of applied linguistics.

The basis of our hypothesis has, after all, been that the Pyramid contains a message based on a code. But if that code is basically architectural and numerical in form, it none the less represents a means of communication between two groups of human beings—both of whom presumably would normally use a form of language more akin to what is usually understood by the term. One would assume, in other words, that the code *is* a code and no more—an arbitrary system of symbols standing, in this case, for ordinary human speech. Consequently we would expect to find in the Pyramid's code (and in the message which emerges from its application) unmistakable signs of linguistic origin—close parallels, in other words, to the universal characteristics of human language.

What are these tell-tale characteristics? To start with the general, and work towards the particular, perhaps one of the first features to look for is the phenomenon of 'linguistic redundancy'. This feature, observable in every known language, is the process whereby the number of signals in an utterance always adds up to *more than* the minimum necessary for communication—the process which ensures that the main ideas of an utterance are expressed simultaneously in a number of different ways.

An illuminating example of this is provided by the use of the telephone. The participants in a telephone conversation have, for a start, to rely entirely on the aural aspects of language: they are denied the use of all facial expressions and other visual signals. This already represents a very serious communicational loss (considerably more so than is often realised) for these features have an extremely important rôle to play in normal speech, and are so ingrained in normal usage that most users of the telephone find it quite impossible to stop producing them, even though in the telephonic context they are quite useless.

What is significant, however, is that linguistic communication via the telephone still succeeds, despite the loss of some of its basic features. Moreover it can be demonstrated that communication can continue to succeed *even when interference results in the loss of up to 90% of the linguistic signals uttered.* Ordinary language seems to be so designed as to permit communication under the most unpromising of conditions—even conditions which result in the loss of a large proportion of the signals produced. The same could not be said, for example, of the conventional postcode, where the loss or substitution of *a single letter or figure* can, in theory, result in delivery to a totally wrong address.

And what of the Pyramid's message and code? It is gratifying to note that exactly the same principle of linguistic redundancy seems to occur throughout the Pyramid's symbolic design. The basic external design, for a start, proclaims in at least half a dozen different ways that the building symbolises the planet earth; and the Pyramid's reduced design, lack of a capstone and six-sided shape, all reveal that that world is in some way to be regarded as imperfect or incomplete.

Again, to consider the Pyramid's internal features, the entrance itself (see page 51) reveals in four different ways that it is the 'gate of the dead' (width, height, horizontal 'death factor' and number of course), in two different ways that it involves a decline or loss of enlightenment (eastward displacement of axis and downward slope) and in two different ways that it involves physical existence (cross-section and nature of masonry). To take other examples, the features of the Well-Shaft indicate in at least three different ways that, entered from the top end, it represents a path of death: the Queen's Chamber speaks in at least twelve different ways of an era of death and rebirth: the Well-Shaft indicates in seven different ways that, for the 'ascending' soul it involves an eventual return to the physical planes: the Grand Gallery lays claim in nine different ways to being reserved for the spiritually perfect enlightened ones and reveals in four separate ways that those who ascend it need not reincarnate during the period it represents. Again, the Antechamber's Granite Leaf identifies itself in nine different ways as symbolic of a returning, spiritually inspired Messianic leader while the Chamber's dimensions suggest in ten different ways that the function of the age in question is to bring perfection to the basis of human life. Finally, the reading on page 128 for the King's Chamber alone reveals at least ten quite separate examples of multiple linguistic redundancy of the type we have been examining; notable among which is the case of the coffer itself. For all three of the coffer's exterior dimensions speak in code of the 'death' which its very form so clearly symbolises—thus providing an excellent piece of evidence that our standard factorisation and decoding procedures have been valid ones.

In short, the multiple symbolisation typical of the principle of linguistic redundancy is to be observed virtually throughout the Pyramid's external design and internal passage-system. And, as in the more normal context, its inclusion can be seen as a logical safety precaution designed to ensure that, even if a number of features were to be in some way altered or destroyed, the vital features of the message would still succeed in getting through.

To turn now to the typical internal mechanics of language, we should expect to find in the Pyramid's code itself a finite, though possibly extensive, 'lexicon' or list of signals, a handful of basic grammatical rules and an almost infinite number of possible permutations of the two. And such, once again, turns out to be the case. The signals are here represented by the various numerical and architectural items listed on pages 37 to 39; and the grammatical rules by the four mathematical functions of addition, subtraction, multiplication and division, by the directional and step-symbolisms outlined on page 37, and by the rules of factorisation referred to on page 41. Moreover, we would expect to find that the words or signals—and even whole phrases—are heavily dependent upon context for the establishment of their precise meaning; and yet again, as I have pointed out on page 41, this does appear to be the case.

Finally, turning to the words or signals themselves, we should expect them to be clearly identifiable, logical in their interrelationships, and constant in their range of meanings—while none the less heavily dependent on context for their *precise* interpretation. And the sections in chapter 2 on tolerances and on self-validation, together with the ensuing read-out, seem to demonstrate that these conditions too are amply fulfilled.

But at this point a most interesting possibility arises. For there is one arithmetical quantity to which we have not so far assigned a meaning—a quantity which none the less makes a number of appearances in the Pyramid's passage-system. And that is the number thirteen. This number has become so firmly associated in the popular mind with the notion of bad luck that it is easy to forget the fact that in the ancient mystery religions the number thirteen was the characteristic number of participants in many an ancient cult group and sacred meal—a fact reflected, of course, in the size of the original Nazarene 'cell' itself.[1] These two apparently contradictory meanings for thirteen clearly made it dangerous initially to assign a code-meaning to the number on the basis of either one or the other interpretation.

But having completed our analysis of the Pyramid's features, we can now compare the various occurrences of the number thirteen and attempt to discover their common denominator. And it soon becomes clear that the meaning suggested on page 98—namely 'soul', or even 'spiritual man'—is at least a possible one. Application of this meaning to the various occurrences of the number thirteen throughout the Pyramid should therefore not only confirm or deny its validity, but also serve as an interesting check on our earlier readings for the passages and chambers in which it occurs. These occurrences are as follows:

Height, sloping passages	52·7452″ (4 × 13)	*The terrestrial (i.e. earthbound) soul*
Depth and width, inset stones in Grand Gallery	10″ × 13″	*The Millennial soul—i.e., the souls of those destined to enter the Millennium*
Length of 2 RC-square King's Chamber Passage (first part)	52·02874″ (4 × 13)	*The earthbound soul 'encased' in mortality*
Length, limestone floor in Antechamber (otherwise mainly of granite)	13·22729″ (13)	*The earthbound soul passing through new spiritual influences*

[1] Note, however, that it is possible to read John's account to mean that there were fourteen, not thirteen, people present at the Last Supper, the 'beloved disciple' being a possible 'extra'.

Width, granite Antechamber ceiling	65·25603″ (13 × 5)	*The soul(s) of the spiritual initiates/Great Initiate overshadowing the soul's path*
Total length, *limestone* floor of King's Chamber Passage	65·25603″[2] (13 × 5)	*The soul(s) of the initiates/ Great Initiate influencing or entering the physical world between* A.D. *1933 and 1999*[3]
Interior width of coffer	Approx. 26″ (2 × 13)	*Productive of the soul*
Sum of coffer's three dimensions	169·7173″ (13^2)	*The 'soul of souls' or ultimate soul*
Width of tunnel to lowest Construction Chamber	26″ (2 × 13)	*Productive of the soul*
Course-number of casing-outlets of Queen's Chamber airshafts	91 (7 × 13)	*The spiritual perfecting of the soul*
Course-number of casing-outlet of King's Chamber's southern airshaft	104 (8 × 13)	*The rebirth of the soul*
Thickness of course 104	26″ (2 × 13)	*Productive of the soul*
N.-S. distance cross top of Well-Shaft	26·7021″ (2 × 13)	*Productive of the soul*

Consideration of these thirteen occurrences of the quantity thirteen in the Pyramid's design[4] not only makes it clear that the reading 'soul' must be the correct one: it confirms convincingly the correctness of the interpretations we have already placed on the various features referred to, while casting some interesting extra light on the manner of the expected Messianic coming.[5] A more fitting—and

[2] Note the exact, and thus clearly deliberate, correspondence between this figure and the one above.

[3] This reference seems to suggest that the Messianic soul or Christos will already be active in the physical sphere long *before* the arrival of the physical man personifying the Messianic presence—and/or that the Messianic kingdom will be prepared beforehand by the returning initiates.

[4] Compare also feature (46) on page 83, which would therefore suggest an association with the soul of an initiate.

[5] In this connection it is interesting to note that the vertical cross-sectional area of the sloping passages (41·21″ × 52·745″, and thus 2RC × 4 × 13″) now has a quite specific symbolism—namely 'the reincarnation of the earthbound soul'. Thus it clearly does *not* represent the discarnate soul between incarnations, which may be thought of as escaping temporarily from the earth-planes (i.e. the Pyramid's limestone) until it finally returns to take up the task again from the point where it

gratifying—check on our earlier conclusion could scarcely be wished for.

In conclusion, then, a critical analysis of the main features not only of the symbolic code provisionally assigned by us to the Great Pyramid, but also of the read-out to which it inevitably leads, suggests that both code and read-out bear, in their essentials, many of the hallmarks of truth. That is not to say, of course, that their reliability and accuracy cannot be improved upon: but it does suggest that the apparent message of the Pyramid merits a great deal of serious thought.

Note: Numerological Parallels in Egyptian Antiquity

Given that the symbolic numerology of the Great Pyramid was as established in the preceding chapters, that part at least of the Pyramid's knowledge seems to have been preserved by the ancient schools of initiates (see Part II of the present volume), and that the Egyptian priesthood in particular is known to have survived into later and better documented dynastic times, it would seem reasonable to examine the ritual number-symbolism of later centuries for signs of affinity with the Pyramid's code.

There was, for example, an enduring tradition among the ancient Egyptians that the soul, at death, is examined not by one, but by forty-two 'assessors'—whose forty-two respective 'questions' are duly set out in a variety of sources.[6] Only when all forty-two questions could be satisfactorily answered by the reincarnating soul could it gain admittance to the glorious after-life.

Now forty-two seems an oddly arbitrary number to choose, either for the assessors or for their questions—unless we interpret the number in terms of the mathematical function 6×7. And this, as we have seen, is the Pyramid-code for the preparation of spiritual perfection—a concept which seems admirably to fit the spiritual function of the assessors in question. Meanwhile we find this same figure reflected in the forty-two settings-up of the Israelites' tabernacle during their exodus through the wilderness—the 'test' which they likewise had to pass. And here also the number-symbolism seems to indicate the preparation of spiritual perfection—in this case

[5] (*continued*) laid it down. (Limestone, as we saw on page 130, cannot symbolically 'contain' spirit.) And this very process of escape and return seems to be specifically referred to by three of the four air-shafts.

[6] See, for example, the concluding chapter of Joan Grant's *Winged Pharaoh.*

the inner process of self-perfection which links a man's initial commitment (i.e. the Red Sea crossing) to its final consummation (represented by the Jordan crossing and entry into the Promised Land).

In his book *The Sphinx and the Megaliths*, John Ivimy points out that a belief in karmic reincarnation seems to have been basic to the original Osirian religion—the process being suspended only in respect of the souls of the exceptionally wicked on the one hand, and of the pharaoh himself on the other. In the case of the wicked, dismemberment and thus annihilation seems to have been foreseen by the accepted system; while in the case of the pharaoh, his very identification with Osiris, the Lord of Eternity, itself ensured the immortality of his soul, and would therefore lead to blissful union with the eternal solar divinity.

It is Ivimy's thesis that Osiris, in his rôle as Judge of the Dead, merely acts as the spokesman, or mouthpiece, for the automatic karmic law. The typical Egyptian representation of the Judgement shows the gods Anubis and Horus weighing the soul against a feather, with Thoth recording the weight, while Osiris himself—holding crook and flail crosswise on his chest—remains aloof from the proceedings, and merely pronounces the appropriate sentence.

But it is Osiris's crook and flail that most intrigue Ivimy. What, he asks with some reason, would the Judge of the Dead want with two implements which are conventionally identified by Egyptologists as the symbols respectively of animal- and plant-husbandry? As Ivimy sees them, they can be nothing less than the symbols of the law of karma. It is the function of the hook-shaped crook to represent the concept of karmic reward, while the flail portrays that of karmic punishment—an Egyptian version of the familiar carrot-and-stick, no less. Hence the fact that Osiris holds the hook in his *left* hand and the flail in his *right*—the natural arrangement for a right-handed man (for Osiris could never, for the Egyptians, be left-handed!) who wished to propel a recalcitrant beast from A to B.

Not that there is anything particularly rewarding, at first sight, about a hook. But just as, in the case of the carrot and the stick, the reaction of the donkey is theoretically governed by its *knowledge* of the existence of both, so in the Osirian context the crook and flail were valid enough symbols for those who took the law of *maat* to be axiomatic to their very existence. It was the prospect of karmic reward that enticed the soul forwards, the fear of karmic punishment that discouraged it from slipping backwards. In such a context the Osirian insignia were apt enough symbols of the karmic process, and one would therefore expect to find a close symbolic association between them and the concept of karmic reincarnation.

As Ivimy points out, the Osirian crook and flail are rarely unornamented. As in the case of the Tutankhamun burial, the crook in particular is generally decorated with alternate bands of gold and blue, and it is very likely that this decoration is not haphazard, but intended to convey a symbolic meaning. Ivimy suggests that the gold bands represent earthly incarnations, while the blue bands of lapis lazuli stand for the discarnate periods between rebirths.

But at this point we can test Ivimy's suggestions against our acquired knowledge of Egyptian numerology. For just as the ornamentation itself is unlikely to be haphazard in its design, so too the number of bands is unlikely to be purely arbitrary. Let us take as an example the dazzling second mummiform coffin of Tutankhamun—that consummate work of art which represents the young pharaoh in full Osiris-regalia, complete with crossed crook and flail. Examination of the crook suggests that it is meant to be taken as having a total of thirteen full-width bands of gold (one being hidden by the left hand), while the flail in the right hand appears to have eight full-width bands of gold, of which the last is contiguous with the lashes of the flail themselves. But eight and thirteen are Pyramid-code respectively for rebirth and soul. The conclusion seems to be that crook and flail, taken together, represent numerologically the rebirth of the soul (the crossing of crook and flail may conceivably refer specifically to multiplication), while the design of the flail in particular suggests that the rebirth in question is specifically connected with the karmic significance of the regalia's design. (Counting the blue bands instead of the gold would give totals of eight and fourteen respectively, and thus would symbolise the notion of rebirth leading to spiritual perfection). Meanwhile the flail terminates in three lashes, each of which bears two groups of seven beads: these seem inevitably to symbolise the preparation of spiritual perfection (6 × 7), while at the same time suggesting that the function of the flail is karmic rather than agricultural, being identical in number-symbol to the forty-two assessors mentioned above.

Again, the head-dress is divided into a number of horizontal gold and blue bands. This time the number of gold bands up to the symbolic uraeus is apparently twenty-five—and once again the number-symbolism appears to fit exactly. For this fact would identify the dead pharaoh, in symbol, with the Great Initiate—in this case the Messianic Osiris—while the appearance of the rearing uraeus or snake-head on the twenty-fifth gold band would signify the awakened wisdom and power of the full initiate (cognate, perhaps, with the yogic *kundalini*). A final twenty-sixth gold band appears to mark the crown of the pharaoh's head, and seems to indicate that his achievement of full initiateship directly permits the final escape of his soul (2×13 = productive of the soul), whose seat is, in oriental tradition, the crown of the head.

The number-symbolism of the second mummiform coffin therefore tallies almost totally with that of the Pyramid's code. And insofar as this splendid artefact probably represents the real *pièce de résistance* of the entire burial we may perhaps adjudge its evidence on the question to be authoritative. But it must be freely admitted that the features of some of the lesser exhibits do not always tally with the figures given above for the second mummiform coffin.

The funeral mask, for example, has twenty-eight gold bands to the crown of the head (here marked by a blue band), as does at least one of the small canopic coffins containing the king's viscera. This would suggest either (a) that the master-design (possibly that of the second mummiform coffin) was inaccurately copied, (b) that exact copying was not considered essential, or (c) that the difference in numbers is symbolically significant and was thus intentional. Yet, on the face of it, (a) seems to fly in the face of the evident high standards of the ancient craftsmen, and (b) seems utterly untypical of the semi-magical priestly mode of thought of the time, so that (c) actually seems the most likely possibility. Could it be, then, that the close connection of the funeral mask and canopic coffins with the pharaoh's physical remains is the reason why both symbolic works of art should seem to speak so clearly of the spiritual perfection of the physical (7×4)? (Or is the number twenty-eight merely a reference to the fact reported by Plutarch that Osiris was held by the Egyptians to have lived twenty-eight years, by analogy with the twenty-eight days of visible moon in each month? In which case why is the same reference not apparent in the second mummiform coffin?)

Again, the numbers of strands in the necklaces of the various representations of the young god-king are likewise not constant, and again we must suspect an intentional symbolism. Could it be that the twelve necklaces of the funeral mask (three of them golden), encompassed by thirteen gold bands, signify that the mask's physical wearer had enshrined the soul of perfect man? Is the ninefold coloured necklace of the illustrious second mummiform coffin, with its ten golden strands, intended to refer to its occupant's Osirian rôle as Lord of Unbounded Eternity? And is it by accident that the fourfold necklace of the humble canopic coffin, with its fivefold gold interstices, seems to proclaim its owner as none the less a true physical initiate?

At all events, let us return to the conjectures of John Ivimy, who next draws attention to the favourite Egyptian ritual symbol known as the *djed*-pillar. Commonly interpreted as a fetish based on a bundle of papyrus, and associated by the Egyptians at one stage with the backbone of Osiris, this symbol too seems, like the Osirian crook and flail, to refer to the concepts of reincarnation and the evolution of the soul.

As Ivimy sees it, the base of the column represents the evolution of the soul through the various elemental kingdoms to the point of self-awareness and the consequent knowledge of good and evil—here represented, once again, by the arms of Osiris bearing their

The *Djed*-pillar. Based on John Ivimy's composite illustration and incorporating details from the *Papyrus of Ani*.

karmic regalia. A number of physical reincarnations follow—represented by a series of flat platforms separated by concave interstices—and then on this basis the *ankh,* symbol of the Life and power of the full initiate, is erected. Finally, from the crosspiece of the *ankh*—which forms as it were, a fifth horizontal platform—a pair of arms reaches up past the egg-shaped *ankh*-head to embrace at last the solar divinity.

On this basis, then, the full *djed*-pillar can be seen as representing nothing less than a kind of 'tree of life'—a beautifully apt *mandala* depicting the upward path of the evolving soul. But at this point we can check the above conclusions once more against our acquired knowledge of Egyptian number-symbolism.

The *djed*-pillar appears in a variety of forms throughout Egyptian antiquity, but in the particular version of it reproduced above it will be seen, first of all, that the base is divided by a number of light-coloured horizontal bands separated by darker layers—similar to the alternate gold and blue bands of Tutankhamun's Osiris-regalia. These light-coloured bands are eight in number, and would therefore seem to symbolise rebirth. Meanwhile, in this particular *djed*-version, the two 'side-skins' of the column each contain a series of dots. These are thirteen in number, and seem (if significant) to symbolise the soul. The number of the horizontal layers completing the column's base varies from representation to representation, and is not always easily discernible. But at all events there seem to be good numerological grounds for interpreting the base of the *djed*-pillar as symbolic of the reincarnating soul—an interpretation entirely consistent with Ivimy's conjectures.

Above the Osiris-arms and regalia we come to the horizontal bars or platforms already mentioned. These are always four in number, and would therefore symbolise the physical—an exegesis once again consistent with Ivimy's belief that they represent earthly incarnations.

The significance of the *ankh* itself is well-known, and the consequent *djed*-pillar notion that the Life-giving power of the full initiate can be attained by physical man is in no way unfamiliar to us. Meanwhile the crosspiece of the *ankh* surmounting the *djed*-column brings the number of horizontal bars above the Osiris-regalia to five—Pyramid-code, appropriately enough, for the initiate, or even the Great Initiate. That it is only from this level that the soul can, in symbol, reach up and embrace the solar divinity, is thus entirely in keeping with our interpretation of the Pyramid's message: it is only the fully enlightened, the true initiate, who can, in the steps of the

Great Initiate himself, finally achieve full union with the Divine (here represented, still in keeping with the Pyramid's code, by the circular solar disc). And the egg-shaped *ankh*-head seems specifically to indicate that the process is ultimately dependent upon some kind of new birth into the spirit-planes. In short, then, the *djed* column constitutes a singularly apt route map for human karmic evolution—and it can thus be seen that the association with the backbone of Osiris is perhaps not quite as far-fetched as it might at first seem to be. Here again one is reminded of the Hindu traditions regarding the ascent of the dormant *kundalini* powers to full wakefulness via the initiate's spinal column.

In a number of typical ancient Egyptian applications, therefore, it is possible to discern distinct echoes of the Great Pyramid's number code. Even the Pyramid's granite-limestone symbolism seems to be reflected in the later dark-light colour symbolism—the dark colours apparently symbolising spirit and the light colours the terrestrial. And we may see all this as evidence that knowledge of the Pyramid's symbolic code may have persisted among the Egyptian priesthood for many centuries after the Pyramid itself was built. And if the code, then perhaps the detailed message, too, survived—possibly even as late as classical times, and at all events long enough to provide the basis for the secret teachings of many of the ancient mystery-schools. This would certainly help to explain a number of remarkable resemblances, some of which are looked into more closely in Part II of this book.

5

Loose Ends—and Evidence for a Master-Plan

IN THE course of the preceding chapters we have unearthed some remarkably conclusive evidence that the Great Pyramid contains a detailed prophecy in mathematical code—a prophecy whose main purpose appears to be the validating of just such a redemptive or Messianic plan for mankind as appears to have been outlined by Jesus of Nazareth.

The Pyramid's whole blueprint for human evolution, in so far as we have deciphered it, is summarised on the chart opposite and the reader is invited to compare the result with the terminology of the Egyptian *Book of the Dead* listed on pages 32–3.

It will be seen that the overall picture is of a broad decline in human enlightenment and spirituality, leading directly to the predominantly materialistic age now in progress and its inevitable result as depicted in the Subterranean Chamber. From this general decline, however, there are two important departures. One is the upward path of mainly physical endeavour systematically forged, against all the odds, by the religious devotees and zealots of as far back as the second millennium B.C., and preserved and persevered with by a variety of highly committed religious groups right up to our own day. The other is the almost vertical path of sudden spiritual enlightenment acquired by men and women of all persuasions (and of none) under the stress of events, and particularly, perhaps, those foretold in the Subterranean Chamber.

It can further be seen that the original upward path of physical endeavour itself bifurcates at a point directly connected with the events in Palestine surrounding the life and death of Jesus of

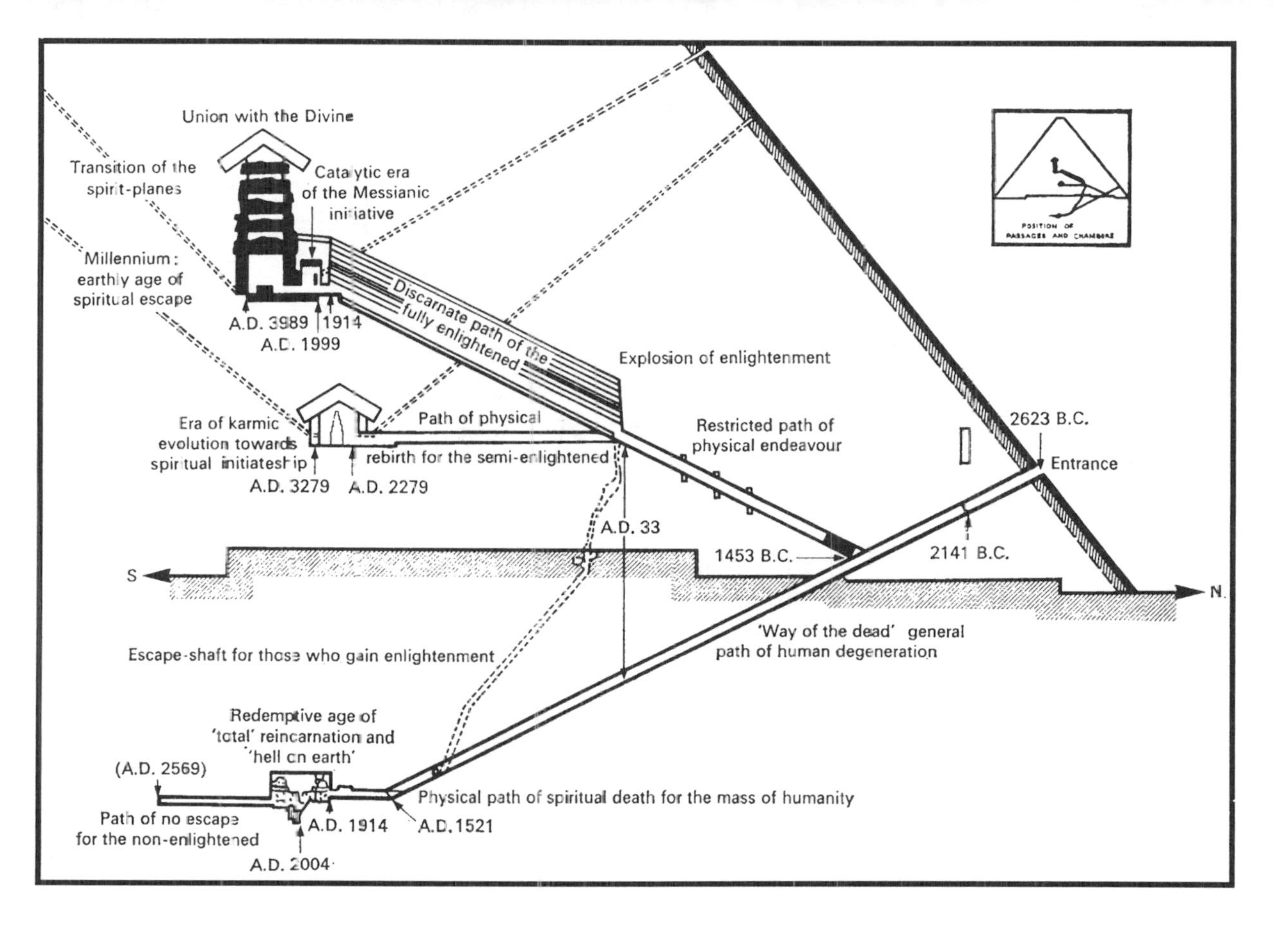

Summary-chart of the Great Pyramid's 'Blueprint for Human Evolution'.

Nazareth. Here, it gives rise to an upper path of the enlightened—involving an extended period of non-rebirth—and a lower path of the semi-enlightened involving a slow, karmic struggle towards the Light. And the point of intersection itself seems to have more than a chance connection with the 'highways and byways' referred to in the biblical parable of the man without a wedding garment (Matthew 22:2–14).

The reader will recall how, in the story, a king holds a marriage feast for his son. But the invited guests refuse to come, and so he sends out his servants to gather anyone they can find from the 'partings of the highways'. Among the latter is one who subsequently arrives without a wedding garment. He is bound hand and foot, and unceremoniously ejected again into the 'outer darkness'.

This strange parable seems to suggest (in confirmation of our earlier interpretation) that entry into the Kingdom of Heaven will be achieved initially only by a few initiates. But those who follow the lower path will also be given a later chance to join the Messianic Banquet of the final age, provided that they have by that time 'found a wedding garment', i.e. made themselves ready. St. Paul's notion of the 'Second coming' springs to mind.

As Maurice Nicoll suggests in his remarkable book *The New Man*, the 'man without a wedding garment' seems to be a reference to those who have learnt and even accepted the Messianic teachings, yet still refuse to put them into effect. Truth has been transmuted into dogma, doctrine into ritual, and even the Life-giving Messianic teachings themselves leave them spiritually dead. And it is in this sense that the Queen's Chamber Passage—the path, above all, of Christianity—may be regarded as the path of mortality *despite* the Messianic teachings.

As for the initiates of the upper path, rebirth during the present era is foreseen as an essential preliminary to their entry into the Final Age of rebirth into the spirit-planes, destined to lead ultimately to union with the Divine. Indeed, the *Christos* itself will also reincarnate during this preliminary period in order that, with the aid of the returned initiates, as many as possible may be guided from the lower paths to join the ranks of the chosen. These efforts will be aided, if not precipitated, by a period of terrible physical upheaval starting in 1914 and reaching its peak in A.D. 2004. All men, including the enlightened, will have to undergo this period of ordeal, but it is those who place the greatest value on physical things who, inevitably, will suffer most acutely from it.[1] By many, however, the need for a complete re-orientation of values will be seen and fully acted upon, and a civilisation of extraordinary vigour and achievement will ensue.

A small minority of men will none the less persist right to the end in their rejection of anything resembling true enlightenment, and for them a path of non-escape is projected by the Pyramid's designer, possibly leading to experience of a further cycle of physical existence.

The reader who has pursued the argument thus far will at this point naturally want to know: 'Where do *I* stand in all this?'—for his or her soul is, by definition, undergoing incarnation at this moment.

Under the terms of our read-out the Pyramid itself cannot directly reveal which level that soul has reached, since *all* souls are predicted as reincarnating at this time. In the absence, therefore, of any conscious memory of former incarnations, our attempted answer to this question has to be a cautious one.

Statistically, the likelihood that the reader is a reincarnation of one of the small and select band of full initiates now returning to the earth-planes is a slight one. It is apparently even more slight if he or she was born before 1914, but very slightly less so, perhaps, for those born after 1933—so, at least, a crude reading of the Pyramid's predictions would suggest. In the event of any doubt, there would in any case appear to be a simple test, for Jesus of Nazareth is recorded as having described very clearly the characteristics of 'those worthy of entering the Kingdom of Heaven', as he habitually described the Final Age of the present earth-cycle.

There is in any case a statistically greater probability that the reader is at present battling his karmic way through the Queen's Chamber Passage. In which case one would perhaps expect to find signs of a considerable innate interest in matters spiritual, and probably, too, a firm and total commitment to some religious orthodoxy—whether Christian or otherwise.

But by far the greatest *statistical* probability must of course be that the reader is one of the great majority who are still imprisoned in the lowest path of all (and the term 'majority' should not delude anybody into thinking that, for 'democratic' reasons, it is anything other than the *worst* path). The fact that he reads books on subjects

[1] In view of the physical nature of the time of the ordeal, it is appropriate that the latter should be depicted as part of the physically orientated Subterranean Inset, rather than in one of the other passages. It is similarly in accord with logic that the predominantly spiritual events of the fourfold Messianic initiative should be portrayed in the equally spiritual upper passageways, rather than in the lower ones.

such as this one—indeed, that he is prepared to persevere as far as this with the present volume—would perhaps argue against this probability to some extent; but that single fact could scarcely be regarded as of any great significance by itself.

None the less it should be remembered that escape is possible, even from this level—an escape which leads directly from the lowest to the topmost path, and which allows the soul in question (potentially at least) to bypass completely the painstaking karmic 'plod' of the Queen's Chamber Passage in a single leap. That escape is one which evidently has close links with the Law of Grace said to have been propounded by Jesus of Nazareth—a law whose operation is apparently dependent upon a total commitment to his teachings. Precisely what those teachings were, however, is a large question beyond the scope of the present volume. Suffice it to say that Jesus himself clearly believed that the Knowledge they enshrined was sufficient to 'set man free'.

On the question of man's overall spiritual state, however, Jesus' views concur completely with the apparent message of the Great Pyramid. 'The gate is wide that leads to perdition,' he is reported as saying at Matthew 7:14, 'there is plenty of room on the road, and many go that way; but the gate that leads to life is small and the road is narrow, and those who find it are few.' And that is the problem which the whole great Messianic Plan was evidently designed to overcome.[2]

So much, then, for the Pyramid's overall message, But we still have a number of loose ends to tidy up. We have still not managed, for example, to attach any firm significance to a number of the features of the passage-system—features whose obviousness suggests that we ought to be able to do so. We have not, for example, discovered any convincing reason for the use by the Pyramid's designer of the principle of horizontal 'insets'; for the fact that there was no apparent attempt to arrange that the various passages should finish in any particular vertical plane; for the 'odd' choice of 1881·2426″ as the distance between the Plane of Life and the Plane of Death; for the various angles of the King's and Queen's Chamber ventilation-shafts; or for the apparent prominence of the massive upright limestone gable above the original entrance, which is today one of the most striking features of the Pyramid's north face (see opposite).

[2] The Pyramid's Descending Passage is built, uniquely, atop an extraordinary sloping limestone platform known as the Great Basement Sheet. This is some 33 feet wide, a fact which perhaps reflects with a special aptness the words attributed to Jesus.

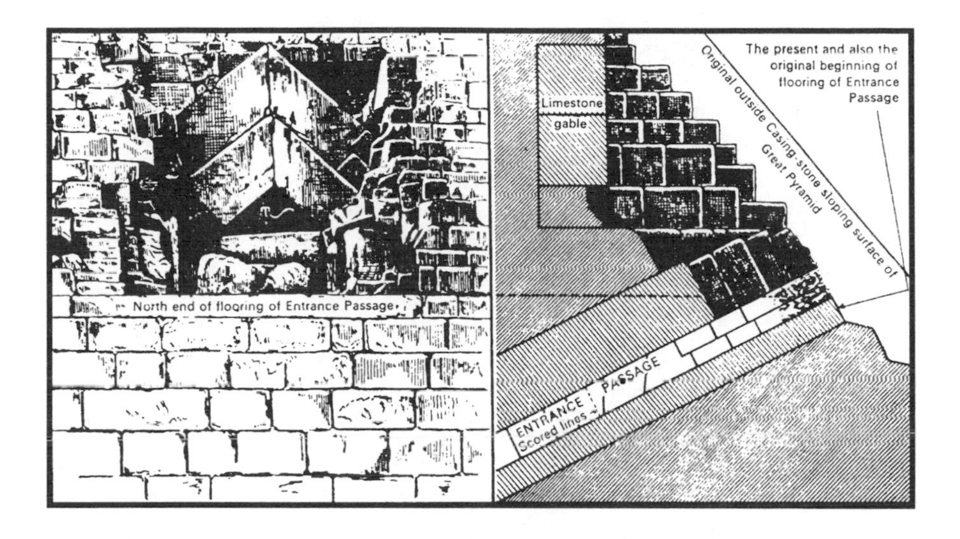

The entrance of the Descending Passage, with limestone gable and Scored Lines.

Yet common sense suggests that there must be some reason for all these features and their apparent idiosyncracies; and the logic of the situation seems to indicate that all these loose ends are the inevitable consequences of some overall geometric and/or symbolic plan to which all the detailed features listed and analysed are subordinate. Somewhere, we ought to be able to find evidence of some overall design or geometric matrix for the passage-system, whose inevitable result would be the appearance of the various features listed above. And in the event it is the massive upright gable over the entrance passageway that seems to provide us with the first vital clue in our search.

Rutherford himself clearly identifies two particular levels in the Pyramid's construction as the Plane of Life and the Plane of Death—corresponding to the levels of the Queen's Chamber floor and the Great Subterranean Chamber roof respectively—and the reasons advanced by him for this symbolic interpretation seem to be reasonably convincing, tying in well with our reading thus far of the Pyramid's message. The two levels are 1881·2426″ vertically apart. But the Plane of Life (or level of potential enlightenment) also passes through the middle of the east-west elevation of the great limestone gable over the entrance-passage, thus giving the clear impression that a further measurement needs to be taken vertically upward from the Plane of Life, since otherwise the top half of the gable would be left, as it were, sticking pointlessly up into the air. This impression is further confirmed in that, from the north, the gable has the clear form of an upward-pointing arrow-head, suggesting once again that the gable is specifically intended to draw attention to something of significance in the geometry of the masonry above.[3]

Now if, acting on this hint, we measure a *further* 1881·2426″ vertically upward from the Plane of Life, we reach a level which is that of the north and south outlets of the Queen's Chamber ventilator-shafts in the Pyramid's 90th course of core-masonry. This level is higher than all the features of the passage-system except the upper portions of the King's Chamber ventilator-shafts (see diagram page 186). The apparent connection of this level with the various ventilator-shaft outlets consequently leads to the conclusion (under the terms of our earlier decoding of the passage-system's geometry) that this level has to do somehow with the notion of 'escape'—escape from the Pyramid's earth-symbolising limestone. We have already noted, for instance, that the Queen's Chamber ventilators are not available for 'instant' escape, but seem to represent an escape that needs to be worked and striven for over a number of incarnations. We have also remarked that the Queen's Chamber seems to symbolise not man's *final* escape, but a discarnate period leading to further incarnations in the context of the King's Chamber; and it is ultimately from here that the 'ultimate' escape needs to be made.

A tentative identification of the level of the top of the Queen's Chamber ventilator-outlets as the Plane of Escape thus suggests itself, for final escape is possible, it seems, only for those who succeed in rising above this particular level. And clearly that escape to pure spirituality must be to the 'life' of the level of the Queen's Chamber floor, as the latter itself is to the 'death' of the level of the Subterranean Chamber roof. Meanwhile we seem to have confirmation that the distance of 1881·2426″ represents the evolutionary gap separating these various levels of attainment.

Now the height above the Pyramid's base of what I have termed the Plane of Escape is 2727·2966″—a distance which appears to be identical with the height of the 90th course-axis, whose notional average value is 2727·7″[4] (see page 377). *But this height is also within three inches of the half-height of the Pyramid's summit-platform*—namely 5448·736″/2, or 2724·368″. The conclusion seems inescapable, therefore:

(a) that there is an intended direct link between all three features

[3] Compare note 17 page 79 and illustrations page 181.

[4] All figures continue to be quoted in Primitive Inches, except where otherwise stated.

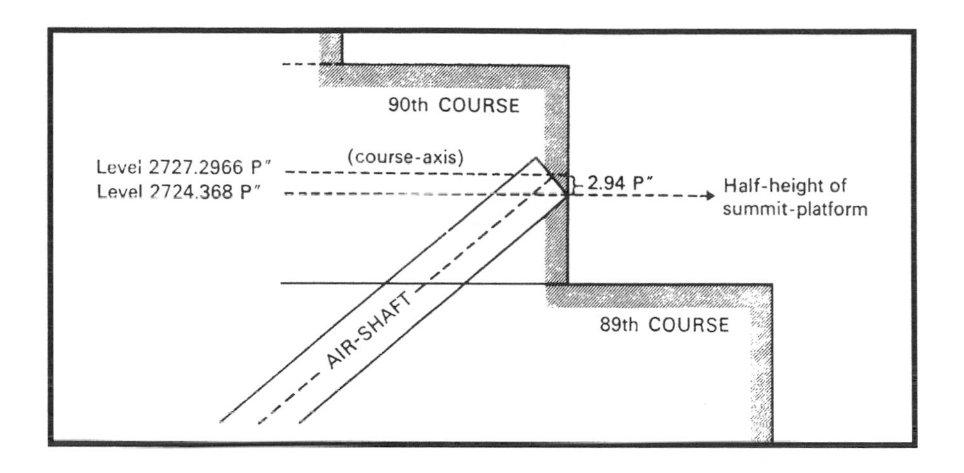

Projected upper termination of Queen's Chamber air-shafts in core-masonry.

(Plane of Escape, axis of 90th course and summit-platform's half-height),

(b) that one purpose at least of the two Queen's Chamber ventilators is to define, in terms of the combined heights above the base of the two core-masonry outlets, the level of the designed summit-platform. In other words,

Height, north outlet + height, south outlet = height, summit-platform.

This would tie in with the fact (see note 63, page 126) that the King's Chamber ventilator-outlets clearly define the course-number of the summit-platform in a similar way. The dual definition of this all-important level was presumably intended by the architect as a precaution against possible later destruction of the upper courses—such as has indeed occurred.

Since, therefore, the difference between the two levels just quoted is apparently quite precise—namely 2·94"—while (b) above would depend upon an extremely fine indication of level, it seems probable that the upper terminations of the Queen's Chamber ventilators were originally as depicted above. Assuming that the angle of slope of the air-shafts was some 40°, the arrangement shown would necessitate a perpendicular air-shaft bore of some 7·67", and would result in the roof of the air-shaft's falling short of the outer surface by some 5"—just as the shaft's lower ends were similarly left uncut by the same amount (see page 80). Meanwhile, the summit-platform's half-height would be indicated with great precision by only the finest of slits in the outer stonework, once the casing had been removed—just such a slit as apparently resulted in the original discovery of the air-shaft's *lower* ends in 1872.

Verification of the above hypothesis will depend largely upon close examination of the already discovered but so far inaccessible north outlet. The south outlet may provide even better evidence, for it has not so far been pinpointed, and may therefore still be in its original closed condition.

The symbolic force of the above arrangement would, as already suggested, be to the effect that the Queen's Chamber air-shafts bring the soul to the Plane of Escape—i.e., the half-height of the summit-platform—but do not yet permit a breakout beyond it. A further five inches of masonry need (symbolically) to be cut away before the soul can finally escape from the earth-planes (symbolised by the Pyramid's limestone and specifically, perhaps, by the *lower* half of the Pyramid), and these five inches, like their counterparts at the inner ends of the shaft, seem to represent the experience of five specific incarnations of ever-increasing spirituality and enlightenment.

If we are correct in this conclusion, we have already gone some way towards accounting for two at least of the hitherto unexplained features of the Pyramid's design. But what of the apparently odd distance of $1881\frac{1}{4}''$ which separates the Plane of Life from the Plane of Death on the one hand, and from what we have dubbed the Plane of Escape on the other?

It needs to be said that this distance (itself incorporated as a feature of our hypothetical code) is the direct code-equivalent of 99 × 19″ and thus presumably signifies the culmination of mortality, i.e. the end of death, and thus the final flowering of the plant which is the physical world. None the less a further significant point about this distance—duplicated in the length of the Grand Gallery's floor—appears to be its unique, and thus possibly deliberate, proximity to 2000″. It is not unreasonable to suggest the possibility that this too may have some definite chronological and/or symbolic significance.

Is there, then, a feature which is not $1881\frac{1}{4}''$, but some 2000″ above the Plane of Death—some feature which may be of real significance? The data considered so far do not appear to offer us any such feature. If, however, there were such a feature, then we should presumably expect it to be, like the Plane of Death, in the nature of a roof, rather than a floor, so that the distance involved could symbolise the corresponding elevation of its counterpart in the human body—namely the head.

Now the roof of the Queen's Chamber Passage itself is only 67·565″ above the Plane of Life—the passage's floor-level—and this is much less than the extra 119″ or so needed to turn $1881\frac{1}{4}$″ into 2000″. Meanwhile, the tops of the Queen's Chamber's north and south walls are already $184\frac{1}{4}$″ above the floor, which is already too much. However, at the point where the roofline of the Queen's Chamber Passage meets the floorline of the Grand Gallery (E on diagram page 58)—what may justly be termed the parting of the ways, both architecturally and symbolically—the roofline of the Ascending Passage (at A) is exactly 52·745″ above that of the Queen's Chamber Passage at E. Point A is thus 1881·2426″ + 67·595″ + 52·745″—*a total of just 2001·5826″*—above the Plane of Death. Could this be the significant point whose existence we speculated about above?

Examination suggests that *point A is nothing less than the focal point of the Pyramid's entire passage-geometry*. For a start, it will be seen from the diagram (page 186) that point A appears to lie horizontally midway between the Scored Lines at the start of the prophecy and the south wall of the King's Chamber at its end—an impression confirmed by trigonometrical calculation.

Moreover, a circle touching the verticals passing through these two points also appears to cut the Plane of Death, pass through the coffer in the King's Chamber and enclose the blocked upper end of the Queen's Chamber's north ventilator. Geometrically, then, the circle circumscribes all the features of the passage-system, from the Scored Lines onwards, except the southern part of the King's Chamber and its ventilator-opening on the one hand, and the lower Well-Shaft opening and Subterranean Inset on the other (the Construction Chambers are also excluded). Symbolically the circle therefore seems to enclose the range of possibilities immediately open to anyone who has attained the level of potential enlightenment, the Plane of Life. These possibilities seem to include direct admission to the final resurrection (the coffer) on the one hand, and descent towards hell on earth on the other.

Meanwhile, the same circle draws attention to the fact that the two verticals in question (NML and OPQ on the diagram) represent the beginning and end of the whole Messianic Plan respectively, if our decoding of the Pyramid's message is valid. Produced to cut the Planes both of Death and of Escape, these verticals can then be seen to complete the rectangle NOQL which, on examination, appears to be of such significance to the overall plan of the passage-system as to merit some such description as the Double House of Redemption.

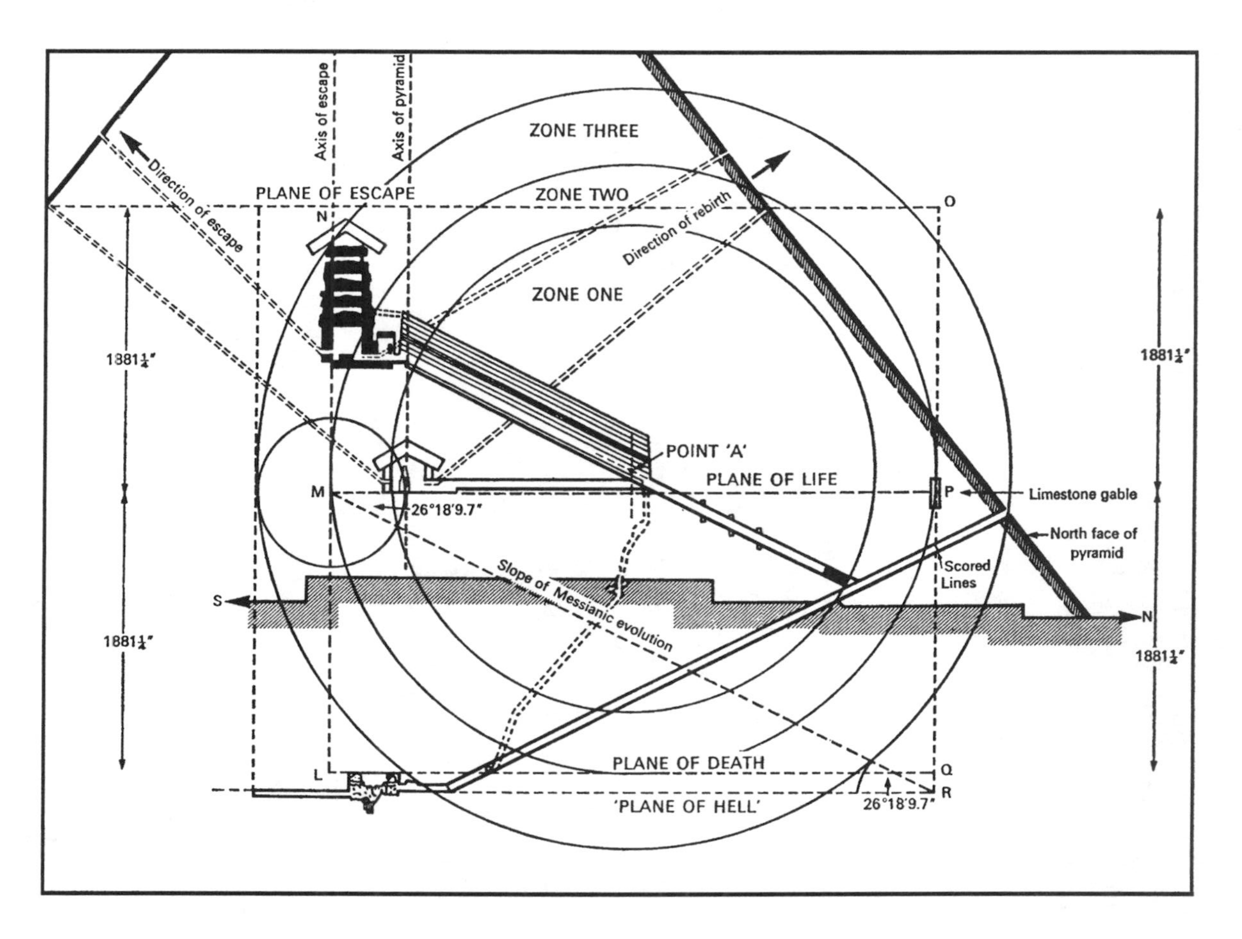

Symbolic disposition of the Great Pyramid's passages and chambers.

For, pivoting about the Plane of Life at MP (and thus, symbolically about the life and teachings of Jesus of Nazareth), this rectangle encloses every significant feature of the passage-system except the Subterranean Inset on the one hand and the King's Chamber ventilator-outlets on the other.

To these exceptions we should also add the blocked south ventilator-opening of the Queen's Chamber, but since this is *below* the Plane of Escape it is in a sense 'hemmed in' by ON produced. Similarly, however, the north ventilator of the King's Chamber is hemmed in by LMN produced—a fact which suggests that this vertical, too, should be regarded as a Plane—or axis—of Escape. Thus, full escape can apparently be attained only *above* the level of ON produced and *to the south* of vertical LMN produced. In other words, the only ultimate escape route from the passage-system (and thus from mortality) is via the south ventilator of the King's Chamber. This notion is confirmed by the fact that the arched entrance at the lower end of the south ventilator is so reminiscent of the pregnant womb (see page 130) as to suggest clearly that it (and it alone) represents the final rebirth into the heavenly planes. The point is emphasised by the shape of the various ventilator-shafts' cross-sections: all except the south ventilator of the King's Chamber are rectangular in cross-section, thus signifying physical rebirth, while that shaft alone is circular in cross-section at its lower end—a clear indication that it represents the path of *spiritual* rebirth.

So far, then, our geometrical speculations have amply confirmed—and thus, in a sense, been confirmed by—our earlier symbolic decoding of the Pyramid's message. The only remaining feature which lies outside what I have termed the Pyramid's Double House of Redemption—namely the Subterranean Inset—adds yet further confirmation to our assumptions. For this too, while it terminates to the south of LMN produced, is still well below level ON—indeed it is even below the base of the rectangle LQ. This fact, plus the further fact that it so clearly comes to a dead-end in the limestone rock beneath the Pyramid, unmistakably identifies the path in question as one of non-escape from mortality and the physical planes. There *is* an escape, it is true—but only outside the Messianic Plan itself as outlined in the Great Pyramid of Giza.

Other circles drawn with centre A lead to further complementary conclusions. If a circle is drawn to touch the south wall of the Queen's Chamber, for example, it will be found to pass through the Granite Plug at a point very close to its lower end, while also cutting the floor of the Grand Gallery at a point just below the Great Step. Its symbolism appears to suggest that the unenlightened 'Old Testament' dispensation, and its modern progeny—including perhaps established Christianity—cannot lead directly to escape at all.

Again, a circle drawn with centre A to cut the lower lip of the entrance to the passage-system appears both to touch the vertical drawn from the extremity of the Dead End Passage, and to pass through the apex of the King's Chamber gable. It encloses within itself every feature of the passage-system, except the two southern ventilators, the Subterranean Chamber and the Dead End Passage. It thus appears to draw attention both to the beginning and to the two alternative finalities of the entire Messianic Plan.

Meanwhile the three concentric circles which we have now constructed (see page 186) give us three distinct zones, each with its own significance. Zone one (the inner zone) is a zone of non-escape, albeit centred on the Plane of Life—and thus appears to bear more than a chance resemblance to the Egyptian notion of the Hall of Truth in Darkness.

Zone two encompasses the entire gamut of spiritual possibilities from direct entry to the final resurrection to consignment to the 'fires of hell'. Within it are portrayed all the Messianic events symbolised by the Passage of the Veil, as well as the entire north ventilator of the Queen's Chamber, which represents the gradual attainment of enlightenment over a number of reincarnations. It might thus be described as the zone of hope or the zone of spiritual endeavour, despite its clear suggestion that, even so, a catastrophic spiritual decline is still possible within it.

Zone three, however, is the zone of escape. Within this zone falls the entrance to the entire passage-system, and thus, paradoxically, the very possibility of redemption in the first place (see pages 85–6). Within it, too, fall both the outer north ventilator-opening of the King's Chamber and its inner south ventilator-opening—thus apparently signifying that the rebirth of the north ventilator will lead to the entry into the spirit-planes signified by its southern counterpart. Even for those who descend to the level of the pit, the lower Well-Shaft opening is ready and waiting for them to effect their escape to the upper regions. Indeed, combining the symbolisms of rectangle NOQL and the concentric circles based on A, it is clear that only the Subterranean Chamber and the Dead End Passage are effectively outside the system.

Finally, it is worthy of note that the diagonals of the two rectangles NOPM and MPQL are so close to the Bethlehem-angle of 26° 18′ 9·7″ as to suggest the application of this gradient to the symbolic geometry of the system as a measure of the vertical distance which man is capable of ascending as a result of the Messianic initiative, as personified in Jesus of Nazareth. If such a slope is constructed, such as to achieve the Plane of Life at M and to cut vertical OQ at R, it is found that the level indicated by R is some 25″ below the Subterrannean Passage floor, i.e., the level of the very rim of the pit. And once again we are struck by the symbolic justness of the traditional claim that the *Christos* 'descended into hell' and is determined to set free even the most wretched of men.

The significance of point A, originally identified by the fact that it lies 2001·58″ above the so-called Plane of Death, thus appears to be considerable. By the same token, therefore, the distance of 2001·58″ ought, like that of 1881·2426″, to be a significant one.[5]

We have already noted that the symbolic, vertical distance of 1881$\frac{1}{4}$″ is directly reflected in the chronographical, sloping length of the Grand Gallery floor. Might there then likewise be some direct connection between the distance of 2001·58″ measured vertically (and thus symbolically) on the one hand, and the same distance measured on the slope (and thus chronographically) on the other? Might this distance mark out some event of special significance?

A diagrammatic representation of the occurrences of these measurements suggests that this is a distinct possibility.

In the diagram (page 190), point A lies on the produced roofline of the Ascending Passage, directly above what we have termed the parting of the ways, and we have already established that point A seems to be the fulcrum of the Pyramid's geometry.

[5] In terms of the Pyramid's code, it is interesting to note that the nearest whole number of inches to 2001·58 is 2002—*which is itself exactly 7 × 286*. It is tempting, therefore, to see this distance as symbolising the spiritual perfection of enlightenment—an interpretation which would square completely, in symbol, with the siting of point A at a point attainable only by the elect who have just succeeded in gaining access to the upper part of the Grand Gallery floor.

On the other hand, it should be pointed out that the 'correcting' of 2001·58″ to 2002″ is apparently contrary to the standard procedure laid down in our hypothetical code (while the 'correcting' of 286·1″ to 286 is quite acceptable). Only by positing a constructional tolerance of up to half-an-inch at this point could 2001·58″ possibly be read as 2002 and not 2001.

Oddly enough, such a reading could conceivably be justified *if the Pyramid's constructional tolerances were to be seen as taking the form not of fixed geometrical quantities* (e.g., ·01 of an inch) *but of fixed arithmetical ratios* (e.g., 1 in 100). The concept would be a reasonable one. And in this case a discrepancy of less than half-an-inch in two thousand would represent a tolerance of better than 1:1000—which is close by any standards. Consequently the symbolic reading suggested above may not, after all, be beyond the realms of reasonable possibility.

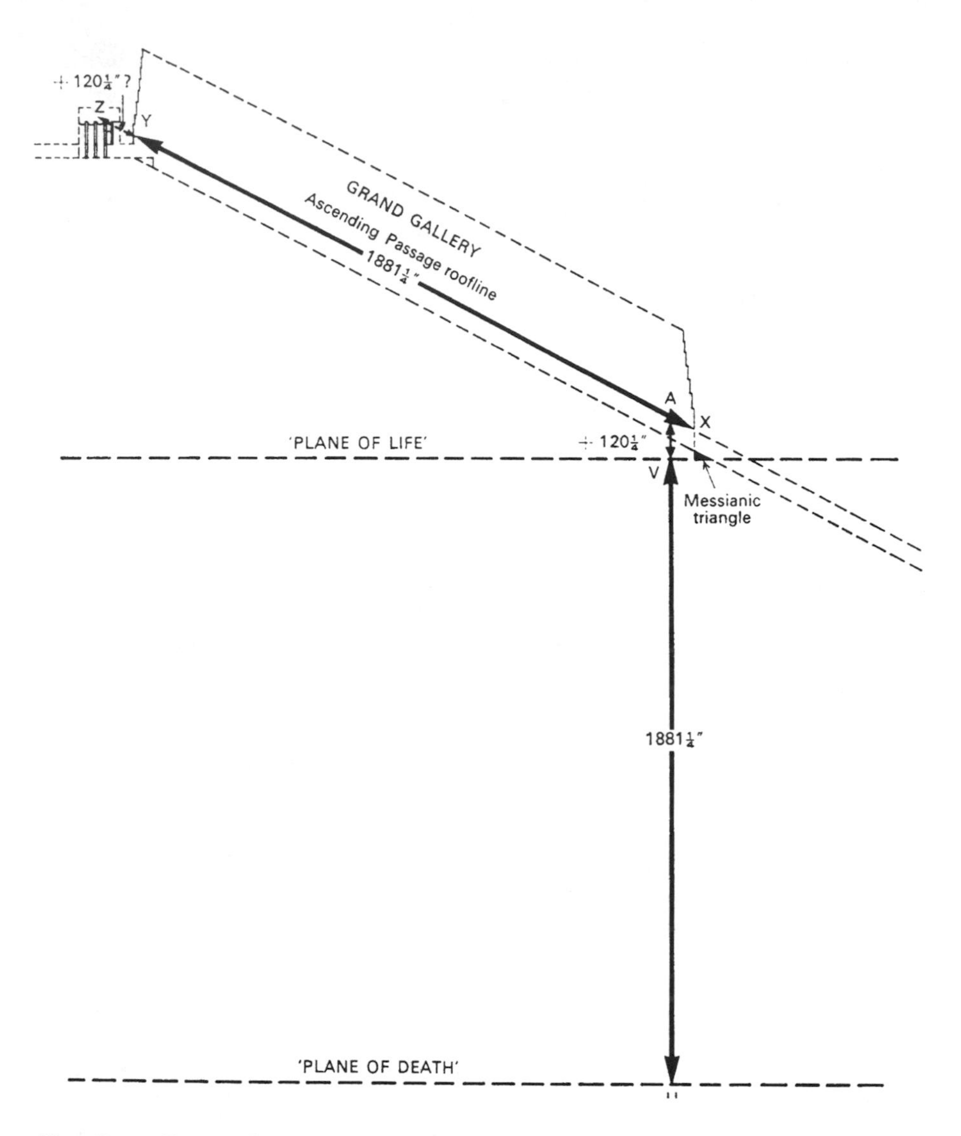

The Great Pyramid: Interrelationship between symbolic and chronographical features.

It may, of course, be objected that A is architecturally non-existent, since neither it nor the intersection of the Grand Gallery floorline and the Queen's Chamber Passage roofline is marked by any tangible feature of masonry. Indeed, the architect seems to have gone out of his way to conceal both points. On the other hand, the projection of the Ascending Passage roofline into the Grand Gallery is affirmed to be a valid interpretative procedure by the mere fact that, throughout its length, its height is exactly 286·1" (a known and significant code-measurement) less than the extreme height of the Grand Gallery's roofline; while the projection of the Grand Gallery floorline and the Queen's Chamber Passage roofline to their intersection is likewise validated by the fact that the point where they meet marks—apparently significantly—a date 153 years after the intersection of the equally intangible Queen's Chamber floorline with the Ascending Passage floorline—i.e., 153 years after the birth of Jesus of Nazareth. Again, the Ascending Passage roofline, produced, appears to pass through the Granite Leaf in the vicinity of the Boss, and to meet the top of the Antechamber's east wainscot at a point between the second and third Messianic portcullises—features which suggest unmistakably that this line is not only significant in itself, *but of direct symbolic relevance in some way to the Messianic return.*

Given, then, that there is a clear numerical link between lines UV and XY in the diagram (in that both equal $1881\frac{1}{4}''$ in length), there would seem to be grounds for thinking in terms of a *geometrical* link between them as well. And if UV needs to be extended upwards by some $120\frac{1}{4}''$ to reach the significant point A, then it would seem logical that a similar upward extension needs to be applied to the sloping line XY to reach the similarly significant point Z. Which leaves us with the question of just what that significance might be.

We have already seen that the point on the floor below point X in the Grand Gallery corresponds to the spring of A.D. 33, so that the point on the floorline below Y necessarily denotes a date $1881\frac{1}{4}$ years later—i.e., the summer of A.D. 1914. But the distance of $1881\frac{1}{4}''$ also denotes symbolically the distance between 'death' and 'life', as we have also seen. Thus we are forced to conclude that, just as the level of the Subterranean Chamber roof is 'as death' compared with the level of the Queen's Chamber floor, so also this level—the so-called Plane of Life—is 'as death' compared with the level represented by the top end of the Grand Gallery. Moreover, the two lines of $1881\frac{1}{4}''$ are not contiguous: only through the Messianic triangle can man bridge the gap (FD on page 58) between the Queen's Chamber floorline and the Grand Gallery floorline. The attainment of the 'higher life' is thus heavily dependent upon the agency of Jesus and the Messianic Plan.

Just, therefore, as the beginning and end of the Grand Gallery are intimately linked with the symbolism of the raising of the elect from death to Life, so also we would expect them to give us some idea of the timing of this process. And, as we have already seen, A.D. 1914 (represented by the top of the Grand Gallery floorline) seems to mark a significant moment in the plan—nothing less than the inception of the Pre-Final Age. In a similar way, just as the lower end of the Grand Gallery marks the physical death of Jesus, so we should also expect to find some chronographical indication, at the top end, of the date of the return of the 'son of man' to physical life, and the actual earthly inception of the higher life so clearly referred to.

At this point we come back to our notion that, since a symbolic distance of $120\frac{1}{4}''$ needs to be added to UV to cut XY at A, so a similar distance (YZ) needs to be added at the top end of XY to reach point Z. For it now becomes clear that, if point Z has any significance at all, then it must be that it determines the chronographical point at which the symbolically-indicated higher life is to be initiated. In conformity with our realisation of the heavy dependence of the Plan's fulfilment on the work of Jesus of Nazareth, we would expect this point to indicate in some way the actual moment of his return—thus presumably serving as a double-check on our earlier calculation of the date of this event.[6]

Just, then, as the inception of the Pre-Final Age appears to come $1881\frac{1}{4}''$ years after A.D. 33, so the event biblically referred to as the 'coming of the son of man' would appear to be timed for a point 2001·5826 years after the same date—namely *the autumn of* A.D. *2034.* This dating, it will be noted, is identical with that already listed.

Indeed, so precise and unmistakable are the measurements involved, that it appears to be possible—fantastic as it may seem—to hazard a well-informed guess at *the actual day of this crucial event* (a fact indirectly made possible for us by our knowledge of historical events presumably unknown to Jesus of Nazareth in his day). For the actual vertical distance from the Plane of Death to point A is, as we have seen, exactly 2001·5826″—which presumably represents chronographically 2001 years and 212·8 days. If we now measure

[6] The Pyramid's design appears to give clear support to this view. As we have already seen, the length of U(V)A—namely 2001·58″—seems to beg identification as the code-equivalent of 7 × 286, thus signifying the spiritual perfection of enlightenment/the enlightened. But there is an obvious parallel, in that the perimeter of the Pyramid's summit-platform's inner (or built) circuit-square similarly measures 7 × 286·1″. Meanwhile this platform is itself designed eventually to support the final capstone, whose designed base circuit-square measures exactly 8 × 286·1″. Moreover the capstone, as its base-perimeter itself confirms (8 × 286·1″ = the rebirth of the enlightened (one), or possibly, the return of the Initiate), is clearly symbolic of the Messiah's return. Thus, this long-awaited event already has, in the capstone's design, quite unmistakable connections with the number 2002 (7 × 286·1 = 2002·7), and this fact may thus well have a chronological, as well as a symbolic, significance.

this period up the floor of the Grand Gallery from the date of the crucifixion—which the Pyramid appears to fix at 1st April A.D. 33 (Gregorian)—then the sign of the Messiah (or the Messiah himself, perhaps) is due to appear in the sky *on 31st October A.D. 2034.*[7]

The aptness of this dating is nothing less than extraordinary, for it coincides with the first day of the traditional and practically world-wide Festival of the Dead—celebrated in the British Isles as All Hallows or Hallowe'en (31st October), All Saints' Day (1st November) and All Souls' Day (2nd November). *Indeed, October 31st is traditionally regarded as none other than the day when the souls of the departed return to earth*—and when, in some countries, candles are even lit on the graves of relatives to guide and welcome the returning spirits. And for good measure, the very day following this Festival of the Dead (3rd November) was held by the ancient Egyptians to commemorate the resurrection of their own Messiah-figure, Osiris.

But that is not all; for it has always been widely held in many parts of the world that this same Festival of the Dead, or its equivalent, commemorates the anniversary of what in Ancient Egypt was called The Destruction of Mankind; in Mexico and Peru, The Destruction of the World; and in Babylonia, Assyria and China, The Deluge. It appears to be nothing less than the world-wide anniversary of the biblical flood—the cataclysmic destruction (apparently through geological upheaval) of that former world whose only known legacy to us seems to lie in the Great Pyramid, in the legend of the watery fate of Atlantis, and in the most ancient religious and mythical traditions of man. And so, perhaps, in the legendary story of Noah and his ark, there may lie an unexpectedly deep stratum of truth—a 'memory of the future', perhaps, centred on the mission of a single, righteous human soul who, having escaped a great destruction, is destined to return once more with his companions, bringing new Life to a desolated world. And in the return of the Messiah—apparently on the very anniversary of the cataclysm—we may yet see that memory's fulfilment[8] (see chapter 9).

One caveat, however, lest the dating given be regarded as more fixed and immutable than it really is. The conclusions arrived at

[7] On the same basis, the later, alternative dating for the crucifixion, suggested in the note on page 62, would put back this Messianic event to A.D. 2039 or 2040—thus conceivably identifying it with the Messianic arrival or assumption of mortality apparently dated by the north face of the Granite Leaf at 21st October 2039 (see page 113).

[8] Note, in this case, the extraordinary aptness of Jesus' own words at Matt. 24:37, '*As things were in Noah's days*, so will they be when the Son of Man comes.'

above are based upon numerical assumptions which are themselves subject to certain tolerances. The length of the Grand Gallery's floor, for example, is not 1881·2426″ (as is the distance from the Plane of Death to the Plane of Life) but 1881·2223″—a difference of about a fiftieth of an inch. If this difference were taken to be significant, then it could have the effect of bringing forward the date of the expected Messianic return by about a week. Again, point A is not 120″, but only some 119″ up the Grand Gallery floor from the point apparently denoting the crucifixion of Jesus of Nazareth. The fact does not affect our calculations above in any way, but if we were to adopt a different approach and measure, not 2001·58″ from point X, but $1881\frac{1}{4}$″ from point A, then the presumed dating of the Messianic advent would have to be brought forward to late A.D. 2033. And finally, our measurements in this connection have been based on a figure of 2001·58″, when it was the figure 2000 that started us off on this line of enquiry—so that a Messianic advent in A.D. 2033 is again not a possibility to be written off entirely, despite the chronograph's own pinpointing of 2034 (see page 112).

And this brings us to the final and perhaps most puzzling loose end—one which we have already discussed in part on a number of occasions. It has become clear that the Great Pyramid was designed and master-planned as a message *to all mankind*, and in view of this it may appear more than a little surprising—initially at least—to find that it apparently has so much to say about the development of Western society in general, and about Christianity in particular. In fact the imbalance may be more apparent than real, and the developments listed may have been more world wide than is generally realised. Nevertheless, the latter part of the chronograph especially seems to spotlight events in Europe and the West in general, rather than those further east.

A little reflection will quickly reveal the reason for this. One has only to ask oneself what historical events were most significant in shaping and moulding the state of today's world to realise that, from the Renaissance onwards, the overriding force in the development of human society has been the influence of European thought and customs. In the arts and sciences, in learning and morality, in social relationships and politics, in almost every sphere of human activity, it has been the West that, increasingly, has tended to 'set the pace' for the rest of the world. Western movements today are world movements, Western crises world crises, Western wars world wars. And meanwhile no effort has been spared during the past few hundred years to sell the idea to the rest of the world that Western society is the best society, Western beliefs the best beliefs, Western dress and life-style the best dress and life-style—ideas whose sheer inanity has started to be questioned only in comparatively recent years. And, oddly enough, the origin of this whole line of thought seems to have had much to do with the characteristic belief on the part of the early Christian missionaries that their *religion* was the best—indeed, the only true—religion, a belief arising directly from the conviction that their founder was God himself in human form. Christianity and 'Europeanity' have, in short, often gone hand-in-hand, and where the 'White Jesus' has led, the rest of the world has followed.

All of which is not entirely surprising when one bears in mind that received Christianity (as opposed to true Nazarenism) is, both conceptually and historically, a basically European religion—a religion, moreover, whose genesis as a 'separate' movement seems to have been foreseen by the Pyramid's designer himself. And at this point we are brought back again to the question of why the Pyramid should apparently have so much to say about the development of Christianity, and so little to say about any of Christianity's 'rival' religions.

The only way in which we can reconcile the apparently partisan claims of Christianity with the avowedly universal message of the Pyramid seems to be to adopt the premise that true Christianity, as preached by Jesus of Nazareth, was *not* partisan and *had* no rival religions. Instead, one suspects, it was simply a logical further development in the world-wide religious and Messianic Plan outlined in the Pyramid and manifested in *all* the world's major religions. Our research into the Pyramid necessitates a view, in other words, which sees Jesus not as an *opponent* of figures such as Osiris, the Buddha, Krishna, Lao Tsu or Quetzalcoatl any more than of Moses—but rather as their logical and ultimate *successor*. Such, indeed, is the view of many a modern Buddhist. Consequently whatever Jesus taught is likely (if the Pyramid's record is to be regarded as true) to have been a logical and consistent development of the core of those earlier beliefs—nothing less than a restatement of what the remarkable James Churchward (see Santesson, page 387) saw as a knowledge that was once universal among the whole of ancient mankind, the true religion from which all established religions have since developed, or rather declined. At which point there would seem to be some justification for taking an entirely new look at Jesus and his teachings from a viewpoint quite independent of the exclusive and partisan exegesis of received Christianity.

None the less many Christians will take the celebrated text: 'I am the way; I am the truth and I am life; no one comes to the Father except by me' (John 14:6) as directly justifying the assumption that Christianity alone, of all religions, has hold of the Truth. It should be remembered, however, that the statement is made by the Jesus of John's gospel, whose author makes it quite clear at the beginning that his *Christos* is an eternal spirit who existed before time began. He is the life-giving Word, or 'man-thought', of God—none other than the same Holy Spirit who in all three synoptic gospels descends on Jesus 'like a dove' at his baptism. From that moment onwards the man Jesus becomes that Spirit's embodiment or incarnation, and what he says is said specifically in its name. Hence the apparently extraordinary claims made for the text quoted above—claims which could clearly never be valid of a mere man, and which have helped to bring about the unhelpful notion that Jesus was in some sense God in human form.

Thus the statement at John 14:6 is a clear affirmation by Jesus that any man who achieves eternal life does so solely through the action of the Messianic spirit-entity known as the Holy Spirit—just as St. Paul later suggested (for example, at 1 Cor. 12:3) that man's every good action is, by definition, necessarily inspired by that same spirit.

The Buddhist and Hindu scriptures and religious literatures (to take only two particular instances) make it quite clear that numerous saints and sages of those religions have in the past succeeded in attaining spiritual immortality—as well as performing a plethora of other miracles more than equal, both in number and in splendour, to the reported miracles of Jesus. There being no a *priori* reason, then, why these oriental writings should be regarded as having either more or less validity than the Jewish and Christian ones, we are bound to assume that the illustrious figures in question, having achieved their immortality, necessarily did so through the agency of the same Holy Spirit.

Now it follows logically from this that the basic teachings of Jesus, who taught in the name of that same Holy Spirit, cannot therefore have been seriously in conflict with those of Hinduism and Buddhism, whatever differences there may have been in the particular words and symbols through which those teachings were expressed. And even here the similarities are often most striking.

'For I am the sacrifice and the offering, the sacred gift and the sacred plant,' says Krishna in the Hindus' *Bhagavad Gita*. 'I am the holy words, the holy food, the holy fire, and the offering that is made in the fire. I am the Father of this universe, and even the Source of the Father. I am the Mother of this universe, and the Creator of all. I am the Highest to be known, the Path of purification, the holy OM, the Three Vedas. I am the Way, and the master who watches in silence; thy friend and thy shelter and thy abode of peace. I am the beginning and the middle and the end of all things: their seed of Eternity, their Treasure supreme. The heat of the sun comes from me, and I send and withhold the rain. I am life immortal and death; I am what is and I am what is not[9]. . .

'Whatever you do, or eat, or give, or offer in adoration, let it be an offering to me; and whatever you suffer, suffer it for me. Thus thou shalt be free from the bonds of karma which yield fruits that are evil and good; and with thy soul one in renunciation thou shalt be free and come to me . . . For this is my word of promise, that he who loves me shall not perish.

'Know thou that whatever is beautiful and good, whatever has glory and power is only a portion of my own radiance . . . Know that with one single fraction of my Being I pervade and support the Universe, and know that I AM.

'When righteousness is weak and faints and unrighteousness exults in pride, then my Spirit arises on earth. For the salvation of those who are good, for the destruction of evil in men, for the fulfilment of the kingdom of righteousness, I come to this world in the ages that pass. He who knows my birth as God and who knows my sacrifice, when he leaves his mortal body, goes no more from death to death, for he in truth comes to me . . . In any way that men love me, in that same way they find my love: for many are the paths of men, but they all in the end come to me' (*Bhagavad Gita* 9, 10, 4, selected, tr. J. Mascaró).

The above argument, of course, proves nothing: it is merely a logical demonstration based on textual premises which are not themselves entirely above suspicion – impressive though the similarities between the texts may be. None the less the point has perhaps been made that Truth—whatever we understand by the term—is ultimately both one and infinite, and that to no man or religion is it given to know all of it. The Buddha once demonstrated this proposition by picking up a handful of leaves from the forest floor: what

[9] Compare Osiris, in the Egyptian *Book of the Dead*: 'I am Yesterday, Today, and Tomorrow; and I have the power to be born a second time. I am the hidden Soul who createth the gods. I am the Lord of those who are raised up from the dead. I am the Great One, Son of the Great One; I am Fire, the Son of Fire . . . I am Osiris, the Lord of Eternity.'

he had revealed, he pointed out, was to what he had not revealed as were the leaves in his hand to those remaining on the forest floor. Jesus likewise is reported by the author of John's gospel as declaring (16:12): 'There is still much that I could say to you, but the burden would be too great for you now.'

The Buddha is said to have further elucidated this proposition in terms of a parable. It seems that a certain Indian ruler once summoned into his presence four blind men (not unrelated in symbol, perhaps, to the blind of the Jewish Messianic prophecies) and, confronting them with an elephant, asked each to say what it was. The first, feeling its trunk, identified the elephant as a pot; the second, encountering a leg, argued that it was a tree; the third, examining its tail, announced that it was a broom; and the fourth, on the evidence of an ear, maintained fiercely that it was a winnowing-fan. And they fell to arguing among themselves, each of them sure that he was right—as, indeed, in a sense he was—and that all the others were wrong. And so it is, concluded the Buddha, when we who know little or nothing of Reality squabble about the nature and definition of the Truth.

And perhaps it is not insignificant in this connection that the present, incomplete, Great Pyramid likewise has four separate faces to north, south, east and west. Like the similarly incomplete (and conceivably cognate!) Tower of Babel of Old Testament legend, it represents divided humanity. But, with the final addition of the crowning Messianic capstone, it will eventually be seen that all four of the facets, in Reality, enshrine but a single diamond of truth, and that all reach their culmination and ultimate unity in the advent of the future Christ.

Historical Footnote

The official history books generally assert that the three main chambers of the Great Pyramid were the results of architectural changes of plan regarding the siting of the alleged burial chamber. However, the clear geometrical relationships between the chambers (see, for example, page 186) render this theory extremely unlikely, while the existence of a complete set of Trial Passages in the rock just to the east of the Pyramid (see map page 300 and diagram opposite)—in which trial-sections corresponding to all the Pyramid's major passages are to be found—finally gives the lie to the whole idea. Indeed, these borings may also have been used to establish the datum for

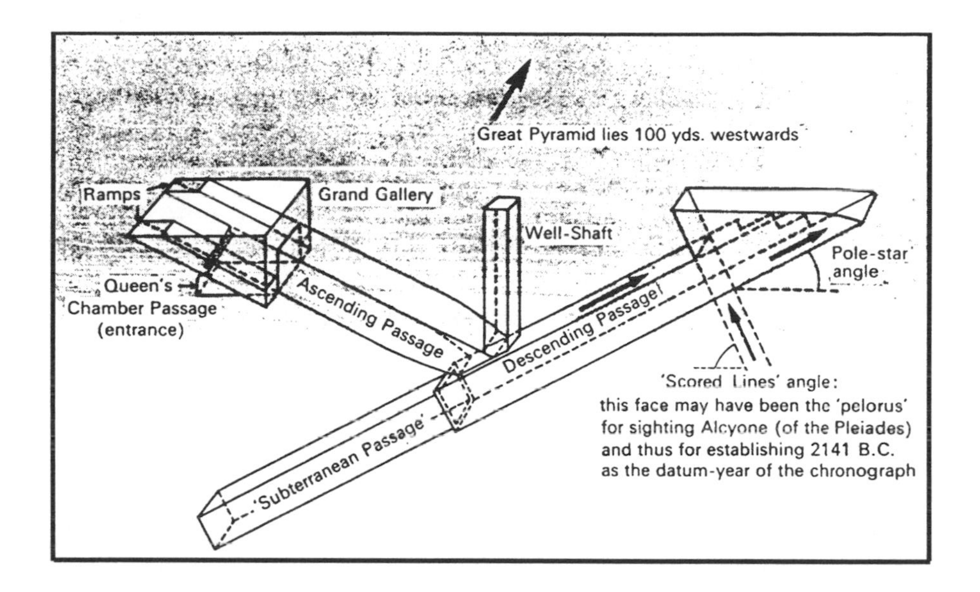

Projection of the Trial Passages. These ancient excavations are to be found some 100 yards east of the Great Pyramid, cut out of the solid rock. Their lateral and vertical dimensions and angle of slope are as for their counterparts in the Pyramid, with the exception that the Subterranean Passage is here cut on the slope, and that the lower part of the Ascending Passage is shown constricted in both height and width (to hold the Granite Plug), whereas in the Pyramid it is only the width that is constricted in this way (see Rutherford p. 965).

That these were Trial Passages, excavated before the Pyramid's construction, is apparent from the displacement of the Well-Shaft and the tilting of the Subterranean Passage (which is horizontal in the Pyramid). Had the passages merely been a later copy of those of the Pyramid, it is unlikely that these 'alterations' would have been made. The tilting of the Subterranean Passage, incidentally, provides good corroboration for Rutherford's conjectures regarding the trigonometrical interpretation of its counterpart in the Pyramid in terms of its projection on to the slope.

the Pyramid's chronograph in terms of the final positioning of the Scored Lines.

Again, accepted historical theory has it that King Khufu was buried in the King's Chamber of the Pyramid (despite Herodotus' and Diodorus' statements to the contrary), and that the Ascending Passage was then blocked by sliding into position the three blocks of the Granite Plug, which until then had been stored in the Grand Gallery on some kind of wooden platform slotted into the walls of the Gallery. The workmen engaged in this task then made their escape, it is claimed, via the Well-Shaft, which was subsequently sealed behind them, as was the lower entrance of the Ascending Passage. Doubts are cast on this theory, however, by the fact that, when the upper passages were first rediscovered by Caliph Al Mamoun in A.D. 820, the coffer in the King's Chamber was found to be empty, uninscribed and lidless, while the passages leading to it remained apparently as firmly sealed as on the day the builders left them. Again, the coffer itself is wider than the lower entrance to the Ascending Passage, so that it could not have been introduced *after* the completion of the building. Moreover, the likelihood that the blocks of the Granite Plug could have been slid down the Ascending Passage after the burial to their present positions is negligible, since there is no vertical clearance; while the insistence of the classical historian Strabo that the lower passage-system could readily be entered in ancient times, merely by opening a pivoted casing-stone, is difficult to square with the suggestion that the Granite Plug was part of a system of elaborate security measures for the protection of a 'royal burial'—when potential robbers could presumably have by-passed the Granite Plug and broken into the upper passageways by the simple expedient of entering by the same path as the original workmen allegedly escaped by. Finally, the ancient Egyptian *Book of the Dead* seems quite clearly to identify the King's Chamber as the Chamber of the Open Tomb—as though that had always been its intended symbolic function, and the coffer had thus always been empty.

None the less, the fact that historians have been convinced by the Pyramid's internal features firstly that there was at least one architectural change of plan, secondly that a royal burial did indeed take place in the King's Chamber before the passages were sealed, and thirdly that the supposed royal mummy and its accoutrements subsequently disappeared—all this may in fact have been an intentional part of the designer's symbolism.

In other words, the Descending Passage represents—as we have seen—the way of the dead, leading inevitably to hell; it is only with the inception, at the beginning of the Ascending Passage, of a *new plan*—which we have termed the Messianic Plan—that the Path of Death becomes potentially transformed into the Way of Life. Again, the passage-system speaks to us of one who, at death, has succeeded in passing up the Ascending Passage, in entering the King's Chamber and—the upward path having been finally sealed behind him—in 'breaking open' the coffer, escaping from the tomb and thereupon achieving full union with the Divine.

The close parallel with the biblical story of the crucifixion and resurrection of Jesus of Nazareth is a clear and uncanny one—even to the sealing of the heavenly path behind him (reflected in the rolling of the stone in front of the tomb, and in Jesus' reported statement: 'Where I am you cannot come' at John 7:34); not to mention the disappearance of the body, the mystery of the empty tomb, the assumption that the body has been stolen, and the notion of a spiritual resurrection. And only with the eventual triumphant return of the King-Messiah in his full power—whether the Egyptian Osiris or the Christian Jesus—can the gates of victory be thrown open to mankind at large, and a righteous people at last be admitted to their eternal inheritance.

Even from the historical misconceptions of the present day, then, there seems to be something of symbolic value for us to learn.

6

Date-Summary of Events Pinpointed by the Great Pyramid

Date	*Event*	*Possible historical identification (updated for the 1996 edition)*
B.C.		
2623 (summer solstice)	(Entrance of passage-system)	Construction (?) of Pyramid begins during reign of Pharaoh Khufu.
2141 (spring equinox)	(Scored Lines aligned with Pleiades)	(Datum-year for Pyramid's chronograph)
1453 (30th March)	Inception of new upward path	Jewish Exodus from Egypt. Formulation of Jewish Law on Mount Horeb. Era of composition of Hindu Vedas.
797–765	Formative period of favourable development	13th Israelite Jubilee (?)
592–559	Formative period of adversity	Early Babylonian captivity of Jews and destruction of Solomon's temple. Decline of Egypt. Era of darkness immediately preceding the sudden dawning of Buddhism, Confucianism, Taoism and, in Europe, the Pythagoreans.
384–352	Favourable formative period	Era of Plato and Aristotle. Birth of Alexander the Great.

2 (27th Sept.)	Birth of central Messianic figure	Birth of Jesus of Nazareth.
A.D. 29 (14th Oct.)	Preparation of Messianic figure completed (?)	Baptism of Jesus of Nazareth (?)
33 (1st April)	Achievement of full enlightenment by Messianic figure: inception of path of the enlightened.	Crucifixion of Jesus of Nazareth.
46 (March)-58 (April)	Mission of lesser Messianic figure	Missionary journeys of Paul to the non-Jewish world. Birth era of Buddhist notion of salvation by faith in a *boddhisattva* (saviour).
58–82	Period of physical death and destruction	Era of Nero, Vespasian, Titus. Jewish revolt and war.
70	Central event of period.	Sacking of Jerusalem by Titus.
152	Separation of discarnate path of the enlightened from the static path of physical rebirth	Final break between Nazarenes, official Judaism, and Christians.
1223–28	Man begins to 'come of age': return to the basics of the Messianic teachings: first steps towards the establishment of hell on earth	Death of Francis of Assisi: Franciscan and Dominican reform. Thomas Aquinas. Tentative establishment of modern scientific principles: Bishop Grosseteste, Friar Bacon.
1440–1521	Crucial events leading directly to the age of hell on earth	Invention of printing-press in Europe; fall of Constantinople; the Renaissance; the Reformation, re-discovery of America; circumnavigation of the globe.
1767–1848	Idealistic time of physical turbulence	American War of Independence; French Revolution; Napoleonic Wars, further revolutions in France, Austria, Hungary, Germany, Italy. Marxism founded. In the arts, the Romantic era; dawn of orientalism in Europe.
1845	Initial steps in the foundation of the Final Age	Spread of effects of Industrial Revolution; railways; invention of telegraph; rapid development in all branches of science.

1914 (summer)	End of discarnate era for the enlightened: beginning of age of hell on earth and of possible universal reincarnation	Beginning of First World War: signs of rapid population increase resulting from improved medical technology.
1918 (± 3)	Materialist low-point	End of First World War.
1921–32 (± 3)	Era of physical progress	
1932–39 (± 3)	Rapid decline in state of civilisation	Lead-up to Second World War.
1933–85	The *Christos* starts to infiltrate the earth-planes	
1933–44	Rise and decline of an anti-Messiah (?)	Adolf Hitler (?)
1935–37 (± 3)	Spiritual decline in mankind at large	
1945 (± 3)	New spiritual influences involving a return to basic Messianic principles	Discovery of Dead Sea Scrolls and revealing of Essenic teachings.
1951–65 (± 3)	Rapid recovery of civilisation	Post-war recovery.
1967 (± 3)	Spiritual/moral decline	The permissive society (?)
1971 (± 3)	Sudden setbacks to civilised societies	1973/74 world oil crisis and economic slump.
1977[1]/78 (± 3) –2004 (± 3)	Partial collapse and continuing decline of civilised societies	World trade-recessions of early 1980s and 1990s; general social and political disintegration.
1985[2] (30th Nov.)	Powerful spiritual influences start to irradiate the enlightened	Halley's comet; first Reagan-Gorbachev summit; new spirit of goodwill.
1999[3] (21st Feb.)	Final establishment of the Kingdom of the spirit—a separate and unique form of	

[1] 1290 years after A.D. 688, the date of the foundation of the Dome of the Rock on the site of the Jerusalem Temple (see Daniel 12:11 and chapter 7 below): but 1290 years after the Mosque's *completion* (in A.D. 691) would give a date some *three years later*—the very correction already suggested (on p. 155 above) as being applicable to the Subterranean Chamber datings shown as approximations above.

[2] 1966, by Rutherford's reckoning.

[3] Beginning of third 'day' of 1000 years after 2 B.C. (but 1979, by Rutherford's reckoning).

	human society based solely on allegiance to the spiritual
2004–25 (± 3)	Total collapse of materialist civilisation
2014–32 (± 3)	Spiritual low-point among mankind at large
2025 (± 3)	Partial re-establishment of civilised societies
2034[4] (31st Oct.)	Appearance in the sky of the sign of the Messiah
2039 (21st Oct.)	The *Christos* incarnates in a physical man—the long-awaited Messianic return
2055 (± 3)	Material progress recovers
2075 (± 3)	Beginning of era of massive physical prosperity and achievement
2076 (± 3)	Beginning of era of massive spiritual expansion; mankind in general raised to new levels
2116 (28th March)	Death or departure of first Messianic leader
2132/33 (± 3)	The path of spiritual escape closed to the totally unenlightened
2134–2238	The *Christos* reincarnates again
2264–2368	Third Messianic reincarnation
2279	Beginning of initiatory age for the reincarnating partially enlightened
2394–2499	Fourth (and last) Messianic visitation
2422–77	Partial effort at reform among the totally unenlightened

[4] Beginning of third 'day' of 1000 years after A.D. 33 (but A.D. 2015, if Rutherford's earlier reckoning is taken as a basis).

2499 (21st Feb.)	Entry of the redeemed into the final path of escape
2569[5]	End of the prophecy for the completely unenlightened (beginning of new cycle?)
2989[6] (2nd July)	Beginning of the true Millennium—the era of final escape for the enlightened
3279	End of initiatory age for the reincarnating partially enlightened
3989[7]	Conclusion of the true Millennium: end of age of human escape into the spirit-planes
Undated	Non-temporal, non-physical progress of the enlightened souls of mankind through five further, non-relative planes of spiritual experience, culminating in full union with the Divine.

[5] 3 'times' of 840 years after A.D. 49 (see Daniel 12:7), when the Messianic prince—Paul—appeared in Jerusalem to explain his missionary journeys, which were dedicated specifically to 'dispersing the power of the holy people' among the Gentiles (see Dan. 12:7). And this same event occurred some seven 'weeks of years' after the birth of Jesus of Nazareth (see Dan. 9:25), dedicated as the latter was to restoring 'Jerusalem', i.e. the Kingdom of Heaven on earth.

[6] 70 'weeks of years' after 2499 (listed above: see Dan. 9:24). Half a 'time' of 840 years after 2569 (listed above: see Dan. 12:7). 2300 years after A.D. 688–91 (the building of the Dome of the Rock apparently alluded to by Dan. 8:14).

These confused and apparently coincidental correlations with the equally confused and obscure time-prophecies of the book of Daniel could conceivably prove to be significant. And in particular the notorious 'time, times and a half' prophecy appears to link two of the dates listed by the Pyramid with a known biblical event, assuming that a 'time' equals 840 years and not 360 years, as is more commonly assumed (in fact the two numbers are directly linked, since 3 × 840 equals 7 × 360).

The erection of 'proofs' on the basis of highly arbitrary readings of the Daniel prophecies is of course pointless; but the similarities are certainly of interest.

[7] At this point it is perhaps worth noting that all three passage-levels terminate in a year ending in the figure 9. Indeed, *no fewer than eight* of the significant dates during the coming centuries do so—a fact which means that they all occur round-numbers of decades after the birth of Jesus of Nazareth in 2 B.C., if its pyramidal dating is correct. These dates (with years since the Nativity in brackets) are: 1999 (2000), 2039 (2040), 2279 (2280), 2499 (2500), 2569 (2570), 2989 (2990), 3279 (3280) and 3989 (3990). These facts could be seen as tending to confirm the potential validity of the time calculations involved.

Dating note

The notional datings for the Queen's Chamber and its Passage (q.v.) are based throughout on the assumption that the Passage's initial downward step marking A.D. 58 was originally vertical, as Rutherford and other authorities assume. On the other hand, it must be admitted that many of the earliest diagrams (e.g. Edgar) show this step in the same broken, and thus 'non-vertical', state in which we find it today. Consequently, the possibility cannot be ignored that it may *always* have been non-vertical in nature.

If we made this assumption, and if we also assumed (contrary to the code as postulated) that a vertical, mid-passage step overrides any other steps affecting the passage in question, then the whole horizontal inset would acquire a new scale of 1*n* per year.

On this basis, if we persisted in dating its beginning at 1st April A.D. 58, the new datings for the Queen's Chamber and Passage would be as follow:

Step in mid-passage:	Midsummer, A.D. 6225	[formerly 1228]
Entry into Queen's Chamber:	Spring, A.D. 7276	[formerly 2279]
South wall of Queen's Chamber:	Spring, A.D. 8276	[formerly 3279]

In other words, the Millennium here represented would not be the age of Aquarius at all, but that of Sagittarius, whose notional dates are A.D. 6330 to 8490 (compare 'Astrological Parallels', pages 322–32). The first half of that age would thus correspond to man's 'coming of age' and 'coming back down to earth' in the latter portion of the Queen's Chamber Passage: the latter half of the age would reflect his eventual escape from the earth-planes as represented by the Chamber itself and its air-shafts—the latter's part-squared, part-rounded cross-section [point (30)] suggesting a spiritual achievement based on physical experience.

In terms of such an 'alternative' reading, we should of course still need to see the upward path of the Grand Gallery and King's Chamber Complex as a totally spiritual path of redemption, reserved for the Nazarene initiates and culminating in the Aquarian age. The Queen's Chamber and its Passage, on the other hand, would need to be partially re-interpreted as the horizontal path whereby the less enlightened souls might slowly learn to encounter spiritual Truth in the heart of the physical, eventually achieving their own escape separately during the later age of Sagittarius.

It must be admitted that such an alternative approach to dating the horizontal inset, while it poses certain problems, nevertheless seems to square rather well in certain respects with zodiacal tradition and astrological symbolism.

PART TWO

Testimony of the Initiates

Surely some revelation is at hand
. . . Somewhere in sands of the desert
A shape with lion body and the head of a man,
A gaze blank and pitiless as the sun,
Is moving its slow thighs . . . but now I know
That twenty centuries of stony sleep
Were vexed to nightmare by a rocking cradle,
And what rough beast, its hour come round at last,
Slouches towards Bethlehem to be born?

W. B. YEATS (from 'The Second Coming')

7

The Pyramid and the Sacred Writings

IN THE COURSE of our investigation of the Great Pyramid of Giza, it has become clear that there are strong and often sustained similarities between the message and symbolism of the Pyramid and those of a variety of the world's sacred writings. Prominent among those writings is, of course, the Judaeo-Christian Bible, and since this is the collection likely to be best known to most of my readers, it is to the Pyramid's specifically biblical links that I propose to devote this chapter.

Quite apart from the fact that the Pyramid's chronograph seems to refer specifically to a number of crucial events which are also reported in the Bible—notably the Israelite Exodus from Egypt and the events surrounding the birth and death of Jesus of Nazareth—there is also, as we have seen, a general similarity of theme. That theme's most obvious and characteristic aspect is its messianism—the firm conviction that man's soul-evolution will achieve its final flowering only under the direct influence of the higher powers during some future time of cataclysmic upheaval. True, the traditional Christian interpretation of the Bible's message has tended, thanks to the writings of the Pharisee Paul, to give the impression that man has little to do but sit idly back, make apologetic noises and wait to be rescued by some *deus ex machina*. But the truth is that the Old Testament and the gospels show, almost as clearly as the Pyramid itself, that there is ultimately no one who can rescue man from physicality and death but man himself—or the 'son of man', to use the orientally inspired term (*ben adam*) still used in Hebrew today to refer to ordinary human beings.

Again, the average Christian may be surprised to learn that the kingdom of heaven referred to repeatedly in the New Testament was not normally thought of by contemporary Jews as some kind of post-mortem state of spiritual bliss at all. On the contrary, it was seen as an uncompromisingly physical Golden Age *on earth*[1]—a future Millennium which the righteous dead would be physically reborn to enjoy. (Christian translators might *call* it 'the world to come'—but the expression in question was actually 'the Reborn World' (Matt. 19:28).) It was a popular misconstruction of this idea of physical rebirth during the Millennium that underlay the whole tradition of the embalming of the dead: for how should the dead return to enjoy that Golden Age, it was argued, if they had no body to incarnate in? 'But thy dead live, their bodies will rise again,' the prophet Isaiah had written (26:19); 'They that sleep in the earth will awake and shout for joy; for thy dew is a dew of sparkling light, and the earth will bring those long dead to birth again.'

It takes an initiate, of course, to realise the extent to which most spiritual truths tend to become corrupted and distorted in the process of popular dissemination. And in this particular case it took a Jesus to point out that the so-called resurrection of the flesh involved not, as was popularly supposed, the re-animation of long-dead corpses, but instead the process described by the age-old doctrine of human reincarnation.

Unpalatable as the more traditional Pharisees such as Nicodemus might find it,[2] the basic fact had to be faced: 'Unless a man *is born again* he cannot see the kingdom of heaven.' Nicodemus might try his hardest—as many Christians still do today—to pretend that Jesus' words meant something other than what they plainly said. But the rebirth of the soul into a succession of earthly human bodies—including that of the Millennium—was absolutely basic to human existence. 'No one can enter the kingdom of God without being born from water *and spirit*,' continued Jesus, referring to a man's dual nature as a physical being with a spiritual identity and inheritance. 'Flesh can give birth only to flesh; it is spirit that gives birth to spirit. You ought not to be astonished, then, when I tell you that you must be born over again. The wind (punning on a word

[1] See, for example, the overwhelming evidence on this question presented by Dr. H. J.Schonfield in his celebrated study of Christian origins, *The Passover Plot*.

[2] As Schonfield points out in *Those Incredible Christians*, the transmigration of souls is known to have formed part of the Pharisees' 'hidden lore', and it may be to the urbane Nicodemus's disquiet at the idea that we owe his initial questioning of Jesus.

which meant both wind *and* spirit) blows where it wills; you hear the sound of it, but you do not know where it comes from, or where it is going. So with everyone who is born from spirit.'

The history of our souls, in other words, is generally unknown to us: so is the pattern of their future rebirths. All we know for certain is that they are with us now. And yet, it seems, there was a time when they were fully spiritual beings ignorant of mortality, and by the same token they could regain that blessed state. 'No one ever went up into heaven except the one who came down from heaven, the son of man whose home is in heaven,' continued Jesus—once again using the customary orientalism 'son of man' for man. But fallen man lacked the self-confidence to make that giant return leap. He needed a leader, an exemplar, and that was the rôle that Jesus saw himself as playing. '*This* son of man,' he went on to explain (this time indicating himself), 'must be lifted up as the serpent was lifted up by Moses in the wilderness, so that everyone who has faith in him may have everlasting life.' The serpent in question (Num. 21:6–9) had been a kind of talisman, the sight of which apparently gave to those who looked at it the will and the strength to recover from snake-bite.[3] Through the knowledge of Jesus' own example, in other words, man, having fallen, could learn to recover from his mortal sickness and rise again.

Nothing magical, no all-conquering saviours, no abstruse theological atonement-dogma. Merely the story of a man who voluntarily undertook the rôle of showing man the way out of his age-old enslavement to the cycle of physical rebirths, and of teaching him the knowledge of the Truth which alone could set him free.

From a traditional Christian viewpoint the interpretation of the third chapter of John's gospel (3:3–8, 13–15) may seem startling. But then few Christians are aware of the extent to which a person's interpretation of the Bible's message depends upon the beliefs and assumptions he 'feeds into it' in the first place. Several of the most basic of Christian assumptions, in fact, are known to have been completely foreign to the minds of the men who wrote the gospels, while it is quite certain that they could never have been believed by that most devout of Jews known to us as Jesus of Nazareth. The traditional Christian kingdom of heaven notion we have already

[3] Moses' symbolism at this point seems to have been related to the Hermetic symbol of knowledge and healing—the staff bearing twin snakes, presumably representative of the Outer and Inner Wisdom respectively. Moses appears to have been saying: 'You are suffering because you have been bitten by the physical snake only: keep your thoughts on the spiritual snake and all will be well.'

referred to as being contrary to the cast of contemporary Jewish thought: the suggestion that the Divine Unity could somehow be neatly compartmentalised into three 'persons' is in flat contradiction to anything that any faithful Jew ever stood for (and, as most Christian scholars are fully aware, owes its genesis mainly to political manoeuvring within the early Church); while the suggestion that any man could be God in human form is not one that is likely to have been subscribed to by Jesus or his immediate followers, simply because, under the basic terms of Jewish doctrine, the very idea would have constituted the most outrageous of blasphemies. True, the Jewish nation regarded itself collectively as the *son* of God, with each of its citizens *a* son of God, but neither idea in any way implied their divinity: and Jesus, for his part, seems to have seen the relationship in terms of affiliation or adoption, not consanguinity (see John 8:42–7, for example). Even the future Messiah (a royal title meaning 'anointed one', and applicable to any Israelite king) was always construed as being quite uncompromisingly physical in nature, whence his specific code-name 'the man' or 'the son of man'. Jesus was indeed crucified for blasphemy, but his crucifixion was the normal Roman sentence for blaspheming *their own god-emperor*. The Jewish sentence for blaspheming Jehovah was stoning.[4]

The wide acceptance in Jesus' day of a number of very different beliefs and assumptions is, however, quite firmly attested by a number of sources. The belief in the advent of a future physical Messianic leader who would lead the world, through the agency of the Jewish nation, to a new age of peace and plenty—this was perhaps one of the most characteristic Jewish convictions of Jesus' day, and one that was in due course to lead perilously close to the wiping-out by the Roman army of the entire Jewish nation.[5] A belief in physical rebirth is also well attested, having been the basis of the religious hopes of most contemporary Jews: and among the Gnostics, Essenes and some Pharisees of the period it seems to have had close affinities with the oriental doctrine of human karmic reincarnation.[6] Meanwhile, the surviving gospel-literature makes it difficult to avoid the conclusion that Jesus too must have taken such a view, and may have based much of his teaching upon it—a fact

[4] For detailed evidence on the above see, once again, Dr. H. J. Schonfield's authoritative research in *The Passover Plot* and *The Pentecost Revolution*.

[5] These events, too, are well documented by Schonfield.

[6] The doctrine of reincarnation, or *gilgul*, is subscribed to even to this day by the Eastern European Hasidic Jews—the direct heirs of that same Hebrew tradition which produced the Essenes.

which must have a direct and radical bearing on the interpretation of such biblical stories and reports of his utterances as have come down to us.[7]

Since, however, these same beliefs (or something very like them) also appear to be basic to the message, or 'gospel', of the Great Pyramid (as well as to most of the world's other major religions), we are faced with the possibility that there may be some direct link between them—a link other, that is, than a mere common spiritual inspiration. Moreover, since the Great Pyramid is much older than the Bible—older, indeed, than the Jewish nation itself—there would seem to be at least a possibility that the Pyramid was *one of the sources* of the apparent knowledge of the Bible's various authors.

If this were so, then we would expect to find in the Bible some hints that an Egyptian pyramid (whether named or otherwise) was considered by those authors to be of some special significance in connection with their sources of knowledge. Perhaps, too, we might unearth some evidence that that pyramid's symbolism was considered by them to have specific links with their own Messianic view of human evolution.

One of the more obvious places to search for clues might therefore be the book of Isaiah, a book riddled from beginning to end with Messianic symbolism and with prophecies relating to the end of the age. One of the book's significant Messianic passages is to be found in chapter 19, where the future Messianic advent is described in terms which might even be taken to refer to a visitation from outer space: 'See how the Lord comes riding swiftly upon a cloud, he shall descend upon Egypt . . .' And then, later in the same chapter, there comes the following extraordinary passage: 'When that day comes there shall be an altar to the Lord in the heart of Egypt, and a sacred pillar set up for the Lord upon her frontier. It shall stand as a token and a reminder to the Lord of Hosts in Egypt, so that when they appeal to him against their oppressors, he may send a deliverer to champion their cause, and he shall rescue them' (19:19–20).

What is extraordinary about this passage is that it describes exactly the site, nature and significance of the Great Pyramid of Giza. For the word here translated as 'altar' had, for the Hebrews, two quite distinct meanings—either the familiar sacrificial stone, or the commemorative altar of witness. The context makes it quite clear that the latter meaning is intended here—whereupon it is interesting

[7] See, for example, the above analysis of the passage quoted from John, chapter 3. Consider, too, the possible implications of Matt. 5:25–6, 8:22, 11:14, 18:3, 18:13–14, 24:34; Mark 10:5; Luke 16:31, 20:34; John 5:24–6 and 29, 8:56–8, 9:1–2.

to note the form which such altars of witness normally took. In the words of Cruden's *Complete Concordance to the Bible*, 'Great heaps of stones raised up for a witness of any memorable event, and to preserve the remembrance of some matter of great importance, are the most ancient monuments among the Hebrews'; and the heaps of stones were of course none other than the traditional altars of witness referred to above—altars such as were allegedly built by Moses, Jacob, Joshua and Gideon, to name but a few. But it should be realised at this point that *exactly the same description* would apply with equal aptness to the Great Pyramid—truly the ultimate in commemorative heaps of stones!

Again, the passage's second sentence uses the word 'it' to refer both to the altar in question and to some kind of sacred pillar. But how can a heap of stones possibly be described as a pillar?

Research reveals that the latter word is perhaps a misleading translation of the word used in the original Hebrew. The word in question is *matstsebah*, which means almost any kind of monument: indeed the word appears to be cognate not only with the ancient Egyptian *mstpt* (funeral bier), but also with the Arabic *mastaba*—a funerary monument which actually seems to have provided the basis for the design of the Egyptian pyramids.

Thus we can now begin to see Isaiah's obviously Messianic monument as taking the form of a commemorative heap of stones bearing a possibly direct relationship to the Egyptian pyramids. The site description which then follows narrows down the range of possibilities even further. For the monument in question is described, paradoxically, as standing both 'in the heart of Egypt' *and* 'upon her frontier'. Yet the apparent paradox is nothing of the kind. For as reference to the map on page 6 will reveal, the ancient kingdom of Lower Egypt comprised, as it happens, the Nile Delta, which forms an almost perfect quadrant of a circle, subtending an angle of 90° at its centre. The geometrical hub of this quadrant could thus be described as the heart of the kingdom, while it would also mark the angle of its southern frontier. And the feature occupying this interesting site (as the United States Coast Survey discovered largely by accident in 1868) turns out to be none other than the Great Pyramid itself.

But Isaiah's description has still more evidence to offer us. Not content with alluding to the monument's nature and site, the author goes on to tell us that it has a specifically Messianic rôle to play—a guarantee of some kind that, when man at last calls out for help, a deliverer of some kind will appear. And we may see this reference too as further confirmation that the building in question is the Great Pyramid.

Yet the author of Isaiah has yet another cryptic clue up his copious sleeve—this time an apparent reference to the building's actual dimensions. It is well-known that the characters of written Hebrew serve a double function in that they represent both letters and numbers. It has long been realised that the passage quoted above seems to have been intentionally cast in such a way as to ensure that the total numerical value of its Hebrew characters is exactly 5449. That, at all events, *is* their value. *It is astonishing, therefore, to discover that this figure is within ·27 of the number of Primitive Inches in the Great Pyramid's height from base to summit-platform (5448·736 P")*—the latter feature (still hidden in Isaiah's day) apparently being designed specifically to refer in symbol to the imperfect world's need of some kind of Messianic deliverer.

The passage quoted from Isaiah 19 therefore comprises, in a number of ways, a strong piece of evidence that *its* author at least was aware of the Great Pyramid's significance and used it as one of the sources of his remarkable knowledge. Moreover, in this he does not seem to have been alone.

The author of Psalm 118, for example, also has Messianic ideas to convey, and like the author of Isaiah 19 he does so in terms which are powerfully reminiscent of the Pyramid and its message:

> 'Hark! Shouts of deliverance
> In the camp of the righteous!
> With his right hand the Lord does mighty deeds,
> the right hand of the Lord raises up.
> I shall not die but live
> to proclaim the works of the Lord.
> The Lord did indeed chasten me,
> but he did not surrender me to Death.
>
> Open to me the gates of Victory;
> I will enter by them and praise the Lord.
> This is the gate of the Lord;
> The righteous shall make their entry through it.
> I will praise thee, for thou hast answered me
> and hast become my deliverer.
> The stone which the builders rejected
> has become the chief corner-stone.'[8]

[8] Psalms 118:15–22.

The first point to notice about this passage is its remarkable resemblance to the words of the Osirian ritual relating to the symbolic entry into the Chamber of Resurrection—an apparent allusion to the King's Chamber of the Great Pyramid:

'Hail thou, my Father of Light, I come having this my flesh freed from decay; I am whole as my Father, the self-begotten God, whose image is in the incorruptible body. Do thou establish me. Do thou perfect me as the Master of the Grave . . .

'I have opened the doors . . . well is the Great One who is in the coffin. For all the dead shall have passages made to Him through their embalming.

'The tortoise dies; Ra lives! O Amen, Amen, Amen, who art in heaven, give thy face to the body of thy Son. Make him well in Hades. It is finished.'[9]

And here in turn we are reminded of the doctrine—and even, in places, of the very words—of Jesus of Nazareth, who is reported (at Matthew 21:42) as having applied *to himself* the extraordinary words, 'The stone which the builders rejected has become the chief corner-stone.' They are extraordinary because, once again, the conclusion seems unavoidable that the original author of the statement was deliberately referring to the Great Pyramid. Any building, of course, can have a corner-stone—most have four or more—but there is only one type of building that can have a *chief* corner-stone, or a 'head-stone in the corner', as one translation has it; and that building is a pyramid. Yet the reference here is specifically to a building whose chief corner-stone *is missing*—and it is further made clear that the final placement of that corner-stone will have a specifically Messianic connotation. The conclusion therefore seems to be that both the psalmist and Jesus of Nazareth were consciously referring to the missing capstone of the Great Pyramid.

And these are not the only biblical passages which suggest that their authors were familiar with aspects of the Pyramid's message. The Messianic significance of the final placement of the capstone on the as-yet incomplete Pyramid also seems to be referred to in the book of Zechariah, for example: 'How does a mountain, the greatest mountain, compare with Zerubbabel?' asks the author in chapter 4. 'It is no higher than a plain. He shall bring out the stone called Possession while men acclaim its beauty . . . . Zerubbabel with his own hands laid the foundations of this house and with his own hands he shall finish it. So shall you know that the Lord of Hosts

[9] From *The Book of the Master of the Hidden Places* by Marshal Adams.

has sent me to you. Who has despised the day of small things? He shall rejoice when he sees Zerubbabel holding the stone called Separation.'

This extract suggests that the building will be completed by the same person who laid its foundations—which, in the case of the Great Pyramid, would necessarily involve his reincarnation. As for the stone's latter name, Separation, this could conceivably be a reference to the Messianic 'creaming off' of the enlightened during the Final Age, as symbolised by the Pyramid's interior features.

Again, there seems to be an oddly pyramidal echo in the 26th chapter of Isaiah, which commences with the words: 'We have a strong city whose walls and ramparts are our deliverance. Open the gates to let a righteous nation in, a nation that keeps faith.' Following on from this almost Osirian passage, we come to verses 4, 7 and 8: 'Trust in the Lord for ever, for the Lord himself is an everlasting rock . . . The path of the righteous is level, and thou markest out the right way for the upright.' If this reference were significant, however, it would suggest an interesting identification between 'the Lord' (Jehovah) and the 'everlasting rock' of the Pyramid—a notion which is further explored in chapter 9 of this book.

Leaving Isaiah and returning once more to the Psalms, we find that the author of Psalm 84 sees life as a pilgrimage with strange affinities to the traversing of the Great Pyramid's passageways:

> 'How dear is thy dwelling-place,
> thou Lord of Hosts! . . .
> Happy are those who dwell in thy house;
> they never cease from praising thee.
> Happy the men whose refuge is in thee,
> whose hearts are set on the pilgrim ways!
> As they pass through the thirsty valley
> they find water from a spring;
> and the Lord provides even men who lose their way
> with pools to quench their thirst.
> So they pass on from outer wall to inner,
> and the God of gods shows himself in Zion.'[10]

The passage refers ostensibly, of course, to the great Temple at Jerusalem. Yet at the same time it seems to make specific reference to the Great Pyramid's Well of Life, to the fact that it is available even to 'men who lose their way'—and to the final entry into the King's Chamber. That the latter is here referred to by the word

[10] Psalms 84:4–7.

Zion—traditionally taken as a reference to Jerusalem—does not necessarily invalidate the possibility of a pyramidal connotation: indeed, it suggests the need to research the possibility that Zion (and perhaps even Jerusalem) may sometimes have been used by Hebrew writers as a code-word for the Great Pyramid.

Meanwhile the author of Jeremiah 6 (16–21) provides a notable further example of a possible link between the Great Pyramid and Bible. In this case the reference seems to be to the Pyramid's Door of Ascent or to the Crossing of the Pure Roads of Life: 'These are the words of the Lord: Stop at the cross-roads; look for the ancient paths; ask, "Where is the way that leads to what is good?" Then take that way, and you will find rest for yourselves. But they said, "We will not." Then I will appoint watchmen to direct you; listen for their trumpet-call. But they said, "We will not." Therefore hear, you nations, and take note, all you who witness it, of the plight of this people. Listen, O earth, I bring ruin on them, the harvest of all their scheming; for they have given no thought to my words and have spurned my instruction . . .' Whereupon a description of some kind of national holocaust follows. Here the reference to watchmen seems to speak of the various prophets, while the consequences of the refusal to heed their instructions once again seem to correspond to events such as those symbolised in the Pyramid's Subterranean Chamber.

* * *

If, then, the links between the Great Pyramid and the biblical Old Testament are as strong as some of them seem to be, it would be surprising if we failed to discover evidence that Jesus of Nazareth himself also knew about them. Indeed, we have already noted the clear pyramidal significance of Jesus' celebrated declaration: 'The stone which the builders rejected has become the chief corner-stone' (Matt. 21:42).

A further report which seems to add fuel to the flames comes at Luke 19:37–40: 'And now, as he approached the descent from the Mount of Olives, the whole company of his disciples in their joy began to sing aloud the praises of God for all the great things they had seen . . . Some Pharisees who were in the crowd said to him, "Master, reprimand your disciples". He answered, "I tell you, if my disciples keep silence the stones will shout aloud." '

The customary interpretation of the last sentence would have Jesus guilty of a measure of exaggeration: but the possibility should not be overlooked that the statement may have been a deliberate reference to the stones of the Great Pyramid—in which case the statement could be seen as a cryptic (and not untypical) way of saying that those stones had a Messianic message for mankind.

Meanwhile the story at Luke 13:6–9 of the fourfold visit of the vineyard-owner seems to be a direct echo of the fourfold Messianic visitation referred to by the Great Pyramid's Antechamber: 'He told them this parable: A man had a fig-tree growing in his vineyard; and he came looking for fruit on it, but found none. So he said to the vine-dresser, "Look here! For the last three years I have come looking for fruit on this fig-tree without finding any. Cut it down. Why should it go on using up the soil?" But he replied, "Leave it, sir, this one year while I dig round it and manure it. And if it bears next season, well and good; if not, you shall have it down." ' The fourth Messianic visitation, in short, is final—as the Pyramid likewise reveals.

Further, in conformity with the Pyramid's message, Jesus himself insists, at John 8:31–2, that it is some form of enlightenment that alone will eventually free man from his physical captivity: 'Turning to the Jews who had believed him, Jesus said, "If you dwell within the revelation I have brought, you are indeed my disciples; you shall know the truth, and the truth will set you free." ' That escape—as the Great Pyramid symbolises in its Construction Chambers—is to the first of a number of spiritual planes, and Jesus seems to have been referring to this same fact when he made the further celebrated comment: 'There are many dwelling-places in my Father's house' (John 14:2).

Jesus, then, seems to have been aware, as were various of the Hebrew prophets, of just such a Messianic Plan for mankind as is spelt out in the Great Pyramid of Giza. Since he is known to have lived in Egypt as a child, it is by no means impossible that he may have been initiated into its priestly mysteries and notably, perhaps, into those of the Great Pyramid—as indeed the late Edgar Cayce (see chapter 8) was repeatedly to affirm. This could in part account for the remarkable knowledge which Jesus was later to display. Moreover, if this supposition is correct, then it seems clear that he must have become aware that he himself was referred to by its chronograph, and that his mission was there spelt out. Two of his most celebrated statements, in fact, appear to be nothing less than direct allusions to the Messianic rôle delineated specifically by the Great Pyramid. With the words 'I am the way: I am the truth and I am life' (John 14:6), Jesus appears to have been identifying himself directly, if cryptically, with the Pyramid's Messianic Triangle, as we saw in chapter 3: and his description of himself as 'that living bread which has come down from heaven' (John 6:51), and of the Son of Man as bearing the seal of God the Father (John 6:28), likewise seems to be a direct symbolic claim to identity with the Antechamber's Granite Leaf (see pages 119–20). Consequently, bearing in mind that the Great Pyramid was already some five hundred years old at the time of the semi-legendary Abraham, it is interesting to speculate further what ulterior meaning Jesus might have intended when he made the extraordinary claim (John 8:58): 'In very truth, before Abraham was born, I am.'[11]

At which point we come to the apparent links between the various pyramidal and biblical prophecies—and especially those which still remain unfulfilled in our own day. How far do the Great Pyramid's predictions for the future accord with those of the biblical prophets?

The Pyramid's predictions are remarkably precise in their timing, though in some cases their content is of a generalised nature. On the other hand, the biblical prophecies are bedevilled by their authors' evident love of obscurity. Many of them seems to have regarded their writings as semi-secret communications for 'those who have ears to hear'—code-messages, as it were, from initiate to initiate. Consequently their identifications of the starting points for their time prophecies are, for the most part, cryptic in the extreme: their calendrical units tend to be ambiguous, often being expressed in terms of each other (a 'day', for example, can apparently represent either a day, a year, 360 years, or even conceivably 840 years—see notes 5 and 6, page 206; and the events referred to are often themselves described in highly symbolic and poetic language which is apparently not intended to be taken literally.

In short, it is possible to read almost anything one likes into the biblical prophecies—as the reader is probably already well aware from the efforts of those who have tried. Rather like reading fortunes in a tea-cup, the result tends to depend largely on what is already in the reader's subconscious mind. Consequently there is little one can do at this point beyond quoting a number of the more celebrated passages relating to future events—notably those surrounding the end of the present age, the Messianic advent and the subsequent Millennium—without attempting any complete interpretation.

Regarding the end of the present age, chapter 24 of Isaiah

[11] Paul, for his part, seems to have inherited from his Pharisaic teachers a picture of the Messianic giant or archetypal man which has distinct affinities to the Pyramid's symbolism. This is particularly noticeable at Ephesians 4:11–16. Compare the diagram on page 159.

contributes the following warning, which might be taken to refer to a shifting of the earth's axis, and to the effects of world-pollution. It also contains echoes of the symbolism of the Pyramid's Subterranean Chamber, Pit and Dead End Passage:

> 'Beware, the Lord will empty the earth,
> split it open and turn it upside down,
> and scatter its inhabitants. . .
> The earth dries up and withers,
> the whole world withers and grows sick
> the earth's high places sicken,
> and earth itself is desecrated by the feet of those who live in it,
> because they have broken the laws, disobeyed the statutes
> and violated the eternal covenant . . .
> The hunter's scare, the pit, and the trap
> threaten all who dwell in the land;
> if a man runs from the rattle of the scare
> he will fall into the pit;
> if he climbs out of the pit
> he will be caught in the trap.
> When the windows of heaven above are opened
> and earth's foundations shake,
> the earth is utterly shattered,
> it is convulsed and reels wildly.
> The earth reels to and fro like a drunken man
> and sways like a watchman's shelter;
> the sins of men weigh heavy upon it,
> and it falls to rise no more.'

Apparently with reference to this time of crisis, Isaiah further claims that 'the earth will bring those long dead to birth again' (26:19), and Jesus of Nazareth is reported (at Matthew 24:35) as having likewise hinted at universal rebirth at the time in question, his reported words being: 'I tell you this, the present generation will live to see it all.'[12]

Meanwhile the book of Daniel provides a field-day for the 'mathematicians of doom' with its famous predictions. Chapters 7 and 8, for example, ostensibly portray the rise and fall of a variety of kingdoms and world-rulers up to the 'end of the age'. As for the timing of the events predicted, the text merely offers the following:

'The one said, "For how long will the period of this vision last? How long will the regular offering be suppressed, how long will

[12] Though study of this statement in context suggests that a more likely meaning is: 'I tell you truly, *this age* will not end until all these things have happened.'

impiety cause desolation, and both the Holy Place and the fairest of all lands be given over to be trodden down?" The answer came, "For two thousand three hundred evenings and mornings, then the Holy Place shall emerge victorious" ' (8:13–14). The significance of the 'evenings and mornings' is anybody's guess, but they are commonly assumed to represent years. (See note 6, page 206.)

The angel Gabriel then says to Daniel: 'Understand, O man: the vision points to the time of the end', and continues:

> 'In the last days of those kingdoms
> when their sin is at its height,
> a king shall appear, harsh and grim, a master of stratagem.
> His power shall be great, he shall work havoc untold;
> he shall succeed in whatever he does.
> He shall work havoc among great nations and upon a holy people.
> His mind shall be ever active,
> and he shall succeed in his crafty designs;
> he shall conjure up great plans
> and, when they least expect it, work havoc on many.
> He shall challenge even the Prince of princes
> and be broken, but not by human hands.
> This revelation which has been given
> of the evenings and mornings is true;
> but you must keep the vision secret,
> for it points to days far ahead' (8:23–6).

But even Daniel himself, it seems, 'was perplexed by the revelation and no one could explain it' (27).

Chapter 9 of the book of Daniel offers us some further mathematical crumbs:

'Seventy weeks are marked out for your people and your holy city; then rebellion shall be stopped, sin brought to an end, iniquity expiated, everlasting right ushered in, vision and prophecy sealed, and the Most Holy Place anointed. Know then and understand: from the time that the word went forth that Jerusalem should be restored and rebuilt, seven weeks shall pass till the appearance of one anointed, a prince; then for sixty-two weeks it shall remain restored, rebuilt with streets and conduits. At the critical time, after the sixty-two weeks, one who is anointed shall be removed with no one to take his part; and the horde of an invading prince shall work havoc on city and sanctuary. The end of it shall be a deluge, inevitable war with all its horrors. He shall make a firm league with the mighty for one week; and, the week half spent, he shall put a stop to sacrifice and offering. And in the train of these abominations shall come an author of desolation; then, in the end, what has been decreed concerning the desolation will be poured out' (9:24–7).

In this passage the 'weeks' could conceivably represent 'weeks of years'. As for the nature of the desolation, chapter 11 spells this out in some detail, after which chapter 12 goes on:

> 'At that moment Michael shall appear,
> Michael the great captain,
> who stands guard over your fellow-countrymen;
> and there will be a time of distress
> such as has never been
> since they became a nation till that moment.
> But at that moment your people will be delivered,
> every one who is written in the book:
> many of those who sleep in the dust of the earth will wake,
> some to everlasting life
> and some to the reproach of eternal abhorrence . . .' (12:1–2).

We may assume, of course, that 'Michael' refers to a Messianic initiate and leader, while it is clear that some kind of mass-reincarnation is also envisaged by the author as coinciding with the period in question—a fact which would, as we have seen, necessitate a dramatic increase in world population. It is also interesting to note that the text seems to assume that those who achieve perfection will do so by their own efforts—as the Pyramid also suggests—rather than by the remote merit of a saviour-figure, for it continues:

'And I said to the man clothed in linen who was above the waters of the river, "How long will it be before these portents cease?" The man clothed in linen above the waters lifted to heaven his right hand and his left, and I heard him swear by him who lives for ever: "It shall be for a time, times, and a half.[13] When the power of the holy people ceases to be dispersed, all these things shall come to an end . . . Many shall purify themselves and be refined, making themselves shining white, but the wicked shall continue in wickedness and none of them shall understand: only the wise leaders shall understand. From the time when the regular offering is abolished and the abomination of desolation is set up, there shall be an interval of one thousand two hundred and ninety days. Happy the man who waits and lives to see the completion of one thousand three hundred and thirty-five days! But go your way to the end and rest, and you shall arise to your destiny at the end of the age" ' (12:6–13).

[13] Generally interpreted as $1 + 2 + \frac{1}{2}$, or 'three-and-a-half-times'. See notes 5 and 6, page 206.

As for the timing of these events, the 1,290 days and 1,335 days in question are generally taken to refer to years (in accordance with Ezekiel 4:6), while the starting-date could conceivably be based on that of the anticipated destruction of the Jerusalem Temple and the building of a Moslem mosque on the site. For the record, the actual construction of what is now known as the Dome of the Rock was started in A.D. 688 and completed in A.D. 691, which would give datings of A.D. 1978–81 and A.D. 2023–6 respectively.

But datings for what? The text seems reluctant to give any clear answer, beyond the fact that some may have difficulty in living and waiting until the end of the period in question. The dates alluded to may thus refer to a time of real world-crisis, during which man's very survival is threatened—and on the whole, therefore, there seems to be a good degree of accord with the Great Pyramid's predictions.

* * *

Finally, we come to the expected Messianic return and the inception of the New Age. The book of Isaiah in particular is so full of allusions to these events that it would be pointless to reproduce them here. The simplest plan seems to be to quote Jesus of Nazareth's own summary of them, as reported at Matthew 24: 4–8, 10–14, 21–3, 27, 29–30, 33, 35–6.

'Take care that no one misleads you. For many will come claiming my name and saying, "I am the Messiah"; and many will be misled by them. The time is coming when you will hear the noise of battle near at hand and the news of battles far away; see that you are not alarmed. Such things are bound to happen; but the end is still to come. For nation will make war upon nation, kingdom upon kingdom; there will be famines and earthquakes in many places. With all these things the birth-pangs of the new age begin . . . Many will fall from their faith; they will betray one another and hate one another. Many false prophets will arise, and will mislead many; and as lawlessness spreads, men's love for one another will grow cold. But the man who holds out to the end will be saved. And this gospel of the Kingdom will be proclaimed throughout the earth as a testimony to all nations; and then the end will come . . .

'It will be a time of great distress; there has never been such a time from the beginning of the world until now, and will never be again. If that time of troubles were not cut short, no living thing could survive; but for the sake of God's chosen it will be cut short . . .

'Then, if anyone says to you, "Look, here is the Messiah," or,

"There he is", do not believe it . . . Like lightning from the east, flashing as far as the west, will be the coming of the Son of Man . . .

'As soon as the distress of those days has passed, the sun will be darkened, the moon will not give her light, the stars will fall from the sky, the celestial powers will be shaken. Then will appear in heaven the sign that heralds the Son of Man. All the peoples of the world will make lamentation, and they will see the Son of Man coming on the clouds of heaven with great power and glory . . .

'When you see all these things, you may know that the end is near, at the very door . . .

'But about that day and hour no one knows, not even the angels in heaven, not even the Son; only the Father.'

Here it should be noted, once again, that the events foreseen seem to include a sudden shift in the earth's axis resulting in violent volcanic and earthquake activity, thick clouds of volcanic dust in the upper atmosphere and the disappearance from the sky of the familiar constellations. Or perhaps the reference to falling stars should be understood more literally to mean that more than one meteor or comet will collide with the earth.

Thus, basing their information on this and other scriptural passages, what might be called 'biblical futurologists' generally cite the following as the definitive 'signs of the times'—the indications of the imminent end of the old world-order and the commencement of the new age:

(a) Widespread wars (Matt. 24:7, Dan. 9:27)
(b) Famine on an unprecedented scale (Matt. 24:7, Isa. 24)
(c) World-wide dissemination of the 'news of the coming kingdom' (Matt. 24:14)
(d) A general spread of lawlessness and lovelessness (Matt. 24:12, 2 Tim. 3)
(e) Preoccupation with pleasure and entertainment (Matt. 24:37–9)
(f) Return of the dispersed Jews to Palestine (Isa. 49 onwards, Dan. 12:7, Ezk. 38:8–9, 28)
(g) Increasing world earthquake activity (Matt. 24.8, Isa. 24:18–20)
(h) Many self-professed 'Messiahs', including a pseudo-religious leader who will attain world-wide power (Matt. 24:11, 23–6, Dan. 8:23–5)
(i) Westward advance of 'kings from the east' (Rev. 16:12)
(j) Invasion of Israel by teeming hordes from the north, and their defeat (Ezk. 38; Dan. 11:40–5)

(k) The appearance *in the sky* of some messianic sign, followed by the Messiah himself (Matt. 24:27–31) and the inception of the great 'sky-kingdom' or 'kingdom of heaven'.

Of these, one of the most significant warning-signs is generally held to be (f)—which is of course actually in process of fulfilment at the present time—while features (a) to (e) seem to have been in evidence for some years already. As for the timing of (k), Jesus himself is reported as claiming ignorance of the subject (see previous page). None the less, he continually associates this event with the expression 'on the third day'—as Matthew 16:21, 17:23 and 20:19 reveal, and as verse 2 of Hosea 6 had already foreshadowed.[14] But the normal 'day' clearly does not apply here. Jesus' alleged posthumous appearances were not his expected physical rebirth to kingship over a new world-order, whatever else they may have been. Nor does the day = year idea put forward in Ezekiel have any validity here: the years A.D. 35 or 36 do not appear to have had any particular significance for the Messianic cause, any more than do dates based on 'days' of 360 or 840 years. 'Go to the people today and tomorrow and make them wash their clothes', Moses is told at Exodus 19:10–11. 'They must be ready by the third day' (i.e., by Hebrew inclusive reckoning, 'the day-after-tomorrow'), because 'on the third day the Lord will descend upon Mount Sinai in the sight of all the people.' And the not dissimilar return of the Great Initiate—the capstone's notional descent upon the Golden Mountain of

[14] Jesus also seems to have offered various cryptic hints of a symbolic dating, if the gospel accounts are to be believed. The Last Supper clearly symbolises, among other things, the 'living bread' or Messianic Banquet of the second Messianic coming: and the upper room in which it is held (the term is extraordinarily reminiscent of the Pyramid's King's Chamber Complex) equally clearly stands for the exalted Final Age. Now the man who leads the disciples to this 'room' is specifically described as 'a man carrying a pitcher of water' (Mark 14:13). But this seems to suggest an intentional symbolic reference *to the zodiacal sign of Aquarius*—for in first-century Palestine the carrying of water was strictly women's work.

Again, Jesus angrily tells the Pharisees that the only Messianic sign they will get is 'the sign of the prophet Jonah' (Matt. 12.39): but this, as Mme. Blavatsky has pointed out, could likewise be a cryptic reference to Aquarius, known to the Babylonians as the 'fish-man', and associated by them with the fish-like creature Oannes, the legendary source of all their wisdom and learning. Either or both of these references, in other words, could have been a symbolic means of dating the Messianic return and the subsequent Final Age as contemporaneous with the Age of Aquarius—the very zodiacal age into which the earth is passing at the present time. If so, then it is perhaps worth noting that the actual point of transition into the New Age is dated by some authorities—including the French Institut Géographique National—at around A.D. 2010.

the Great Pyramid—is clearly destined to be one of the most momentous events in world history. As such, the 'resurrection of the Christ (= Messiah)' has not yet occurred. What, then, is meant by the third day?

Psalm 90:4 contains a clue: 'A thousand years in thy sight are as yesterday.' The point is taken up at 2 Peter 3:8—part of a letter on the subject which appears to have been produced by the Jewish Christians at Jerusalem, who, after all, should have known what they were talking about. 'And here', writes the author, 'is one point, my friends, which you must not lose sight of: with the Lord one day is like a thousand years and a thousand years like one day . . . because it is not his will for any to be lost, but for all to come to repentance.' Further, verse 10 refers specifically to the 'day of the Lord' which is to come—and this day is itself described as a thousand years at Revelation 20:2 and 4. The reference is clearly to the expected Golden Age, which will culminate in 'a new heaven and new earth'—a new plane of human existence entirely.

We may perhaps conclude from this that 'on the third day' means at the beginning of the third millennium. But after what? The beginning of the third millennium after the birth of Jesus of Nazareth (if the Pyramid's dating of that event is right) is 1999—a date associated by numerous seers from Nostradamus to Edgar Cayce with the beginning of the New Age. The third millennium after his crucifixion, on the other hand, begins in 2033. Perhaps, therefore, it is at this time that men should expect to see the 'man coming on the clouds of heaven'—conceivably, indeed, arriving literally out of space. And certainly the dating is one which the Great Pyramid seems to confirm.

If Isaiah is right, the foundations of a new and extraordinary civilisation will then be laid—an age of peace, plenty and justice amply described in the later chapters of that book, if in largely symbolic terms. Meanwhile the gospel reports of Jesus' teaching make it clear again and again that the eventual flowering of that civilisation will be marked by man's final escape from the eternal cycle of rebirths and redeaths, and his entry into the eternal life of the planes of the spirit. And this, too, the Pyramid seems already to have foreshadowed some twenty-eight centuries before him.

* * *

In view of the foregoing, then, there does seem to have been a deep awareness among many of the ancient Hebrew initiates that the great Messianic Plan for human evolution was somehow bound up with the significance of the Great Pyramid of Giza. Not only the knowledge of its exterior symbolism, but also that of its interior detail seems to have been preserved into a later age by the guardians of the sacred mysteries. The author of John's gospel in particular seems to have been well versed in the Great Pyramid's message and symbolism.

As a case in point, the curious story of Jesus and the Samaritan woman (recounted in John 4) shows us Jesus on a journey from Judæa to Galilee via Samaria. Sitting down at noon by a well at a town called Sychar, he tells a woman drawing water that his own 'living water', unlike hers, is 'an inner spring always welling up for eternal life'. When she asks for a drink of it, however, Jesus apparently demurs, telling her that she has had five husbands but is now living with a sixth man who is not her husband. Nevertheless the woman, convinced of his Messiahship, spreads the news of his coming. Meanwhile Jesus and his disciples decide to break their journey there for two days before setting out again for Galilee, where a warm welcome awaits them from the Galileans who had earlier been with them in Jerusalem.

The whole story seems to display a veiled symbolism reminiscent of the message of the Great Pyramid. Even the geography is symbolic. We should first note that, in the New Testament, 'Galilee' (the 'country', it will be remembered, where Jesus promised to rejoin his disciples after his expected return from the dead) often seems to represent the eventual Golden Age on earth (Jesus' promise, in other words, was really of physical reunion in the Millennial Kingdom which was to precede man's eventual attainment of full spirituality). The same name was often formerly given to the porch, or penitentiary antechapel, of a church—symbolic, perhaps, of the same Millennial gateway to the eternal mysteries as is represented specifically by the Pyramid's Antechamber. Meanwhile Judæa appears to represent the ancient Jewish dispensation centred on the Torah; while the despised 'Samaria', traditionally regarded by orthodox Jews as well 'beyond the pale', naturally symbolises the non-Jewish or gentile world.

Reading between the lines, then, the story goes as follows. The Great Initiate is on his reincarnatory way from the orthodox Jewish dispensation to the New Order of the Millennium. However, he interrupts his journey at the well of Sychar—an unidentified town whose name could be a corrupt form of the Hebrew *schichah*, meaning pit or corruption. Not only these facts but also Jesus' own words seem to indicate that the place somehow signifies the experience both of physical death and of spiritual Life. But that spiritual Life is denied the woman (who seems to represent the gentiles) because she has rejected 'five husbands' (the true Messianic initiation?) and has instead chosen to live with an impostor (conceivably the 'leaven of the Pharisees' or the Pauline teachings).

Jesus and his disciples now decide to 'break their journey' for 'two days' at this place of 'Life-in-death', while the woman's testimony serves to spread the Messianic message among her compatriots. In other words, we may perhaps see Jesus' 'cell of initiates' as enjoying a 'two-day' period of suspended reincarnation while the Messianic teachings are spread among the gentiles. Thus Jesus finally arrives 'in Galilee' on the 'third day', where the 'Galileans' give him a warm welcome, having themselves been with him 'at the festival in Jerusalem'. And there he once again visits 'Cana-in-Galilee', the scene of the 'changing of water into wine'. There can be little doubt that these final details constitute a direct reference to the expected Messianic return during the Golden Age, when Jesus' reincarnated contemporaries are expected to join him in the Millennial Messianic Banquet.

As in the Great Pyramid, then, Jesus and the other initiates are seen to 'rest' by a Well symbolic both of Life and death, while the teachings are given time to spread throughout the gentile world. After 'two days'—presumably representing the familiar two thousand years—the journey is resumed and the final earthly Kingdom itself reached amid the plaudits of Jesus' own reborn contemporaries, who are heirs to the long-awaited Golden Age.

This episode from John's gospel seems to be nothing less than an allegorical representation of the events specifically symbolised by the Great Pyramid's Well-Shaft, Grand Gallery and Antechamber. Bearing in mind the other patently Pyramidal references in John's gospel, it is difficult to avoid the conclusion that one at least of the gospel's authors had been trained in the Pyramid's mysteries.

But in this, as we have seen, he was merely continuing a tradition common to many of the ancient Hebrew prophets, whose writings likewise reveal a familiarity with the message of the Great Pyramid. To it they may all have owed, at least in part, the basis of their message to mankind and their vast visions of the shape and flow—and even the timing—of crucial future events. And to the extent that this is true, the Great Pyramid must inevitably be seen as the house, *par excellence*, of their secret knowledge.

8

A Third Eye on the Future

IN THE preceding chapters we have seen how the Pyramid's apparent predictions for the future show a high degree of precision in their timing of events and tendencies affecting mankind in general. The chronology of the complementary biblical prophecies, on the other hand, is less easy to decipher, even though some of the events in question are described in more detail than is the case with the Pyramid predictions. Moreover, the terminology used often seems to have been left intentionally vague—added to which, many of the terms and concepts necessary to describe the events of the twenty-first century A.D., say, simply did not exist in the contemporary language and thought of the authors. (If Ezekiel's famous visitation *was* a spaceship he could not have said so,[1] even if he had understood the concept in the first place.)

If, therefore, we are to gain a more precise insight into the future events referred to by both Pyramid and Bible, we need to turn to sources of prophetic gnosis rather nearer to our own times. Among the best known of these sources are the various 'seers' whose predictions are summarised below. The reader is invited to compare them with the Pyramidal and biblical predictions already listed and with known historical events thus far, as well as with the various extrapolations into the future currently being derived from observed present-day tendencies by 'experts' in a wide range of specialist fields.

[1] Ezk. 1 and 10. Compare *The Spaceships of Ezekiel* by J. F. Blumrich.

St. Malachy (Irish, 12th Century)

St. Malachy predicted all the popes from his day until the fall of the Vatican. The last four are described as follows:

De Medietate Lunae (of a half-moon), during whose reign the Roman Church will be persecuted, and at the end of which he himself will fall victim to the persecutors. His 'label' could have referred to his physical features, his coat of arms, the length of his pontificate . . . (*Possibility No. 3 fulfilled in 1978: John Paul I reigned for just a month.*–P.L.)
De Labore Solis (from the toil of the sun). His name could have suggested a descendant of former Negro slaves. *Suggestion unfulfilled: instead, John Paul II worked with his bare hands in the stone-quarries of Nazi-occupied Poland! P.L.*
Gloria Olivae (the glory of the olive), whose reign will be a glorious one, uniting humanity in the Christian faith. The 'Monk of Padua' names him as Leo XIV. His Latin name suggests either a period of peace or a black skin.
Petrus Romanus (Peter of Rome), who will be the last pope, and whose reign will conclude with the burning of Rome at the end of the 20th century.

Nostradamus (French, 16th Century)

Most of the predictions of this celebrated physician are, if anything, even more obscure than the Old Testament ones—apparently because of his fear that he would be accused of witchcraft if their predictive message were made too overt. He himself admitted to having deliberately garbled the order of the cryptic quatrains comprising his *Centuries*, while both datings and events are expressed via a Nostradamian code that is still barely understood. Even the names of people and places are sometimes expressed as anagrams. Consequently it has in the past proved difficult to decipher his predictions other than in respect of events that have already happened: none the less the correspondences, once established, are often impressive, being particularly detailed and accurate in respect of French history. And it turns out that there are few major events of subsequent world history that cannot be shown to have been predicted in the quatrains of Nostradamus.

Perhaps the best known of his as yet unfulfilled prophecies apparently predicts the appearance of a 'King of Terror' in 1999. His 27-year reign, it seems, will be one during which heretics will die, be taken captive or exiled; a period of 'blood, human corpses, reddened water, hell on earth'; a period during which grim invaders will sweep across Europe, and Paris will be destroyed by air attack before the tide of invasion is eventually turned.[2]

According to Jean-Charles Pichon's decoding in *Nostradamus en Clair*, this King of Terror will be only one of a number of world tyrants, several of whom will be in evidence around the year 2000. One of them will be an American overlord of Germanic origin, given power on the strength of his earlier reputation as a peacemaker. Nostradamus gives his name as 'Chiren'.[3]

In another context, Nostradamus offers the following: 'La grande étoile par sept jours brûlera, Nuée fera deux soleils apparoir. Le gros mâtin toute nuit hurlera, Quand grand pontife changera de terroir.' The reference is apparently to a comet which will approach the earth so closely as to resemble a second sun for several days; and this event appears to be linked with the journey of some great spiritual overlord. In another allusion, apparently to the same event, Nostradamus refers to: 'Castor, Pollux en nef, astre crinite.' In this case, 'ship' and 'comet' (astre crinite) seem to be directly linked—as though in reference to some kind of spaceship. And the expression 'Castor, Pollux' itself seems to be a direct reference to the ancient legend of the immortal who foreswore his immortality in order to rescue his mortal 'twin-brother'—as Messianic an allusion as one could wish to find. Pichon places this intriguing event 'shortly before A.D. 2164',[4] but it must be said that his dating-theory, though ingenious, is in some respects unconvincing.

My own work on Nostradamus since this book was first published in 1977 has – as reported in the titles listed on p. iii – produced other conclusions. A massive Asiatic Muslim invasion will, it seems, start in 1998, be halted in the Middle East and then (headed by the antichrist in person) resume in 1999, overrun Italy in 2000 and reach south-eastern France by 2005 (compare the Pyramid's predictions for this time). Further hordes of marauders from North Africa will enter south-western France via Spain and the Balearics in 2007 and, using a succession of chemical, biological and possibly nuclear weapons, reach the English Channel coast by 2022. It will take until around 2037 for a great European counter-invasion to chase the grim invaders back to the Middle East. A great era of peace and prosperity, presided over by the almost Messianic Henri V of

[2] Compare the passage quoted from Isaiah on p. 223.

[3] Another writer on the subject, Roger Frontenac, in his *La Clef Secrète de Nostradamus* (1950), credits Nostradamus with predicting a great era of peace starting in A.D. 2080.

[4] Compare page 113 (points 63 and 64) and page 324.

France, will then ensue. Following apparent extraterrestrial interventions (whether strictly 'Messianic' or not), the Last Judgement (whatever its true nature) is timed for 2828 – though time itself will continue until 3797 and well beyond. – P.L.

Coinneah Odhar Fiossaiche (Scottish, 17th Century, 'The Brahan Seer')

Almost the only prophecy of this seer which has not yet been fulfilled is that 'a dun hornless cow' will appear 'in Minch' and will make a bellow which will knock the six chimneys off Gairloch House (which in the seer's lifetime *had* no chimneys, but now indeed has six!). The whole country will become so desolated and depopulated that 'the crow of a cock shall not be heard', and wild life generally will be exterminated by 'horrid black rain'. (The explosion of a nuclear submarine might conceivably fit these details, though the reference to 'black rain' sounds more like oil.)

Edgar Cayce (American, 20th Century)

This remarkable and saintly clairvoyant, who died in 1945, appears to have had access, while in hypnotic trance, to almost unlimited knowledge. This gift was freely used, mainly to diagnose the illnesses of those named to him (they did not have to be present) and to prescribe suitable remedies—with recommendations which ranged from the common sense to the frankly unorthodox, often placing considerable stress on right *mental* attitudes ('Mind is the builder'). The conscious Cayce himself was often unable to understand the medical terminology he had used while in trance.

Investigated by official medical circles, Cayce's 'methods' were adjudged to be basically sound, and in 1932 a research foundation (now known as the A.R.E.) was set up at Virginia Beach, Virginia, to preserve and collate the stenographic records of over fourteen thousand of his trance-readings.

In the course of time Cayce (whose work is today gaining considerable respect, especially in America) turned his trance attentions to fields of inquiry other than medicine: he affirmed and explained the reality of reincarnation (to his own waking consternation as a Sunday-school teacher), shed a great deal of light on ancient Egypt, on the mission of Jesus of Nazareth, on dreams, and on the ancient

Atlantis legend, and made a number of important predictions for the years up to 1998, many of which have already been fulfilled.

To summarise his outstanding predictions, Cayce saw 1958 to 1998 as a period of intensive geological upheaval and of the shifting of the earth's axis.[5] The bed of the Mediterranean would rise and fall, Japan would become submerged, the 'upper portions' of Europe would be submerged 'in the twinkling of an eye', Los Angeles, San Francisco and (later) New York would all be destroyed, and parts of Atlantis would 'rise again' from the waters.[6] The period would be heralded by a major eruption of Vesuvius or Etna, while the major risings and sinkings would occur towards the end of it, contemporaneously with the predicted change in the earth's axis. These years would also be marked by the reincarnation of many souls whose last incarnation took place in Atlantis, with a consequent rapid advance in science and technology through what will, in fact, be the re-learning of old skills. And at some date associated with 1998 the 'Great Initiate' would return to earth.[7]

Meanwhile the message of the Great Pyramid would be finally decoded (see the passage quoted at the beginning of Part I of the present volume), and an ancient 'hall of records' would be discovered 'at the proper time' somewhere between the Sphinx and the Nile. The building of the Great Pyramid, Cayce insisted, was started in 10,490 B.C. and took a hundred years.[8] Within it were recorded—in terms of passage-angles, types of rock etc.—the future

[5] See pp. 223 and 226–7.

[6] See p. 318, and especially the research of Dr. J. Manson Valentine at Bimini.

[7] Compare pp. 110 and 113, points (37) to (58).

[8] This dating accords with extraordinary exactness with the date of the last known reversal of the earth's magnetic field, as determined in 1971 by the Swedish scientists N.-A. Mörner, J. P. Lanser and J. Hospers on the basis of geological core-samples (*New Scientist*, 6th January 1972, p. 7). The end of the reversal period was calculated by them to have occurred some 12,400 years ago—and thus, notionally, in around 10430 B.C. If, then, we accept that the construction of the Great Pyramid (as dated by Cayce) may have been associated with some cataclysmic event, the possibility arises of a direct link between that event and the magnetic reversal referred to. And one possibility in particular which springs to mind is that the magnetic reversal may have resulted not from changing currents within the earth's core, *but from a geologically or astronomically induced 'flip-over' of the earth itself* (producing a reversed relationship between the dynamo-'rotor' of the spinning earth and the magnetic 'stator' of its cosmic environment). Such a 'flip-over' would of course have produced tidal waves and geological upheavals of unbelievable magnitude—and certainly more than enough to merit the title cataclysm. It would also accord with the ancient Egyptian priesthood's insistence to Herodotus that such a 'flip-over' had indeed happened *at least four times* during the period to which their records referred.

rise and fall of the nations and the evolution of world religious thought.

As for the hall of records (sometimes also referred to by him as a 'pyramid'), Cayce described its site as follows: 'As the sun rises from the waters, the line of the shadow (or light) falls between the paws of the Sphinx, that was later set as the sentinel or guard, and which may not be entered from the connecting chambers from the Sphinx's paw (right paw) until the time has been fulfilled when the changes must be active in this sphere of man's experience. Between, then, the Sphinx and the river.'[9]

The fact that the sun is described as rising from the waters indicates that the ancient time of High Nile is being referred to—and this annual flooding occurred only from late June onwards. The bearing of the sun at sunrise during the summer should therefore indicate without much difficulty the bearing from the Sphinx of Cayce's hall of records, which would presumably be marked by some sort of low mound, sufficient only to cast a shadow between the Sphinx's paws at sunrise.

Assuming that sunrise is defined as the moment when the sun's lower limb sits tangent on the horizon (the definition normally used by the ancient megalith-builders) the bearing of sunrise would formerly have reached E.$23\frac{1}{2}$°N. (the angle of the Earth's tilt, and thus of the sun's maximum declination) on 20th May, subsequently increasing to E.$27\frac{3}{4}$°N. on midsummer's day, before declining again to its former value on 24th July. The two bearings in question would thus seem to mark the likely northern and southern limits of the site of Cayce's hall of records.[10]

Reference to the map on page 300 will reveal that these bearings do indeed mark out a low mound between the Sphinx and the Nile—one which is at present used as a midden by the inhabitants of the nearby Arab village of Nazlet-el-Samman. Indeed, it is the *only* piece of elevated ground between the Sphinx and the Nile on the bearing in question.[11]

If Cayce is to be believed, the hall of records will be found to take the form of a sealed 'time-capsule' containing ancient records and artefacts left by the Atlantean refugees and colonists who founded

[9] *Edgar Cayce on Prophecy* by M. E. Carter (Paperback Library), p. 106.

[10] As calculated for the third millennium B.C. Adjustment of a day or so may be necessary in respect of a date 10,000 years B.C.

[11] Visitors will find the mound just beyond the low wall diagonally across the road from the Sphinx, and directly opposite the Son-et-Lumière seating.

the Egyptian civilisation and built the Great Pyramid.[12] Its likely form might be a pit sealed by a stone pyramidion. And if that pyramidion subsequently turns out to be some 30 feet 8 inches high and 46 feet 11 inches along the base, with an angle of slope of 51° 51′ 14·3″, then of course this crowning 'cultural bombshell' will be, fittingly, the missing capstone of the Great Pyramid itself.

Cayce's predictions for the future, despite their apparently unorthodox nature and uncertain datings, certainly open up some intriguing possibilities.

Mario de Sabato (French, still living)

A professional clairvoyant with a long list of fulfilled prophecies already to his credit, de Sabato, writing in 1971,[13] has made a series of predictions of world-wide import. (My original 1977 reports follow, with my 1996 comments appended in italics—P. L.):

(i) A long war-cum-peaceful-exodus involving the Chinese on the one hand and a united East and West on the other, and starting with incidents between China and India. The Chinese will spread throughout the Eurasian continent, often unarmed, eventually being halted in the eastern part of France. Grave economic problems, the reorganisation of Eurasia and an intermingling of races will follow, and after the Chinese in exile have arranged a 'surrender' Peking will be destroyed. A short period of world harmony will then immediately precede a Golden Age.

(ii) Prior to the Chinese conflict, a variety of events will occur, of which the following are perhaps the most notable:

Italy: grave political and economic crises, whole periods without any government, a semi-revolution, a left-wing but non-Communist government, floods, earthquakes . . . *Largely fulfilled since 1977.*
France: political and economic reforms, a revised constitution, the abolition of the Senate, an increase in national power; the franc will strengthen and become the exchange-currency of the United States of Europe; Paris the 'capital' of Europe, retirement at 60, compulsory birth-control.

[12] Compare page 16, and bear in mind that Thoth/Hermes was credited by the ancients with being particularly skilled at the art of *sealing airtight containers*—whence our modern term 'hermetically sealed'.
[13] *Confidences d'un Voyant*, Hachette (1971).

Belgium: conflicts over the monarchy and the succession, constitutional reform.
Germany: a Treaty of Berlin giving the city international status, continuation of the East-West division, East Germany transformed into a 'shop-window' for Communist ideas in the West. (*Largely incorrect!*)
Holland: new techniques of winning land from the sea will enlarge the country; important commercial links outside the Common Market.
Spain: restitution of the monarchy, commercial self-enrichment (*fulfilled from 1975 onwards*); neutral during the Asiatic conflict.
Portugal: an *avant-garde* republic established by a 'leader of the opposition'. (*Fulfilled by left-wing Costa Gomes, 1974.*)
Britain: a 'difficult' member of the European Common Market on account of actions in favour of Commonwealth members; new political and economic directions, grave problems in the sphere of international politics; independence for Northern Ireland after a conflict with Eire over the question; a research accident, probably nuclear, with catastrophic results (compare 'The Brahan Seer'). (*Partly fulfilled.*)
Yugoslavia: an important rôle in preventing disaster at the time of the Asiatic conflict, through the suppression of 'a treason'.
Sweden: a campaign against violence; important scientific discoveries.
U.S.S.R.: centre of a new movement for peace and equality, economic expansion, a 'capitalist' country with communist doctrines, revolutionary scientific theories. (*Largely fulfilled from 1991.*) After the Asiatic conflict, co-operation with Christianity over the reorganisation of Eurasia.
Israel: persecution of Jews in Arab lands, periodic wars with Arab countries, intervention of Great Powers. (*Fulfilled repeatedly.*)
Iran: fall of empire and royal family (*fulfilled 1980*), admission of foreign military forces which will in time help to stem the Chinese advance.
Turkey: a 'Cuba' affair, nearly precipitating a world war. (*Cyprus, 1974.*)
Vietnam: an end of fighting, and then a new war brought about by China; eventually completely communist. (*Fulfilled 1976.*)
Japan: considerable scientific pre-eminence and commercial expansion (*fulfilled*); a new and surprising military technique.
Pakistan, Formosa, Korea: all centres of crises.
American Continent: U.S.A. will take over all central and southern

America as far south as the equator (excepting Cuba, which will remain Communist); south of this line, a United States of Latin America, with its capital in Brazil, living in peace.
Canada: independence for Quebec, under a pan-Canadian 'super-government'.
U.S.A.: grave economic crisis; joined by other states; collaboration with U.S.S.R. during the Asiatic conflict; one man's folly may endanger peace.
Africa: A Saharan war, during which oil wells will be set alight (*Gulf War*). Troubles in Congo, Cameroons, Angola and Mozambique.
Egypt: two revolutions, extending across the Red Sea.
Ethiopia: a revolution over the succession, resulting from the pride of one man (who will be assassinated). (*Largelyfulfilled.*) A general transformation of the country, based on government by experts and scholars.
Dakar: upheavals and explosions, both aerial and underwater.
Eurasia: following the Asiatic conflict, a mingling of the races, commercial prosperity, a federal system of government without customs barriers and with a unified military organisation—though there will be no further wars after the Asiatic conflict. The East reorganised by Russia and Japan.

(iii) World-wide developments will include:

Eventual acceptance of government by experts rather than politicians.
Eventual overcoming of the hunger problem through use of the sea's resources.
New evolution of man in every sense resulting from submarine events.
Birth-control, probably compulsory.
Possible local wars from time to time—but not world-wide after the Chinese war.
Severe atomic pollution originating in North America and Britain: effects throughout Europe, South-East Asia and North America.
A new kind of leukemia affecting the spinal marrow. (*HTLV-3/Aids [?], 1981.*)
Genetic mutations.
Medical advances capable of dealing with the above.
Popular revolts against those responsible for the above atomic pollution, further discrediting politics and politicians.
Defeat of cancer in 1975, following the discovery of an

'intermediate' serum which will preserve sufferers long enough to benefit from the final cure. Applicable also to the leukemia mentioned. (*Unfulfilled*)
Heart-transplants successful from 1974. (*Fulfilled.*)
Increase in longevity as a result of scientific discoveries.
International scientific co-operation universal.
Eventual interplanetary voyages: a 'jumping-off' station on the moon.
Complete revolution in the interpretation of religious doctrines, and in morality: collaboration between Communism and progressive Christianity.
End of papacy, after a new rule permitting abdication of popes: the Vatican the centre of a new peace commission by A.D. 2000.

(iv) The Golden Age will start between 1993 and 2021, and will last 730 years—i.e., a 170-year 'Progressive' period, a 370-year 'Prophetic' period within 14 years either way of 2177, and a 190-year 'Apocalyptic' period starting within 14 years either way of 2547 and finishing around A.D. 2737 (compare with the Pyramid's prophecies above).

During this time winter will be largely abolished, manual work will virtually cease to exist, and the seas will be exploited for the benefit of man.

At some time between about A.D. 2163 and 2191 an astonishing event will happen: 'The arrival of a man on earth! A man from another planet older than ours and which will already have undergone its apocalypse. The mysterious arrival of this man will settle the problem of good and evil and also the whole system of the after-life, the discarnate and reincarnation. The arrival of this man will be all the more mysterious for the fact that it will take place near Palestine. The physical appearance of this being, of white race, will strongly resemble that of the Jewish race . . . The Jews will claim the arrival of the Messiah, covered with glory and bearer of riches and marvellous techniques. Moreover his language too will be similar to Hebrew . . . The whole world will be evangelised on new bases . . .

'Once they have got over their astonishment, the inhabitants of earth will welcome other personages similar to the first, who will bring us the fruit of their magnificent civilisation. Thanks to them we shall evolve to the extent of several centuries in just a few years . . .[14]

[14] Compare Jalal'ud-Din Rumi, the thirteenth-century Sufi mystic: 'From these stars like inverted candles, from these blue awnings of the sky, There has come forth a wondrous people, that the mysteries may be revealed' (Tr. R. A. Nicholson).

'Thanks to a revolutionary new device man will cross the universe as easily as he now does the seas . . . Wretchedness will no longer exist . . .

'At this time the future of the universe will be written, including the destiny of earth with relation to the other planets. This message could be seen as a kind of Third Testament . . . Included in it will be the "end of the world"—which in fact will merely be a kind of purification: the rest of the universe will live on.'

One characteristic of the Golden Age, according to de Sabato, will be the eventual achievement by man of 'complete freedom of conduct'. A given area of the globe will be set apart for the practice of each particular type of life-ethos, and each person will thus be free to choose how he wishes to live. None the less, at a later date, the areas inhabited by what de Sabato terms 'the wicked' will be destroyed by some kind of natural atomic disaster. This event will in fact be a delayed result of the interaction of present-day nuclear experiments with the earth's own 'polar forces'.

With the inception of the last part of the Golden Age the sky will finally turn bright orange. And around the year 2800 the whole universe will embark on an entirely new era.

Mrs. Jeane Dixon (American, still living)

Like de Sabato, Jeane Dixon (by profession an estate agent and by persuasion a Roman Catholic) has for many years been justly famed in her own country for the accuracy and reliability of her predictions. These have ranged from the deaths of John Kennedy, Robert Kennedy and Martin Luther King on the one hand, to innumerable 'minor' predictions for friends and acquaintances on the other, taking in various important incidents in the American defence and space programmes on the way. She distinguishes between irrevocable predictions arising from semi-trance 'visions' and premonitory telepathic experiences which enable her subjects to alter dangerous courses of events. Her only self-doubts appear to concern her own ability to interpret reliably the contents of her 'visions'.

In many respects Mrs. Dixon's predictions for the world mirror de Sabato's. The complete overhaul of religious doctrines, the dawning of a new age at around the turn of the present century, far-reaching political changes in Great Britain, the solving of the Irish politico-religious quarrel, the economic advance of Japan, the resumption of hostilities in Korea and Vietnam, the continuance of the Arab-Israeli conflict until 'the great earthquake that will hit Jerusalem', the invasion of the Middle East and Europe (as far as the German border) by Chinese forces, the development of new propulsion systems permitting greatly simplified space travel, gigantic advances in medicine, the harvesting of the oceans—all of these are predicted by Mrs. Dixon and de Sabato alike.

In addition to these Mrs. Dixon, writing in 1969 (*My Life and Prophecies*), foresees a woman president of the United States, the mysterious loss of a British submarine (compare 'The Brahan Seer'), a germ war, and an increasing ability in mankind at large to use psychic powers (a tendency also predicted by Edgar Cayce). She also predicts, like St. Malachy, that there will be only three or four more popes. Of these one will be injured in office, a second will be elected but then replaced in favour of another by the cardinals to whom he will have given increased powers. However, the same figure will in due course re-ascend the papal throne to become the last pope, and will eventually be assassinated. These details naturally throw some interesting light on St. Malachy's predictions. She also expects the U.S.S.R. to become a Christian country, claims that the Jews will eventually accept the Christian Messiah, and foresees the undermining of Christianity by a 'false oriental religious philosophy'.

Mrs. Dixon's most dramatic prophecies, however, cover the expected Chinese war and the advent of several major politico-religious leaders, two of whom she sees in terms of the biblical Antichrist and his prophet. Her relevant predictions and revelations may be summed up as follows:

1962 Birth in the Middle East of a baby whom Mrs. Dixon identifies as the Antichrist—subsequently taken by his parents to live in an Egyptian city (possibly Cairo).

1973–4 Important development in the life of the Antichrist, leading to the eventual formation of a 'cell' of close followers.

1979 A world food shortage may start to be felt.

1985 A comet hits the earth in the area of one of the great oceans, resulting in tidal waves and earthquakes.

1991–2 Beginning of the Antichrist's world mission to capture the youth of the world—described in terms of the Pied Piper of Hamelin—from his headquarters in Jerusalem.

1999 Sudden destruction and war. Appearance of a great flaming cross in the eastern sky which will lead to Christian unity.[15]

[15] It is intriguing to speculate on the significance of this extraordinary celestial phenomenon. It would appear to correspond directly to the 'flaming sword' set up

2000	Chinese troops defeated in the Middle East.
2020–30	Invasion of Europe by an 'oriental religious philosophy'.
2025	The conquering Chinese march into Russia, eventually reaching the German border.
2030	A peacemaker promoted in America (?) as overlord to defeat the Chinese menace, subsequently himself becoming a military tyrant. Possibly the prophet of the Antichrist, who will make full use of mass communications media for the replacement of religion with scientific atheism.
2037	End of the Chinese war.

Possibly the most interesting and contentious aspect of the above is Mrs. Dixon's identification of the figure born in 1962 with the Antichrist. In her initial vision on this subject, she 'saw' an Egyptian queen whom she took to be Nefertiti, accompanied by the pharaoh whom she took to be her husband Ikhnaton, and bearing a baby in soiled, ragged, swaddling-clothes whom she proceeded to 'present to the people'. Over the baby's head a small cross gleamed, and subsequently grew until it covered the whole world, and the baby was enveloped in the light of the rising sun. The pharaoh subsequently disappeared from the scene, and the queen herself was unexpectedly stabbed in the back. Meanwhile the child, now a young man, was receiving the adoration of all peoples, while emanating love, knowledge and 'serene wisdom'—yet 'the channel that emanated from him was not that of the Holy Trinity'.

Eventually, says Mrs. Dixon, this figure will obtain the backing of the United States propaganda machine (contemporaneously with the prophet of the Antichrist), the youth of the world will place the entire world in his hands (compare de Sabato's prophecy of the discrediting and ousting of the traditional politicians), and a fusion of Christianity and Eastern religions will come about. Eventually, however, he will retire from the scene. Mrs. Dixon sees him leading the world's peoples forward towards their goal, *but then taking a turn to the left*: the majority will follow him, but a small 'faithful remnant' will continue up the ever-steepening path to their final goal.

Mrs. Dixon's identification of the 'babe' with the Antichrist

[15] (*continued*) in the Genesis creation-story to prevent the return of Adam and Eve to their Garden of Eden. Mrs. Dixon's 'flaming cross' may in other words signify nothing less than man's eventual and definitive entry into the promised Golden Age—that same 'kingdom of heaven on earth' that he is said to have forsaken so many ages before. Or perhaps it may simply be that the Genesis story is a kind of 'prophecy in reverse'—in which case the idea might more properly be considered alongside other cases of this type in our next chapter.

springs from (a) the uncanny and, she feels, sinister resemblance between his own life and that of Jesus of Nazareth, and the 'unnatural planning' which is bringing this about, (b) her feeling that his 'channel' is not 'that of the Holy Trinity', and (c) her identification of the Egyptian queen with Nefertiti. Yet (a) by itself would seem to be an odd reason for condemning anybody and (b) is perhaps no more than one should expect in view of Mrs. Dixon's traditional Roman Catholic outlook, if our earlier suspicions regarding the pedigree of 'received Christianity' are at all valid. Much would therefore appear to hang on (c).

Yet there seems to be good reason to doubt her interpretation of the Egyptian vision. Mrs. Dixon's identification of the queen with Nefertiti is of course natural enough, since as a result of the discovery of the famous bust of Nefertiti this queen has become, for us, the best-known female figure (apart from Cleopatra) in the whole of Egyptian antiquity. Thus, in view of this, one wonders whether any *other* similarly featured and adorned Egyptian queen of the same dynasty might not equally well have been identified by Mrs. Dixon as 'Nefertiti'—for a woman of our generation would presumably see far more similarities than differences between any two such ancient figures. Again, if the pharaoh was Ikhnaton, then it is surprising that Mrs. Dixon noticed no peculiarities about him—for Ikhnaton was a man of most unusual build, having a strange, elongated face and head, and extraordinarily bloated thighs.[16] Moreover, so far as is known, Nefertiti barely outlived Ikhnaton—which makes the latter's early disappearance from the scene somewhat puzzling—while both of Ikhnaton's successors (Smenkhare and Tutankhamun) may have been the children of Ikhnaton not by Nefertiti, but by his own mother Queen Tiye.

But at this point we come to the 'babe's soiled, ragged, swaddling clothes'. Why soiled and ragged? one feels bound to ask. And at this point the key to the whole riddle becomes apparent. For it is clear that the babe must be none other than the child Moses, newly rescued from among the bulrushes by the pharaoh's daughter Hatshepsut—probably the most remarkable woman in the whole of Egyptian history. Indeed, she shared in her father's rule over Egypt from the age of eighteen, was the *de facto* ruler during his successor's reign, and even continued to exercise considerable power for the first sixteen years of the reign of Tuthmosis III—under whose successor's reign the Hebrew Exodus subsequently took place. She

[16] See Velikovsky's absorbing study, *Oedipus and Akhnaton*.

it was who, having rescued the child, named it apparently after the pharaoh himself (Tuthmosis I), and was subsequently 'stabbed in the back'—long after her father's death—by Moses' assumption of the leadership of his people against the rule of his own foster-mother's family. It should be noted that Nefertiti was a *direct descendant* of Hatshepsut, who died some half a century before she was born—and, in view of the deliberate policy of 'in-breeding' adopted by the Egyptian royal family, some similarity of physical features might of course be expected. Any confusion between the two women on Mrs. Dixon's part could thus be forgivable.

True it is that Ikhnaton tried to substitute a monotheistic solar religion for the earlier worship of many gods, but Mrs. Dixon's argument that this was a *bad* development seems less than convincing: indeed, it seems merely to have been a development of a wider monotheistic movement which was also affecting the Israelites at around the same time. Meanwhile the fact that Mrs. Dixon's cross-crowned babe manages to radiate love, knowledge and serene wisdom hardly seems typical of a *false* Christ—especially when one remembers that he is seen to lead the world's youth *in the right direction,* albeit only a part of the way. And in this we once again see a reflection of the story of Moses, who was likewise to 'turn to the left' (fall prey to mortality) before actually reaching the Promised Land. The last part of the Old Testament journey, it will be remembered, had to be achieved by the surviving Israelites under the leadership of one Jesus (Joshua in Hebrew)—a figure who will presumably be represented by the expected Messiah himself.

There therefore seems to be a strong case for identifying Mrs. Dixon's Antichrist as none other than 'the prophet of the Messiah'—a latter-day Moses-cum-John-the-Baptist—and for interpreting her various predictions relating to him in a similar light.

Vladimir Soloviev

The Russian Soloviev, author of *War, Progress and the End of History* (1899), is reported by George Every in *Christian Mythology* as making in that book some remarkable predictions of future events.

Every identifies a sermon by the fourth-century St. Ephrem as providing some of the inspiration behind Soloviev's book. Notions included in this sermon include those of an Antichrist who will be attractive, modest, humble and a popular fighter for just causes, as well as possessing remarkable powers, but who, on becoming world-king and destroying the opposition, will turn into a sadistic monster laying waste the earth for 'three and a half years' before the coming of the true Messiah.

Soloviev, for his part, appears to have foreseen the results of the Russo-Japanese war and the collapse of Czarist rule, together with the Japanese take-over of China. He then expected the Japanese (apparently confused in this instance with the later Communist Chinese) to invade Russia and sweep into Europe where they would eventually be defeated. Meanwhile the revival of a measure of Christian unity—albeit somewhat muted—would coincide with the formation of a kind of United States of Europe, presided over by a former businessman and ballistics expert (the son of a notorious mother, it seems), who is destined subsequently to become world-emperor. This figure appears to correspond to St. Ephrem's Antichrist, and Soloviev sees him as actually regarding himself as a kind of 'autonomous' Messiah. He will be an offerer of specious panaceas, based on impressive displays of scientific gimmickry combined with oriental mysticism, and his purposes will be furthered by the propaganda media of the Anglo-Americans. Finally he will preside over a World Congress of Religions in Jerusalem, as a result of which most of the established churches and religions will agree to accept his political patronage. Only the newly elected Pope, Peter II, (appointed while at Damascus on his way to the Congress), together with a former Russian bishop described as 'the Elder John' and a German theologian called 'Professor Ernst Pauli' will oppose this fateful move, and will continue to lead the 'faithful remnant' who will choose to remain 'outside the fold'. Pope Peter and the Elder John, it seems, will then 'die'—the Pope being replaced by a collaborator of the imperial Antichrist—before eventually 'rising again'.

It is perhaps worth noticing that the three 'leaders of the elect' described here appear to be thinly disguised reincarnations of the apostles Peter, John and Paul. Meanwhile this account clearly invites particular comparison with the extraordinarily similar predictions of St. Malachy, Nostradamus, de Sabato and Jeane Dixon—predictions which are none the less in sufficient disagreement with each other to scotch any suggestion of complicity.

Buddhism and Hinduism

The typical Hindu conception of history is based on an overall view of the universe which is truly stupendous in its scale. Man acts out his brief evolutionary struggle against the background of an oscillatory universe which undergoes (complete with its creator) an explosion/implosion cycle with a period of 8,640,000,000 years (the celebrated 'day and night of Brahma'). Within this cosmic framework, further planetary cycles of up to 84,000 years are in operation, notable among them being the 26,000-year cycle of the precession of the equinoxes. According to some traditions, it takes seven groups of twelve such cycles—or some 2,100,000 years—for man to complete any given phase in his development. And within each of these cycles in turn his progress is governed by successive 'ages' of 2,160 years, each having its own characteristics, and corresponding to the sun's zodiacal traversal (compare chapter 9).

If the apparent knowledge which permeates the vast library of Hindu scriptures seems extraordinary in its cosmological aspect, it is no less extraordinary in its view of past and future events on earth. In particular, the great Hindu epics paint a vivid kaleidoscope of a race of ancient 'gods' with access both to flying machines and to weapons of apparently nuclear proportions—explosive devices 'brighter than ten thousand suns', whose gruesome after-effects are described in all their now familiar detail.

At the other end of the scale the overall astrological matrix is used to foretell the destiny of future ages. On this basis the current precessional cycle—that of Kali—is said by some sources to be now in its eleventh 'age' of 2,160 years. The twelfth age (corresponding to Aquarius, and thus destined to begin around the end of the present century) will complete the cycle. The solar chariot-wheel illustrated on page 23 symbolises just such an age.

The latter part of the cycle of Kali (so the *Vishnu Purana* suggests) will be marked by a progressive degeneration of human values, a spread of dishonesty, materialism and violence. And then, shortly before the end of the cycle, new spiritual or divine influences will bring about a complete transformation of the human psyche, in time for the commencement of the succeeding 'age of purity'. Some traditions (see next section) expect this to start as early as A.D. 4000.

These 'spiritual influences' are traditionally linked with the reappearance of one or other of the great avatars, or divine reincarnations. As Krishna states in the *Bhagavad Gita* (4:5–11):

> 'I have been born many times, Arjuna, and many times hast thou been born. But I remember my past lives, and thou has forgotten thine.
> 'Although I am unborn, everlasting, and I am the Lord of all, I come to my realm of nature and through my wondrous power I am born.
> 'When righteousness is weak and faints and unrighteousness exults in pride, then my Spirit arises on earth.

'For the salvation of those who are good, for the destruction of evil in man, for the fulfilment of the kingdom of righteousness, I come to this world in the ages that pass.'

Many of the traditions of Hinduism are paralleled in the teachings of its offshoots, Mahayana Buddhism, though in this case the reappearing avatars are replaced by the reincarnating Bodhisattvas—generally regarded as enlightened beings who have voluntarily deferred their ascent to higher planes of existence in order to assist struggling humanity in its painful upward climb. Indeed, in some schools of Buddhism, the Bodhisattva actually takes on the rôle of vicarious saviour (Amida), much after the style of the Jesus of received Christianity. And eventually, it is said, this process will lead to the appearance of the next human Buddha—though that event still lies some hundreds of thousands of years in the future.

T. Lobsang Rampa (d. 1981)

This controversial figure, author of the best-selling *The Third Eye* and other popularisations of Tibetan Buddhism, makes a number of predictions, notably in his *Chapters of Life*.

Rampa claims that a Messianic leader is born into each of the twelve ages of the zodiacal traversal. The approaching age of Aquarius is the eleventh of the present cycle of Kali, and the Golden Age proper will thus begin with the inception of the twelfth age in around A.D. 4000.

As for the coming Messianic leader of the eleventh age, several of his 'disciples' have already been born (the first of them in 1941), while the Messiah himself will be born in 1985. After specialised training (claims Rampa) this leader will deliver some severe shocks to the world in the year 2005. As in the case of the baptism of Jesus, the same figure will be 'taken over' by a higher power, with dramatic results for subsequent world history. During the course of the next two millennia, a completely new world order will then emerge, while man will evolve in remarkable and unexpected directions.

There seem to be a number of similarities between Rampa's datings and those of the Great Pyramid—notably, perhaps, in respect of the events of A.D. 1985 and 2005. As for Rampa's dating of A.D. 4000, this would appear to correspond to the Pyramid's dating of A.D. 3989 for the completion of the Messianic Plan.

Incidentally, Rampa also predicts an increase in world heat in 1981, together with a reduction in rainfall, as well as cataclysmic geological upheavals at a later date.

Criticisms are frequently levelled at Rampa as a result of doubts about the accuracy of his autobiographical claims. The issue appears to be irrelevant, however, to the validity or otherwise of any prophecies he may have made. Rampa's predictions, like those of all other sources mentioned (including the Great Pyramid itself), are valid to the extent that they prove to fit actual events—no more, and no less—and therefore time alone must be the judge of their accuracy.

9

Memories of the Future?

DURING THE course of chapter seven, we unearthed what appears to be a series of surprisingly overt links between the message of the Great Pyramid, the writings of the great biblical prophets, and the recorded teachings of Jesus of Nazareth. Exploring further the implications of the various ideas involved, we found in the last chapter that many of the prognostications offered by all three sources are largely corroborated by a number of more recent sources of prophetic gnosis, and that the latter have the advantage of being, in many cases, quite clear and specific about the course of future events, in a way which the earlier sources often were not.

What we have not so far considered in any detail, however, is the possibility that much of what is recorded in the world's sacred writings as past history may instead *itself* be a form of concealed prophecy for the information of initiates—or conceivably both at once, if it is accepted that history may take a basically cyclic form.

Some of the most ancient and basic traditions of the biblical Old Testament, in particular, consistently bear the most astonishing and sustained of resemblances to the predictions apparently outlined by the Great Pyramid of Giza. Of these traditions, the following are perhaps the most striking.

The Creation Story in the Book of Genesis

Taken at its face value, the familiar biblical creation story of the first chapter of Genesis is not untypical of the creation myths widely current among primitive societies. On the other hand, the strict and dogmatic delineation of the activities of the seven days of creation suggests that hidden beneath the literal meaning lies some kind of allegory with a much deeper meaning. Comparison with the Pyramid's revelations suggests quite strongly, in fact, that it is to the Messianic Plan itself that the creation story actually refers. Indeed, when one thinks about it, there is little obvious *religious* reason why the sacred scriptures should commence with a highly speculative account of the world's origins at all—unless the account is meant to be read as a religious allegory rather than as a historical treatise: whereas the fact that it can be read as a kind of blueprint for the Messianic Plan would, if true, be ample justification for giving it pride of place over the rest of the scriptures, whose purpose may then be seen as the detailed historical revelation of the stated Plan in action. In this case the creation story itself would be seen to be little more than a chronologically highly plausible 'blind' concealing the 'key' to the knowledge of the meaning of the scriptures which follow. But what evidence is there to support this view?

The first 'day' of the Genesis story is devoted to the creation of 'light' in opposition to the all-pervading darkness. Before this, we are told, there was only the spirit, or wind, of God, moving upon the face of the waters. But in this case it is clear that the waters already existed. The idea seems an odd one, until one realises that the text may be referring *not* to the obvious subject—the creation of the 'opposites' of the physical world out of the undifferentiated 'spiritual ether'—but to a much deeper and religiously more pertinent one; none other, in fact, than the coming into the world's 'darkness' of the Messianic 'Light', as proclaimed by the author of St. John's gospel in the 'creation story' with which his own contribution to the scriptures likewise commences. 'The real light', he calls it, 'which enlightens every man' (1:19); and he goes on to make it clear that the Light in question came to find its physical manifestation in the person and teaching of Jesus of Nazareth.

Thus we may see verse 1 of the Genesis story as an allegory of the Divine spirit, or wind, breathing life upon the fathomless uterine waters of (re)birth and mortality to bring into the world the Light personified by Jesus of Nazareth. We may take the first 'day', then, as referring to the original Nazarene or Gospel Age—as represented in the Pyramid by the Grand Gallery. It is appropriate, too, to remember at this point that The Light was the ancient Egyptian name for the Great Pyramid itself.

Further evidence for this interpretation of the initial act of the 'week of creation' can be seen in the fact that the Lord's Day, as celebrated by Christians ever since the original apostles, has never been the Jewish sabbath, contrary to what one might perhaps have expected, but the *first* day of the week—the Mithraic 'day of the conquering Sun'. The fact that Jesus is reported to have given proof of his victory over death and the powers of darkness on this same day of the week makes even more inevitable the identification of Jesus and the Nazarene dispensation with the sun and the coming of light into the world—even if one ignores the fact that December 25th, traditionally celebrated as the birthday not only of Jesus, but also (among others) of the mythical Adonis, Horus, Dionysus and Mithras himself, was the date of the winter solstice on the Julian calendar, and thus symbolised, once again, the victory of the sun's light over the powers of darkness.

Moreover it is not only in the New Testament that the Messianic advent is identified with the sun and thus with the coming of light into the world. The Pyramid's capstone, symbolic of the Messiah's return, is closely associated with the sun, as we have seen, and so, too, is the angle of its sloping passageways (the Bethlehem-angle). Moreover, in the Old Testament, Malachi (4:2) likewise refers to the coming Messiah as the 'sun of righteousness' which 'shall arise with healing in his wings'.

The second day of creation sees the separation of the 'water' above the vault of heaven from that beneath it. Mythologically the imagery may seem odd to Western minds, but the idea that rain is the result of leaks in the vault of heaven is not, on the whole, an entirely illogical one, bearing in mind the obvious analogy between the rainy sky and early man's supposed difficulties in waterproofing the roofs of his primitive dwellings.

None the less, a deeper meaning is again probable. For 'water' we may once again read 'fruit of the womb', so that the separation in question may be seen as the final separation between those human souls redeemed to heavenly life in eternity and those condemned by their own acts to continuing death and rebirth—or, to use the imagery of Matthew 25, between the sheep and the goats. In the Pyramid the second day is thus represented by the King's Chamber Complex, with its promise of an age for 'judgement' and deliverance.

The third day of the Genesis story is primarily that in which all life bears fruit 'according to its kind'. (We should bear in mind at this point the great significance given throughout the scriptures to the 'third day' as the day of action and fulfilment, following two days of preparation—see, for example, Ex. 19:11–15, Num. 19:11, Hos. 6:2 and the various New Testament references to the same notion in connection with the return of the Messiah.) In terms of the Messianic Plan, then, the reference must clearly be to the age in which the redeemed finally gain entry to the lowest spirit-planes, while for the 'lost' the redemptive path of rebirth is finally closed—the age when, the crops having borne fruit 'after their kind' (i.e. good or bad), the harvest is finally gathered in. In the Pyramid, this age is represented by the lowest of the five Construction Chambers whose name, in the event, may be seen as oddly apt in view of their symbolic significance.

The fourth day sees the creation of the sun and moon 'to separate light from darkness'—which seems an odd feature in the story when one considers that it is the *second* time that this has been done (see above). In terms of the Messianic Plan, however, the notion is perhaps less confusing. On the first day the Light was sent into the world to separate the light from the darkness, or to separate the wheat from the chaff: its purpose, as Jesus himself pointed out, was essentially a divisive one, as well as—in another sense—an atoning one. On the fourth day, however, the process is complete: the final eternal barrier is placed between the lost and the redeemed, whose path henceforward must lie irrevocably upwards.

The fifth day is the day of the 'living creatures'; in it they are made to 'be fruitful and increase'. Here we see the continued, triumphant, upward growth of the Messianically redeemed: already they are 'living creatures', i.e. they have been restored to their former status as fully-fledged spiritual beings. And appropriately enough we are told of their rediscovered freedom via the symbolism of 'birds' and 'fishes'—creatures of three dimensions as opposed to the human animal's meagre two. Both, we may infer, stand here for liberated or redeemed souls.

The sixth day is of course the day of consummation: in it man is finally made 'in(to) the image of God', i.e. he becomes one with the Divine 'eikon' or spirit. Having attained the summit of his spiritual evolution, man finds that 'all things are in subjection under his feet'. He realises that the physical world and all things in it, from the lowliest grass to the most evolved form of animal life, constitute no more than a chain of earthly phenomena designed expressly to culminate in his own spiritual apotheosis—for he is its crowning glory, the spiritual harvest of the planet earth, the ultimate flower of the whole process of earthly evolution.

The continual struggle for perfection has now reached its glorious conclusion, and man can 'lose himself in heaven above'. At last all the striving can cease, and man is once more united with the Divinity from which he issued and to which he has for so long struggled to return. Once more God is in man and man in God, and at last—to use a Buddhist analogy—'the dewdrop slips into the shining sea'. This is the seventh day, the day of rest, in which God 'ceased from all the work he had set himself to do' (Gen. 2:3). For the Messianic Plan, having reached the topmost of the five Construction Chambers, and thus the last of the seven Messianic ages, is complete.

Seven ages—seven days: the very week around which the life of our society revolves seems, in short, to have been devised as a direct symbolic memorialisation of the Messianic Plan. A plan which, as the week's number of days suggests, has as its prime object the spiritual perfecting of man.

The Garden of Eden

The familiar story of Adam and Eve (from two Hebrew words which mean 'Man' and 'Life' respectively) needs no introduction in its essentials. That it conflicts in numerous fundamental respects with the details of the immediately preceding creation story is self-evident, a fact which of course suggests that it needs to be treated as an entirely separate issue from the creation story itself, whose deeper significance, as suggested above, is almost certainly prophetic. This suggests a pseudo-historical, or even cosmological, significance—and it has long been widely accepted that within it are to be discovered the causes of man's 'fallen' condition, from which the Messianic Plan was designed to save him.

The picture of man which emerges from this account is of a non-sexual being living in contentment and able to eat of the 'tree of life', which confers immortality upon him. 'Knowledge of good and evil' is, however, denied to him. He thus appears to be a denizen of the spheres of non-differentiated matter—there are no opposites, no good or evil, hot or cold, matter or anti-matter—so that, reading between the lines, we may regard Adam and his world as basically spiritual in nature. This is also apparent from man's obvious intimacy with God, and from the complete absence of the need for any form of religion (of 'binding-back-together') in the garden, just as in the eternal heavenly city of St. John's Revelation religion is once again entirely absent. At the end of the Bible, in other words, man's state is once again a spiritual one, as it was in the beginning. 'And I saw no temple therein,' says the familiar Authorised Version (21:22), 'for the Lord God Almighty and the Lamb are the temple of it.' Indeed, by the same token, even the light of the first day of creation is no longer necessary: 'And the city had no need of the sun, neither of the moon, to shine in it: for the glory of God did lighten it, and the Lamb is the light thereof'. Or, as the *Bhagavad Gita* puts it: 'There the sun shines not, nor the moon gives light, nor fire burns, for the Light of my glory is there. Those who reach that abode return no more.'

Such, then, is man's state in the beginning also, but from this idyllic state he eventually falls. It is the 'snake' which first lures the 'woman' to tempt man to eat of the fruit of the 'knowledge of good and evil'; as a consequence, both man and woman become aware of their sexuality and are expelled eastwards from the garden.

Thus the union of 'Man' with 'Life' is seen as coinciding with man's first experience of sexuality and of the physical world of differentiated forms. And from that moment man is sent out of his paradise to the east, the direction of (re)birth and thus of mortality.

On this basis, then, we may regard the proverbial fall as a loss of spirituality incurred by true man—an essentially immortal and spiritual being—as a result of becoming ensnared by the meshes of physical existence (to this extent we are here in agreement with one of the basic tenets of the so-called 'Gnostic heresy'). Man is, as it were, imprisoned in the cells of the body, and it therefore becomes the prime function of the Messianic Plan to 'set the prisoners free', in response to appeals such as that of Psalm 142: 'Set me free from my prison, so that I may praise thy name'. Of all the rôles assigned by the Old Testament prophecies to the Messiah, none is more persistent than that of 'releasing the captives', as reference to the appropriate passages will confirm. As Malachi has it (4:2): 'The sun of righteousness shall arise with healing in his wings, and you shall break loose like calves released from the stall.'

The above notion of loss of spirituality agrees well, of course, with the displacement of the Pyramid's passage-system 286·1" to the east (left) of its centre-line—but precisely how this loss of spirituality came about is a matter for conjecture. Bearing in mind, however, that other men and women must have existed at least as early as the biblical Adam[1] and that chapter 6 of Genesis makes mysterious reference to the 'sons of the gods' marrying 'the daughters of men', the vague outlines of a possible answer begin to take shape amid the mists of time. In particular, the idea starts to emerge that the Adam and Eve story relates to some kind of special creation, a

[1] The Bible suggests that the 'sons of Adam and Eve' somehow managed to find wives to marry.

spiritual incursion into the physical world long after what we refer to as the creation was completed (if indeed it is *ever* completed).

In other words, we may assume that the appearance of life on earth was already a matter of ancient history; plants, animals, insects and all known forms of life had already been through the long process of evolution, and the humanoid ape to which we attach the name 'man' had long since made his appearance. But the process of evolution did not stop there. All forms of animal life—perhaps even all forms of life—have some kind of 'mind', a partly neurological phenomenon which also appears to partake of the spiritual world, existing as it were in two different sets of dimensions at once. Each living mind, in other words, creates a kind of disturbance in the spiritual 'ether' during its lifetime—a disturbance which continues to exist after its death as a phenomenon which those familiar with the occult refer to as a 'thought-form'. And this process would naturally apply equally to the humanoid apes already referred to. Indeed, so evolved had their brains become—and so developed, in consequence, were their minds—that the thought-forms produced by them had almost succeeded in evolving into fully-fledged spirits or souls in their own right. Such beings were potentially capable of escaping entirely from their material environment and of achieving direct reunion with the Divine powers which had brought the physical world into existence in the first place. Physical creation, in other words, had almost come full circle and was within an ace, after long aeons of painful evolution, of rendering back to its Creator a remarkable harvest of spiritual fruit.

Yet somehow that final evolutionary step (I surmise) continued to elude, for the most part, those unimaginably remote ancestors of ours. And it was at this point that a group of spirits or souls felt moved to act as midwives for the hoped-for new birth, by themselves infiltrating the minds and bodies of what now became 'mankind' as we know it. This is a notion which, oddly enough, Central American legend still preserves in its concept of 'heavenly beings' being sent down as the earliest kings to help man's early development. Spiritual man, in other words, plunged into the waters of death and re-birth in order to help his drowning brother, well aware that he might lose his own Life in the process. And so it proved. For spiritual man—the 'sons of the gods' as personified by Adam—only too soon became asphyxiated and weakened by the polluting influences of the physical world, to the point where neither he nor the creature he had intended to save—the 'daughters of men' as personified by Eve—could muster enough strength to regain unaided the receding shores of eternity. Just as the drowning man is deprived of breath, so the 'sons of the gods' became increasingly deprived of their spiritual nature. In the words of Psalm 69 (1–3): 'Save me, O God; for the waters have risen up to my neck. I sink in muddy depths and have no foothold: I am swept into deep water, and the flood carries me away. I am wearied with crying out, my throat is sore, my eyes grow dim as I wait for God to help me.'

The 'fall' of man, then, may well have been a *voluntary* act, a kind of 'induced psychic mutation', undertaken with the noblest of intentions but with disastrous results, and it is to this event that the Adam and Eve story may obliquely refer.[2]

Fortunately for man, there are (as the psalmist realised) other more powerful spiritual influences in the universe, and in response to his cries for help the Messianic rescue-plan was instituted by the spiritual world. As part of it, the man Jesus showed how the attempted rescue mission could be achieved, despite apparent physical death, *without* loss of spiritual immortality—a demonstration which involved allowing himself, from the moment of his baptism, to be 'taken over' by a further emanation from the world of pure spirit known by some as the Holy Spirit. His return as Messiah is destined to provide the final proof and culmination of the demonstration.

Thanks to this spiritual intervention in man's affairs—one which, clearly, can come about only with the consent of man himself—the day will eventually come when man will at last be saved from the waters of mortality, along with a goodly proportion of the souls he himself had tried to save. The harvest may be great and the labourers few, but in the prophetic words of Psalm 126: 'A man may go out weeping, carrying his bag of seed; but he will come back with songs of joy, carrying home his sheaves.'

[2] This notion is supported and elaborated by a fairly late Jewish exegesis of the statement at Genesis 6:4 that 'the sons of the gods had intercourse with the daughters of men' in the time immediately before the 'Flood'. In this interpretation the sons of the gods are represented as loyal angels—as opposed to the alternative rebel-angel tradition—attempting a last-ditch pre-flood mission of human redemption by Divine permission. The attractions of the flesh, represented by the daughters of men, prove too great for them, however, and the result is their own fall and the production of the legendary Nephilim, or giants.

The Story of Noah's Flood

The familiar story of Genesis 6–8 starts amid an age of wickedness when, according to the biblical account, 'the sons of the gods had intercourse with the daughters of men' and the Nephilim, or giants, were on earth. Forewarned, because of his exceptional righteousness, of an impending world-flood, Noah, in his 600th year, having built an ark of reeds to a divinely prescribed specification, enters it with his wife, his three sons and their wives, together with one pair of each species of living animal. Oddly enough, Noah seems to have disobeyed his instructions, which were to take with him *seven* pairs of most of the species involved—but no matter.

After the best part of seven days, forty days of rain begin, while vast subterranean waters also pour forth, and for 150 days the flood waters rise, until the mountains are covered to the extent of fifteen cubits—some thirty feet, if the measurement is meant to be taken seriously. After a further 150 days the floodwaters have receded to the point where the ark is aground on a mountain in Ararat and the first mountain tops are visible.

Now Noah releases a raven—a bird commonly used in the ancient world for shore-sighting purposes—but the raven merely flies to and fro, an activity normally interpreted by mariners of old as an indication that they were at about the mid-point of their voyage (but none the less an odd activity, when one considers that the mountain tops were allegedly already visible and therefore clear of the water). After seven days he releases a dove, which promptly returns to the ark, being unable to find any spot dry enough to settle (another unlikely story, on the face of it). After another seven days he again releases the dove, and it again returns, this time bearing an olive-branch. Finally, after a further seven days, he releases the dove a third time, and this time it fails to return at all, whereupon Noah and his companions are at last ordered to disembark with their living cargo, and the divine command goes forth for all living things to be fruitful and increase—the language being reminiscent of the events of the fifth 'day' of creation (see page 254). Moreover Noah receives a divine promise that never again will all creatures be killed by any flood, and the promise is sealed with the heavenly sign of the rainbow.

The story is, of course, full of inconsistencies, both of fact and figure. On the other hand it is also full of symbolism, both factual and numerological. Once again, one suspects, a typical explanatory myth—whatever its basis in historical fact—is being used as a vehicle for some much deeper message. The flood narrative itself is strongly reminiscent of the Babylonian *Epic of Gilgamesh*; but the symbolism is equally strongly Messianic. Indeed, Jesus of Nazareth himself seems explicitly to have acknowledged the story's Messianic significance at Matthew 24:37.[3]

To start with, the word 'ark' is generally used in the Bible to refer to a chest or coffer for keeping things safe or secret—principally in connection with the much later Ark of the Covenant. In fact there is only one exception to this general interpretation—Noah's ark itself.[4] But in this case why suppose Noah's ark to be an exception in the first place? Could we not, in other words, read Noah's entry into the ark with his flock in terms of the entry of the world's one righteous man and his followers into a secret place of safety?

But safety from what? What is signified in this case by the all-engulfing waters? Once again the obvious symbolism is of the 'water' of the womb,[5] and thus of (re)birth and human mortality. By extension we may interpret it as also signifying human wickedness—with which indeed the story itself explicitly links it.

Already, then, the Messianic significance of the story starts to become apparent. In Noah we can see a Messianic figure who leads his immediate followers to a secret place of safety at the beginning of an age when a great tide of human wickedness is to lay waste the whole earth. The imagery already seems to foreshadow the rôle of Jesus of Nazareth, who in the Pyramid's symbolism is shown as bearing the souls of the initiates upon his back and protecting them from further reincarnation until they are resurrected with him in the Final Age. And the fact that Noah is instructed to take with him seven pairs of each kind of 'bird' and 'clean beast' suggest that the latter represent a special group of souls whose function is to produce spiritual perfection (2 × 7).[6]

As for the account's chronology, the period of the flood, from Noah's entry into the ark until he eventually steps forth on to dry land again, appears to last 'the best part of seven days', plus 150 days, plus another 150 days, plus forty days, plus (3 × 7) days—which gives a figure remarkably close to a year of 365 days (the text

[3] 'As things were in Noah's days, so will they be when the Son of Man comes.'

[4] A different word is used for this in the Hebrew, but not in the Greek.

[5] Compare, for example, Job 38 :8, 'Who watched over the birth of the sea, when it burst in flood from the womb?'

[6] It is perhaps worth noting that the Hindu equivalent of Noah's ark—that of Vaivasvata Manu—likewise contains seven Rishis, who are the progenitors of the various terrestrial life-forms.

itself comes up with a total of a year and ten days). We may see this fact as tending to confirm our earlier assumption that the flood signifies an earthly 'time' or age of death and destruction.

Eventually the ark comes to rest with its precious cargo upon a mountain in Ararat (the word itself means 'high peaks'). Yet Noah refuses to venture out of the ark and set foot upon the earth until all traces of the flood have gone. And at this point he sends out his four feathered emissaries to the drying world. In the light of the 'birds' of the creation story (see page 254) we might at once suspect that these shore-sighting birds again represent beings who have succeeded in attaining spirituality—beings who in this case are returning to the scene of their former endeavours. The fact that Noah waits forty days before sending them out at seven-day intervals would even seem to identify them numerologically: they are reborn initiates (8 × 5) whose task is to re-establish spiritual perfection (7) on earth—and thus identical in symbol with the initiates whose discarnate period before eventual rebirth is symbolised by the forty roof-slabs of the Pyramid's Grand Gallery.

Let us consider the 'birds' in more detail. The first—a raven, and thus black in colour—continues flying to and fro until the water on the earth has dried up. The second—a dove, the symbol of peace and presumably white in colour—returns, unable to find a secure footing. The third—the same dove of peace again—also returns to the ark, but this time bearing an olive-branch, which seems to represent a reciprocal gesture of peace from the drying earth. The fourth emissary—the dove again—finds itself so much at home on the new earth that it is happy to stay there, whereupon Noah at last descends with his followers upon the dry land, offers sacrifices to his divine protector, and is made what appears to be a promise of an end to universal death.

The symbolism of the last part of the flood epic, then, seems to refer to the return of the initiates from their discarnate place of safety after the worst of the age of death, destruction and wickedness has passed. The descent of the ark upon the mountain in Ararat corresponds closely in symbol to the 'descent' of the final capstone upon the incomplete Pyramid, and thus signifies the return of the Great Initiate himself to physical existence at the onset of the Final Age. The first Messianic emissary is not entirely happy with things at first—represented by the raven, he 'flies to and fro', i.e. he is repeatedly incarnated—until the Messianic Age reaches its predestined fulfilment with the removal of death, destruction and wickedness from the earth. And the black 'raven' corresponds closely to the dark granite of the Pyramid's Granite Leaf, which likewise represents the first Messianic resurrection. Even the species chosen—the raven—is appropriate, for the raven is by inclination a scavenger or gatherer, whose function is to salvage what is still edible even from the debris of death and destruction.

The second Messianic visitation is symbolised by the white dove of peace, but as yet the world is not ready for its peaceful overtures. Its light colour is reminiscent of the Pyramid's white limestone, and may perhaps signify that the figure in question will, like his two successors, be a physical man of normal earthly origin. He will allow himself to be 'taken over' completely by the spiritual entity manifesting in the person of the first Messianic figure—a case of spiritual 'possession', in fact. (In this case we might assume that all three vertical 'slabs' suggested by the grooves in the Pyramid's Antechamber would have been made of limestone, unlike the Granite Leaf which appears to represent a Messianic figure of extra-terrestrial origin.)

This spiritual possession is exactly parallel to the reported descent of the Holy Spirit upon Jesus at his baptism—the 'spiritual anointing' which seems to have sparked off his whole ministry. And it should be noted that this spiritual anointing of the earthly Messiah is described in all three synoptic gospels *specifically in terms of the descent of a dove*—a fact which provides powerful support for our interpretation of this aspect of the Noah story.

A third Messianic initiative follows, but this time there are some signs of reciprocated goodwill and of peaceful intentions, so that by the time the fourth Messianic emissary appears he is welcomed with open arms and is happy to remain, now that the earthly environment has become, as it were, attuned to the Messianic purpose.

Finally, with the non-return of the last Messianic figure—the last dove—at the end of his allotted physical life-span, it becomes established that conditions are ripe for 'Noah's disembarkation'—the inception of the great earthly Millennium. With that event heaven and earth are, in symbol, finally reconciled, and universal death is abolished. The Messianic Plan has reached its fulfilment, man is reunited with his true spiritual nature, humanity is saved, and death is at last swallowed up in victory. And the seal of the new covenant (compare de Sabato's 'Third Testament') is in the shape of the rainbow—a shape with distinct affinities to that of the Seal or Boss on the Pyramid's Granite Leaf, itself likewise symbolic of the Messianic return and the culmination of the Messianic Plan.

There is one further point to be considered before we leave this remarkable biblical episode. Jesus, for his part, apparently saw the Messianic rôle as quintessentially symbolised by what he called 'the sign of the prophet Jonah' (see Matt. 12:39, Mark 8:12, Luke 11:29, 30). This apparent reference to Jonah's symbolic three-day sojourn in the 'fish's belly' and subsequent return to the land of the living may also be an esoteric reference to the Messianic significance of the 'age of the fish-man' (Aquarius), as Madame Blavatsky has suggested. On the other hand, the Hebrew name Jonah *in fact means 'dove'*—so that Jesus' own reference to this particular 'reappearing dove' may conceivably be taken as evidence that he was aware of a symbolic link between what may be called the 'legend of the returning dove or homing-pigeon' (originally exemplified in the Noah story) and the pattern of the Messianic initiative. Indeed, he appears to have hinted as much at Matthew 24:37, 'As things were in Noah's days, so will they be when the Son of Man comes.' Meanwhile the Pyramidal dating of that next Messianic reappearance seems to mark the very anniversary of Noah's flood (see page 193), while Noah's alleged age at death of 950 years (Gen. 9:29) corresponds exactly to the Pyramidal period between the Messianic reassumption of mortality (A.D. 2039) and the inception of the final Millennium (A.D. 2989). The evidence seems little short of overwhelming that the Noah story, like that of Jonah, is intended to be read as nothing less than a prophetic allegory of the coming Messianic age.

* * *

The notion of a *multiple* Messianic visitation is not, admittedly, one that is familiar to modern Christians—though a Messianic trinity of 'prophet, priest and king' was certainly one of the possibilities envisaged by certain of the Jewish religious sectarians at around the time of Jesus. None the less there is at least one group of Gospel accounts which may perhaps contain the germ of such an idea. I refer to the story of the supposedly miraculous stilling of the storm at Mark 4:35–41 and Luke 8:22–5, together with the possibly associated incident of the walking on the water at Matthew 14:22–34.

Whatever the factual pedigree of the latter story, it would, after all, be hard to find a more apt symbolic allegory for the workings of the Messianic rescue-plan. Here we are given, in symbol, an unforgettable picture of Jesus as alone able to rise above the waters which immediately beg identification as the waters of mortality. Even the foremost of his followers, attempting to emulate him, cannot do so unless Jesus himself 'helps him up'. There could scarcely be a clearer picture of the redemptive function of the *Christos* (or Holy Spirit).

But if the walking on the water story is a patently symbolic allegory, might not the story of the stilling of the storm likewise have an allegorical aspect—connected, perhaps, with the predicted Messianic rôle of bringer of peace? And in this case might not both accounts have originated as symbolic parables *once told by Jesus himself*?

I should at this point confess to having long inclined to the view that many of the ostensible miracle accounts which have come down to us in the gospels may in fact have originated as parables told by Jesus himself—parables which have subsequently been adopted and adapted by Christian chroniclers as factual accounts of the deeds of Jesus, primarily because of their aptness to the thesis that Jesus *physically* fulfilled all the various Messianic prophecies. The evangelists, indeed, make no secret of the fact that their whole object is to demonstrate this thesis, since upon its veracity hung (in the contemporary view) any man's claim to be the promised Messiah. The possibility that the Messianic prophecies of Isaiah (say) might instead demand interpretation on a deeper, more spiritual level does not seem to have carried much weight with them. None the less the development of this particular argument will have to await the appearance of a further book devoted specifically to the life and teachings of Jesus.

At all events, if both accounts referred to did have a common source in stories told by Jesus himself, then the vital elements in the stilling of the storm story were presumably: the crossing of a lake; the falling asleep of the Messiah; the rising of a storm; the reawakening of the Messiah, and the restoration of peace at his behest. The possible relevance of these symbolic events to the notion of a multiple Messianic visitation, involving the repeated reincarnation of a single personage, may already be apparent to the reader. To proceed to the story of the walking on the water, we then have to add the following elements: the Messiah withdrawing to a hilltop to pray; his companions going on without him; the return of the Messiah, walking upon the water; the Messiah saving the leader of his followers from drowning in the water.

My own hunch is that we have here a memory of two different parables told by Jesus at different times, but on the same theme—namely the coming disappearance and return of the Messiah. Now this, of course, is not a surprising notion: we have records of his having warned his disciples many times of his imminent departure and eventual return as King-Messiah. On the other hand, the theme on this occasion could equally well have been the less familiar notion that even the expected *future* Messiah might from time to time during his final reign withdraw from earth for short periods—as though to recharge his batteries, so to speak, or to 'seek fresh instructions'. Matthew's reference to Jesus' withdrawal to a hilltop to pray seems to add specific weight to this latter theory, since it recalls the similar withdrawals of Moses to the top of the Holy Mountain and his subsequent return, each of them (see page 280) apparently being symbolic of a Messianic departure and return.

Moreover, Mark's account—perhaps the oldest and most reliable of those we have—places the stilling of the storm story directly after a parable by Jesus in which he likens the coming Kingdom to the great tree eventually produced by a mustard-seed—a tree so great 'that the birds can settle in its shade.' Now this expression is perhaps not a little odd, since most small birds can settle quite comfortably in the shade even of a tomato-plant: why, therefore, bring in birds at this point in an expression apparently quite irrelevant to the notion of a tree's size?

If, in fact, the juxtaposition of the mustard-seed parable and the stilling of the storm incident is intentional and significant, I would offer the following explanation. The birds of the mustard-tree story are none other than the successive visits of the dove of the Noah story currently under discussion: the dove of the Holy Spirit will not settle finally on earth until it can find a tree large enough to shelter under—in other words the Messiah will not stay permanently on earth until the expected Kingdom has finally blossomed into its full glory. Until that time 'Foxes have their holes, the birds their roosts; but the Son of Man has nowhere to lay his head' (Matt. 8:20). Thus, from time to time he will again depart from earth—will appear to fall asleep while crossing the waters of mortality—and for a while his followers will be left without him. In his absence things will naturally start to deteriorate (see page 155 and compare the associated notion in the story of Moses on page 275) until his followers start to become seriously threatened and alarmed; but, at their call, he will return to the aid of the faithful, triumphant over the waters of mortality, and will restore the world to rights.

A further scriptural passage which seems to have possible connections with the walking on the water story, and with the notion of a multiple Messianic visitation, is the last chapter of John's gospel, where Peter again 'goes overboard' to meet Jesus—this time by swimming to his master on the shore ahead of his companions, who are dragging behind them the miraculous 'draught of fishes' (probably itself cognate with the similar 'miraculous catch' at Luke 5:1–

11). Once again, however, it seems more than likely that these miraculous catches of fish are symbolic, and probably represent memories of a parable once told by Jesus himself about the nature of the coming of the Kingdom of Heaven—the fish being men, some of whom will eventually become evolved enough to be 'collected up' by the 'angels of the Son of Man', while the rest will have to be 'thrown away'.

But if the idea of the fishes is purely allegorical, how factual is the whole story reported in John 21 likely to be? There seems to be good reason for regarding it—rather like the beginning of the same gospel—as almost entirely allegorical in nature. The posthumous appearances of Jesus reported in John 20 may conceivably be of dubious historicity, but at least they are reported in fairly straightforward, down-to-earth terms. But the whole atmosphere of chapter 21 is quite different: it is somehow dreamlike, almost surrealist—a ritual and idyllic celebration, it seems, of the very end of time itself.

And, moreover, its theme appears to be highly relevant to our present discussion, for the basic elements of the story are as follows. Seven disciples go fishing all night on Lake Tiberias. They catch nothing. At dawn they see Jesus, standing on the beach: they do not recognise him. Jesus tells them to cast their nets to starboard (to the right). They do so, and catch 153 big fishes without breaking the net. Peter, who is naked, wraps his coat about him, plunges in, and swims some 200 cubits to the shore, ahead of the others. They then bring the fish ashore to where Jesus (whose identity they are still not sure of) has some bread and fish for them to eat. The author stresses particularly (v. 14) that this is Jesus' *third* appearance to the disciples since his resurrection. Jesus asks Peter three times whether he loves him, and three times commissions him to look after his flock (in terms which suggest that the flock is getting older all the time). The chapter concludes with a highly esoteric passage on the relationships between the 'beloved disciple', the author of the gospel, and Jesus.

These various elements seem to invite interpretation as follows:

The seven disciples represent (in terms of the Pyramid's code) the spiritually perfect—or even the angels of the Son of Man—going about their work of attempted redemption during the long night preceding the dawning of the Kingdom.

The appearance of Jesus on the shore at dawn is Jesus' return during the Final Age: the difficulty in recognising him suggests that he has been resurrected/reincarnated *in a different body*. There may be some allusion here to the alleged ancient Egyptian term for the

Pyramid's Antechamber, the Chamber of the Triple *Veil*—but in the event it is his actions that confirm his identity.

The great catch is made when they cast to the right—on the 'good' side of the boat—and the 153 fishes seem to be directly symbolic (again in terms of the Pyramid's code) of the redeemed enlightened ones.

Whether the boat itself is intended to be symbolic of Noah's ark is a matter for conjecture. But its seven occupants may have more than a chance connection with the seven pairs of each kind of creature that Noah was commanded to take into the ark with him, in order that they (symbolising the 'producers of spiritual perfection') might be saved from the waters of destruction.

The significance of Peter's strange act in putting on his coat before plunging in seems, astonishingly, to be a direct reference to the Pyramid's symbolism. For Peter (Cephas)—the Rock—seems here to symbolise nothing less than the Pyramid itself. He, like the present Pyramid, is 'naked'—stripped of the normal 'casing'. But when the time comes for him to lead the faithful to where Jesus is waiting for them on the shores of eternity, he 'puts his coat on again'—the Pyramid is symbolically re-completed to its full dimensions. He then swims some 200 cubits to the shore. This distance sounds as though it may somehow be deliberately symbolic of the total distance to redemption—and, sure enough, we find, on examining the Pyramid, that the total of the floor-length of the whole Messianic Plan, from the entrance to the triumphant south wall and escape-shaft of the King's Chamber, *is just 200·17 Sacred Cubits*.

The bread and fish seem here to symbolise the harvest of land and sea—the whole spiritual harvest of planet earth.

Meanwhile the author's emphasis of the point that this is Jesus' third appearance since his resurrection seems to contain more than a hint that the Final Age may be marked by *three further appearances* of the Messiah after his initial one—as well as by certain initial difficulties of identification on each reappearance.

Jesus' threefold injunction to look after his flock makes sense only in terms of a 'shepherd' who indeed proposes to forsake it on three separate occasions. The impression is strengthened by the fact that the flock seems to 'get older' in the process, starting as lambs and finishing up as sheep.

In the final chapter of John's gospel, then, we seem to have found further biblical backing for the notion that the future Royal Messiah will be succeeded by *three more* Messianic figures, whose advent will immediately presage the establishment of the true Millennium—a notion which seems to find a strong Old Testament echo in the destruction of the biblical Sodom and Gomorrah immediately after the visit of *three angels* and the consequent escape of Lot and his family at the behest of the first two of them (Genesis 18–19). But we should remember that we gleaned the notion originally from the Pyramid's Antechamber, and substantiated it via the story of Noah's flood.

* * *

The historicity of Noah's flood is of course very much a matter for conjecture at the present time. Much depends upon its alleged extent and dating, and neither is easy to determine exactly from the available accounts. Certainly the folk-mythologies of virtually all races attest the reality of some sort of primeval deluge which allegedly destroyed an earlier human civilisation. A similar claim is made by Plato in his *Timaeus* (see footnote page 14), apparently on the authority of the Egyptian priesthood, who held that such cataclysms occurred at regular intervals as the natural result of irregularities in the movements of the heavenly bodies. Nor do such claims seem inherently unreasonable from the scientific point of view.

The geological record makes it quite clear, in fact, that the eleven-thousand-year period starting in around 15,000 B.C. was characterised by world-wide flooding. For, as a direct result of the earth's 26,000-year precessional cycle, solar radiation is known to have increased considerably during this period, with the result that world sea-levels rose by an incredible 350 feet. This would certainly have inundated large areas formerly inhabited by man. Moreover, the final peak was reached in around 4,000 B.C., a date which appears to fit in extremely well with the approximate chronology of the Babylonian, Egyptian, Mayan and Hebrew flood-accounts.[7] What is more, the published data indicate that this final 250-year upsurge raised world sea-levels by some $9\frac{1}{2}$ metres. Translated into Hebrew measurements this represents some fifteen cubits—and we find that a distance of fifteen cubits is mentioned specifically by the Genesis flood-account in association with the height of the floodwaters (7:21).[8]

[7] R. W. Fairbridge, 'The Changing Level of the Sea' (*Scientific American*, May 1960, Vol. 202, No. 5).

[8] The text associates this distance with the water's height *above the mountains*; but a moment's thought will reveal the untenability of the notion. Mountains are not uniformly high. Unless, then, we take the extraordinary view that the text is somehow referring to Mount Everest, we have to assume that the reference to the mountains is a later intrusion—a piece of exaggeration, in fact—which has crept into the earlier account of a simple fifteen-cubit rise in sea-level.

Again, it is becoming increasingly clear that the earth has been, for over two million years now, subject to periodic climatic changes of considerable violence—one at least of which seems to have been responsible for the sudden disappearance of many of the great prehistoric animals. And the natural cause of these changes (as also of the frequent changes in the earth's magnetic field observed by geologists and used to justify the theories of continental drift and of plate-tectonics) might conceivably have been sudden periodic changes either in the inclination of the earth's axis or in the planet's orbital environment. And that, of course, is without going into the cataclysmic world-wide effects of the periodic large-scale meteoric impacts which occur, it is estimated, at least once every 10,000 years, probably causing, in the case of the necessarily more frequent oceanic impacts, 'tidal waves' *up to four miles high.*[9]

Perhaps the most striking account of the overwhelming of a civilisation by watery catastrophe is to be found in Plato's description of the fate of the semi-legendary Atlantis, apparently in around 10,000 B.C. (a period known geologically to have marked an especially rapid rise in world sea-levels).[10]

It is perhaps worth recording that Edgar Cayce, the celebrated American healer and visionary referred to in chapter 8, affirmed the reality both of Atlantis and of its eventual destruction by flood in around 10,000 B.C., following the misuse of scientific powers no less potent and advanced in their own way than ours today. While some of the survivors headed west to Central America, and others to the Pyrenean area of Europe, a large group of them, bearing the knowledge of these advanced techniques, eventually settled in Egypt. Hence, among other things of course, the remarkable similarities between the ancient Egyptian and Central American civilisations. (Further discussion of this topic will be found under 'The Atlantean Tradition' later in this chapter.)

According to Cayce, the 'Egyptian' group arrived in Egypt at about the time when that country was being taken over by a Caucasian invader named Arart, assisted by his son Araaraart and a priest by the name of Ra-Ta (conceivably Ra-Ptah?). It was with the cooperation of these that the Atlantean refugees then planned

[9] R. S. Dietz, 'Astroblemes' (*Scientific American*, August 1971, Vol. No. 2).

[10] Compare Fairbridge, op. cit. If Plato's dating is right, and if we ignore the possibility of major geological upheavals, then the published figures for ancient sea-levels suggest that the bulk of any Atlantean remains would now be submerged under anything up to 150 feet of sea water. Actual soundings in the area, however, may have been reduced by later depositions of sediment.

and commenced the building of the Great Pyramid in order that it, together with other records and artefacts yet to be discovered, might bear witness to a world history very different from what we have hitherto supposed it to be, and to a Messianic Plan for human evolution whose glories are virtually unknown at present to most of those whom it is designed to benefit.

It is even held by some that the former Atlanteans had already developed some kind of space-technology which allowed some of them to escape from this planet entirely and to colonise some other part of the universe—conceivably the moons of Jupiter, say. And thence, it is said, they will one day return. This notion, fantastic as it may seem, is nevertheless remarkably consistent both with the symbolism of the Noah story and with the Messianic expectation. It also accords well with the predictions of Mario de Sabato (see chapter 8) and the theories of von Däniken, Blumrich and others. How big a grain of truth the idea contains is anybody's guess at present.

At all events, it is no less than extraordinary that the culminating phase of the Messianic Plan—biblically described as the 'sky-kingdom' or kingdom of heaven—should be predicted as commencing with a Messianic arrival (apparently from outer space) which is symbolised in the Great Pyramid by the Noah-like descent of the culminating capstone upon what, in Cayce's terms, may well be described as 'the mountain of Araaraart'.[11]

The Story of Abraham

In view of Abraham's rôle as the semi-legendary founder of the Jewish nation, one would expect to discover rather more Messianic parallels in the story of his life than in fact are to be found.

True, his rôle is to set up a new nation of 'sons of God' in a distant 'promised land'. True, his initial problem is a lack of sons—a

[11] The biblical account seems to hint directly at such a link. The height to which the floodwaters—and thus Noah and his ark—rise above the mountains during the deluge is put at some fifteen cubits. By the same token, this is also (apparently) the distance the ark descends before it comes to rest. Thus in symbol, fifteen cubits is the distance which spiritual perfection is capable of rising above the 'mountains' of the physical world, and also the 'distance' the Messiah will 'descend' in order to re-contact that world. But the height of the designed capstone of the 'mountain of Araaraart', symbolic of that descent, is similarly 364·28 P", *or 14·57 Sacred Cubits*. It is difficult to avoid the suspicion, in other words, that the apparently arbitrary height of fifteen cubits may have a direct Pyramidal connotation.

problem destined to be amply overcome as from the date of his ritual circumcision and consequent commitment to the son-of-god concept. And in both these respects there are perhaps foreshadowings of the later Jewish Messianism.

But in many respects the events of Abraham's life seem to be random enough to suggest that a real human figure stands at the basis of the legend, and that a basically true story has been embroidered with only a few Messianic embellishments in the course of time.

Among the more obvious of the few allusions relevant to our researches in the story of Abraham are perhaps the following:

The year of his birth as Abram would appear to be almost identical to the date of the Great Pyramid's Scored Lines—the apparent starting-point for the whole Messianic Plan.

He later apparently attained a position of some power in Egypt—a fact which poses the question of whether he received any kind of priestly training during his years there, as a man of his obvious religious leanings might well have done. If so, he would inevitably have become acquainted with a number of Messianic ideas.

At the age of ninety-nine, Abram's name is changed to Abraham. The biblical text suggests in explanation a change of meaning from 'High Father' to 'Father of a Multitude'. Be that as it may, the reported addition of an 'h'-sound—also applied simultaneously to Sarai his wife, who thus became Sarah—is widely recognised as having a special esoteric connotation. For the sound in question is esoterically associated first with breath and then with spirit (many languages, indeed, fail to distinguish between the two): moreover, the corresponding Hebrew character also signifies the number five, symbolic of an initiate. Consequently, reading this curious episode 'between the lines', we may perhaps deduce that Abram—who was in any case always prone to oracular 'visions'—underwent at this stage some kind of special initiation into the secret mysteries.

The horrific Sodom and Gomorrah episode comprises perhaps one of the more strikingly Messianic allusions in the story of Abraham. Informed beforehand by three messengers or angels of the impending destruction of these two vice-ridden cities, Abraham pleads for mercy for any righteous men still living there. His nephew Lot and his family are duly rescued by the first two 'strangers' just in time to escape the predicted holocaust, which is presumably unleashed by the third 'stranger'—the one whom Abraham is reported as addressing as 'the Lord'. Only Lot's wife, unable to resist a backward glance, fails to make good her escape.

The extraordinary resemblance of the reported holocaust itself to the effects of a nuclear explosion has often been pointed out. But what ought to concern us mainly here is the fact that the development of the story apparently corresponds with some exactitude to the Great Pyramid's view of the coming multiple Messianic initiative. There are four potential saviours of the people of Sodom and Gomorrah—namely Abraham himself and the three extra-terrestrial visitors. On the one hand a man on the side of the angels; on the other, three angels on the side of man. And in this respect Abraham seems to correspond to the Pyramid's Granite Leaf, his function being essentially redemptive. Like the Granite Leaf, his rôle is to rescue, not to judge or cut off—in short, to 'let the people pass through', after the style of Moses at the later crossing of the Red Sea.

The first two of the visitors are less forgiving, however, and their demands more strenuous; and although they are still prepared to rescue those who are single-minded enough to deserve it, they are insistent that the destruction must be carried out. To the third visitor apparently falls the task of 'pushing the button', thus finally destroying those who are unworthy for the sake of those who have escaped and are prepared to 'refuse to look back'. And in all this we may see the three visitors as symbolising the function of the three further portcullises in the Pyramid's Antechamber—portcullises which, unlike the Granite Leaf, are designed to descend as far as, and even below, floor level, thus finally cutting off those single-minded ones who have succeeded in passing through to the King's Chamber from those who have not.

Thus, the story of Sodom and Gommorah serves as a direct allegory of the time of testing foretold in the Pyramid's Chamber of Ordeal, and of the multiple Messianic visitation which is foretold as accompanying it.

Finally, the story of Abraham's would-be sacrifice of his own son Isaac, and his last-minute substitution of a ram, has long been accorded a deep symbolic significance. For a start it is an obvious mirror-version of Jesus' reported new covenant, which involved the substitution of human self-sacrifice for the atoning blood of the traditional Passover lamb. Be it noted, however, that in the earlier version the sacrifice of the son *is avoided at the last moment*—a feature which may have given Jesus himself some pause for thought immediately before his arrest and crucifixion. But then another Jewish version of the story has Isaac duly sacrificed, *and subsequently returning from the dead after three years.* And this legend too has considerable affinities with the Messianic tradition. The person of

Isaac has long been regarded by many Jews as symbolic of the Jewish nation itself, whose claimed basic function as 'collective world-Messiah' thus supplies a further link between the story of Abraham and the Messianic traditions.

The story of Abraham, in short, is not as overtly Messianic as that of Noah, still less than that of Moses (q.v.)—possibly because of the strength of the real man's enduring legend. But there are none the less sufficient Messianic hints and episodes to suggest that his life story, like theirs, has been to some extent overworked by unknown hands to deliberate prophetic effect. It is therefore not too surprising to discover that his age at death is reported as 175 years—a figure which in Pyramid-code would indicate his attainment of the spiritual perfection of the Great Initiate (7×5^2).

The Story of Moses and the Exodus

Of all the Old Testament stories, none contains more Messianic parallels and portents than the story of Moses. The book of Exodus records how he was born at a time when the Israelites in Egypt were becoming so numerous as to incur the wrath of the reigning pharaoh: as a result they were subjected to harsher and harsher treatment, until eventually it was decided to drown all male Israelite babies at birth. The young Moses himself only escaped this fate by being placed in a watertight cradle of rushes and hidden among the reeds, whence he was rescued and brought up in the pharaoh's own household (the reference seems, in fact, to be to the traditional Egyptian ceremony of Nile baptism—the offering of the new-born infant to the sacred waters).

Already, then, Moses had survived, Noah-like, an ordeal by water in a boat of reeds. Moreover, he shared with the much later Jesus of Nazareth the distinction of having survived an attempt at mass infanticide while still in the cradle—though Herod's massacre seems unhistorical, and could conceivably be merely a literary 'borrowing' from the story of Moses himself, undertaken for symbolic effect.

Having murdered an Egyptian official—apparently under severe provocation—Moses goes into exile, where the cries of his people for deliverance from Egypt eventually reach him on Mount Horeb via the voice of his God. Given detailed divine instructions—rather like Noah—for bringing the Israelites out of Egypt to the sacred mountain, and thence back to their Promised Land of Canaan, Moses sets out to meet his brother Aaron, and together they start to put the plan into effect. Of these two, Aaron, the mouthpiece, is to do the speaking, while Moses is to be 'the god he speaks for' (Ex. 4:16). After a long series of distinctly unpleasant prodigies performed by Moses before the horrified Pharaoh with the aid of a special 'staff' with which the Lord equips him (some believe that this may have been the Egyptian *ankh*, or staff of life, thus: ☥),[12] a mysterious death strikes all the Egyptian firstborn—a death avoided by the Israelites only by dint of keeping a newly prescribed feast (the Passover) which involves the eating of an unspotted lamb with unleavened bread and bitter herbs. Then Moses at last succeeds in leading his people out of Egypt, after 430 years of captivity, and the great Exodus back to the Promised Land begins, guided by the flames and smoke of the erupting Mount Horeb.

In a last attempt to retain his rapidly disappearing labour force, the pharaoh and his army set out in pursuit in the teeth of a strong east wind, which apparently causes the volcano to create a smoke-screen between them and their quarry. Scarcely has contact been made when an early nightfall causes the running engagement to be broken off. Under cover of darkness, the Israelites, perilously embayed on the shore of a branch of the Red (or rather 'Reed') Sea, somehow contrive to cross it virtually dry-shod, thanks to a local ebbing of the waters—associated by the text (Ex. 14:21) with the east wind already referred to (some sources would even associate it with the historical eruption of the Greek island of Santorini) but traditionally ascribed to the semi-magical agency of Moses' staff, which he holds up over the waters. In the light of dawn the Egyptians, suddenly realising that their prey has given them the slip, set off in pursuit and attempt to emulate the feat, but their chariots are quickly bogged down and the waters, unexpectedly returning, engulf the Egyptian army and at the same time cut off the Israelites from further pursuit.

There follows a long period of desert wanderings, during which the army of refugees is sustained by its leaders through a number of further prodigies. A tainted well is made drinkable by casting in a log; a flock of quails inexplicably flies in and supplies the travellers with meat; miraculous 'bread from heaven' appears on the ground each morning at daybreak, and continues to do so for forty years;

[12] Compare Acts 6:22, where Stephen claims that 'Moses was trained in all the wisdom of the Egyptians'—a statement which adds support to the theory that Moses was a full initiate of the Egyptian mysteries, and was thus thoroughly familiar with the knowledge enshrined in the Great Pyramid. This would help to explain the extraordinary Messianic symbolism of the Exodus.

Moses strikes the rock with his miraculous staff and a spring of water gushes forth.[13] And eventually they come to Horeb, the sacred fire-mountain.

Here, in the course of five ascents of the erupting volcano, Moses receives the divine promise: 'You shall be my kingdom of priests, my holy nation', and is given precise instructions for bringing this about. The divine laws for Israel, designed to qualify it for its redemptive, priestly function, are twice inscribed in stone—the repetition being necessitated by a temporary rebellion (connived at by Aaron) among the impatient tribesmen who, during Moses' long absence, revert to the worship of a 'golden calf' (i.e. the ancient bull-cult). Moses brutally stamps out the incipient revolt among his faithless people by arranging for the slaughter of all those who do not publicly side with him and proclaim their willingness to submit to the divine will. Finally an Ark of the Covenant is constructed to house the divine 'tokens', with a tent to cover the ark and to house the divine presence. Strict rules of ritual are set up with a priesthood to administer them, and an equally strict—and surprisingly advanced—social code is laid down.

Then the Israelites set out once again on their long wanderings, eventually arriving many years later on the borders of the Promised Land of Canaan—the land which their ancestors had left so many years before to 'go down into Egypt'. And here, at the age of 120 and within sight of his goal, the great prophet Moses eventually dies and is buried, and his grave is unknown to this day. Thereupon the hosts of the chosen people, led by one Jesus (or in Hebrew, Joshua), pass miraculously through the waters of Jordan, the final river which separates them from their divinely ordained inheritance.

The whole story, as I have said, is riddled with Messianic symbolism, and one has the feeling that either the events or the biblical account, or perhaps both, were deliberately arranged with this end in view—by whom and to what end one can only surmise. As Paul was later to put it (1 Cor. 10:11): 'All of these things that happened to them were symbolic, and were recorded for our benefit as a warning.'

To start with we have the purely linguistic evidence surrounding the very names Moses and Aaron. According to the biblical account in the first chapter of Exodus, the child rescued from among the

[13] As Keller has shown in his *The Bible as History*, most of these are well-attested natural phenomena, still observable today. Indeed *manna* (an exudation from the tamarisk plant) is nowadays one of the commercial exports of the region in question.

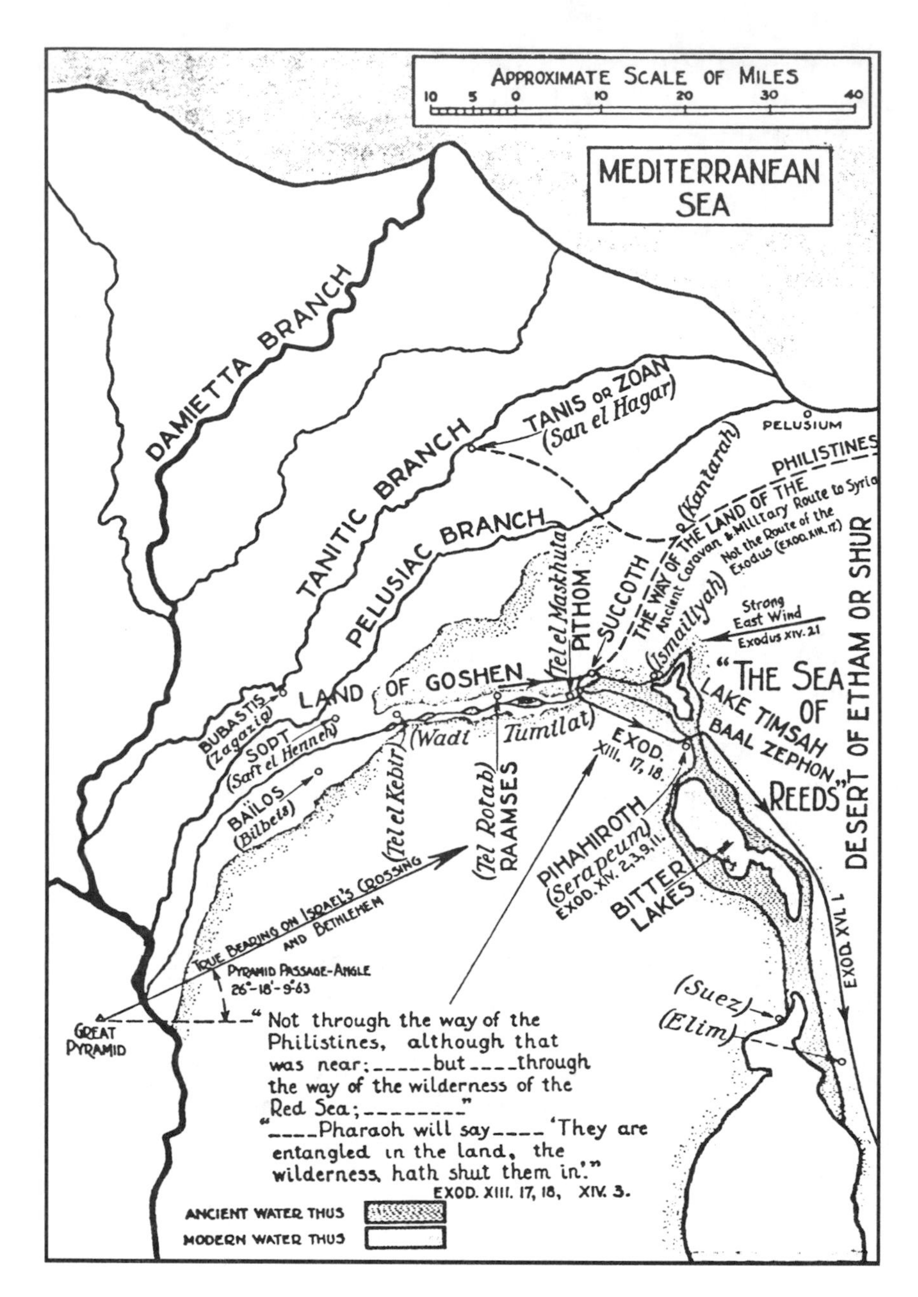

Sketch map of the probable outline of the Eastern Delta and Gulf of Suez at the time of the Exodus. The map shows the possible location of the crossing of the Reed Sea as defined by the Pyramid's passage-angle, and also the function of the east wind in providing first of all a pillar of smoke ahead of the Israelites, and later a smoke-screen behind them, finally holding back the Nile waters in the Wadi Tumilat during the crossing.

reeds is called 'Mosheh' by the pharaoh's daughter ostensibly because she 'drew' him out of the water—*mashah* being the Hebrew verb 'to draw'. Which might symbolically suggest the distinctly Messianic notion that the child was one who had previously succeeded in escaping from the waters of mortality and had now returned to help rescue his fellow men.

On the other hand, the biblical Old Testament is so full of obviously dubious etymological explanations of ancient proper names that this explanation could clearly be equally dubious. Why, to take the present case as an example, would an *Egyptian* princess base the name of the child on a *Hebrew* verb in the first place? It seems far more likely, in fact (as we suggested in chapter 8), that the princess, who appears on chronological evidence to have been none other than the celebrated Hatshepsut herself, simply named the child *after her own father Tuthmosis*.

And there we could quite happily leave the matter, but for the fact that, as adapted into Hebrew, the child's name (Mosheh) and especially its alleged Hebrew derivation (*mashah*) bear an extraordinarily close similarity to the Hebrew verb *maschah* (to anoint) and thus to its direct derivative *maschiach*, meaning 'anointed' or 'Messiah'. The conclusion seems virtually unavoidable that the link must originally have been a semi-overt one, designed to refer directly to Moses' basically Messianic rôle. The Egyptian loan word, in other words, was probably adapted to fit a pre-existing Hebrew symbolism.

The case of Aaron is almost as intriguing. Ostensibly, the name means 'enlightened'—thus marking out Aaron as an initiate whose task is to speak for the almost godlike Messiah figure personified by Moses. As such, his interpretative rôle is clearly that of all initiates throughout the ages. But, more than that, his name also appears to be directly related to the word *aron*—the very word used by the text to refer to the sacred coffer or ark which the Israelites are to bear with them on all their travels. The ark, in other words, seems to be marked out as a source of enlightenment—hence, no doubt, the extreme respect with which it is always treated. Indeed, the very success and survival of the people as a whole seems to be directly proportional to their devotion to the ark's safety and the dictates of the divine will promulgated by its chief guardians and by the mysterious tokens housed within it. Meanwhile it should be remembered that the Pyramid's King's Chamber coffer—its interior volume identical (according to Rutherford) to that of the Israelites' ark—is also essentially a 'coffer of enlightenment'.

Not only the name of Moses, then, but also those of Aaron and the sacred ark itself, speak directly of a leadership of powerful Messianic initiates which alone is capable of bringing the people back to the promised land of the forefathers—in symbol, of course, man's long-lost spiritual inheritance.

The Exodus story itself begins at a time when the enslaved Israelites in Egypt have reportedly become so numerous as to constitute a veritable blight on the face of the land. Their consequent ill-treatment by their Egyptian overseers causes them to cry out for deliverance amid their increasing woes and hardships. The parallel with the Messianic prophecies is plain, for we have already seen that the age in which the Great Initiate at last returns must by definition see a world population explosion of unparalleled proportions if it is to be accompanied by the universal reincarnation predicted by the Pyramid and other sources of gnosis. In the drowning of the firstborn with which the story commences, we may even see a foreshadowing of modern attempts to control population increase by means of birth-control and particularly abortion—the drowned firstborn representing those infants who are never allowed to escape alive from the waters of the womb. We have also seen that both Bible and Pyramid agree that the second advent will be preceded by a time of unprecedented death and destruction, in which the Higher Powers will intervene only when there is a sufficiently general wish among the people that they should do so (compare the olive-branch-bearing dove in the Noah story).

The almost incredible parallelism continues. The recall of Moses to Egypt and the appointment of his brother Aaron as his mouthpiece immediately suggests a spiritual intervention in world affairs at the onset of the final age of escape through the medium of a physical man. The semi-magic staff or *ankh* with which he is equipped appears to correspond to the sceptre of the expected Messiah described in Psalm 110 (v. 2). The prodigies are reminiscent of the 'marvellous new techniques' with which de Sabato invests his expected visitor from outer space (q.v.). The death of the Egyptian firstborn foreshadows the succumbing to mortality of those who choose the way of physicality in the Pyramid's Subterranean Chamber ('Egypt' is traditionally understood as standing for the physical world in the prophetic scriptures); while the Israelites who escape that way of death do so only by virtue of eating the untainted meat of a pure lamb and the unleavened bread of penitence. The symbolism is one that was later to be adopted by Jesus of Nazareth as one of the specifically Messianic concepts basic to his teaching. For him, the eating of that self-same unleavened Passover bread was to symbolise the unconditional acceptance of his self-abnegatory teachings—an acceptance which alone would permit man access to the 'promised land' of eternal Life.

And so it is that the great Exodus from Egypt—i.e. from the physical world—finally gets under way, guided by the fire of the holy mountain—a reflection, conceivably, of the spiritual light symbolised by the Great Pyramid's Messianic capstone. But the battle is not over yet. The physical world (personified by the Egyptian Pharaoh and his army) still has its claims on the chosen ones, the souls of the enlightened. The community of the faithful is still pursued by the earthly powers but, led by the Great Initiate, they escape by passing unscathed through the waters of death—as symbolised by the 'Red Sea'. They thus qualify for rebirth into the Messianic era which is destined eventually to lead them to the Promised Land of Earth's Golden Age, and during which their function will be to act as a 'kingdom of priests and holy nation'—a kind of collective Messiah, or midwife to the New Age. Meanwhile their pursuers, attempting the same feat, fail to make good the crossing, and thus in symbol remain bound to utter mortality.

The Israelites' forty years of desert wandering, which now follow, would seem to represent the progress of the enlightened through a still hostile environment, and the choice of 40 (8 × 5) years, rather than any other number (the period seems an extraordinary length of time for the *historical* journey in question, even via the longest route), may well have some connection with the Pyramid's symbolism. If the latter is applied here, it would suggest that the period in question is to be a time during which the Messiah will be reincarnated (or resurrected), a notion which of course squares well with the Pyramid's predictions. During this time there are hints in the Exodus-story of pollution and its cure, and of the discovery of new and unexpected sources of food. Meanwhile the *manna* or bread from heaven finds a clear echo in Jesus' Messianic words: 'I am the bread of life. Your forefathers ate the manna in the desert and they are dead. I am speaking of the bread that comes down from heaven, which a man may eat, and never die. I am that living bread which has come down from heaven; if anyone eats this bread he shall live for ever' (John 6:48–51). At the same time the miraculous spring, spurting from the rock, recalls Jesus': 'Whoever drinks of the water that *I* shall give him will never suffer thirst any more. The water that I shall give him will be an inner spring always welling up for eternal life' (John 4:14). The symbolic function of the Pyramid's Well-Shaft also comes to mind.

Meanwhile we come to the holy volcano, or fire-mountain—its red granite virtually identical to that of the Pyramid's Granite Plug, Granite Leaf and entire King's Chamber Complex. The identification of the sacred mountain with the Pyramid, crowned with its shining, golden capstone—as well as the holy Mount Zion of the Old Testament prophecies—becomes a notion difficult to avoid. It is here that, during the course of five ascents of the mountain (reminiscent of the Pyramid's five Messianic 'ascents', starting with that of Jesus) the faithful flock is shown its role as a kingdom of priests, i.e. its function as a sort of communal saviour of mankind, or Messiah. This involves the giving of the divine law—the Messianic standard which men must eventually either follow or die, as both the Pyramid and the Exodus story unequivocally suggest.

We should note at this point that the law is duly inscribed *in stone*, a medium which inevitably reminds one again of the Great Pyramid, in whose stone the natural and divine laws are likewise (it seems) indelibly inscribed. The parallelism is made even more striking by Moses' symbolic breaking of the tablets of the law on finding the Israelites worshipping the golden calf, and by the laborious production of new tablets to replace them. For here we see, in symbol, not merely the destruction of the old law and the establishment of the new Messianic dispensation, but also the physical process which specifically symbolises it in Rutherford's view—namely the partial dismembering of the original, slightly reduced 'Pyramid of the Divine Law' (i.e. the observable fact that its original casing-stones have, in the course of time, almost all been removed) and its eventual reconstruction to its full, Messianic, dimensions—the 'growing up' of mankind 'into Christ' described by Paul in Ephesians 4:11–16. And even the precise historical reason for this reconstruction and re-expression of the divine law is carefully enshrined in the details of the Exodus story—namely the long absence of the people's Messianic leader, as a result of which they revert to worshipping the golden calf, an activity which must clearly be interpreted in terms of an obsessive pursuit of wealth and affluence which typifies above all, perhaps, our own twentieth-century Western society. And let it be noted that it is precisely when these degenerate activities have reached their climax that—in the story—Moses, with Joshua (= Jesus) his assistant, at last unexpectedly returns, just as Jesus himself predicted the return of the Messiah 'at the time you least expect him' (Matthew 24:44).

Moreover, there is further prophetic significance in what follows—for, as the story makes plain, the utter unpreparedness and degeneracy of the people—actually connived at by Moses' appointed mouthpiece, Aaron—results in the repudiation of the new Law, the destruction by Moses of the golden calf and the bloody slaughter by his agents the Levites of a large number of the Israelites. These events would appear to foreshadow symbolically the rejection of the coming Messiah's message by much of mankind, the destruction of the whole monetary basis of capitalist society, and a consequent period of unprecedented death and destruction. Meanwhile one wonders whether the earlier connivance of Aaron in the establishment of the worship of the golden calf, in an effort to respond to the people's clamour for a new god, might not be a direct foreshadowing of the historical pandering of official Christianity—the supposed mouthpiece of God—to the demands of wealth and worldly power, rather than to the spiritual ideals of its founder.

Eventually Moses retires once more to the summit of the Holy Mountain for forty days and nights, where without food or drink he makes intercession for the people, and once again the eternal law is inscribed on stone tablets. Then, on redescending at last from the mountain, he demands that the Israelites devote their gold and silver and other precious possessions to a new and better use—the construction of the Ark of the Covenant to the divinely ordained specification, a device designed to assist and protect the Israelites during their long journey to the Promised Land.

We may see in this part of the story unmistakable indications that the Messiah-to-come will in due course depart for the spiritual planes and then once more be reincarnated. The '*forty* days and nights without eating and drinking' clearly suggest a spiritual sojourn culminating in the rebirth of the Messiah, and incidentally recalls directly the departing Jesus' parallel statement that he would 'no more drink of the vine until I drink it new with you in my Father's kingdom'. Following this, a new demand is made for men single-mindedly to devote all their energies and wealth to the urgent need finally to escape from physicality. And at long last the kingdom of heaven is put *first* among human priorities, while even man's technology (it seems) is at length subordinated entirely to assisting him in his ultimate quest for full spirituality.

At last, then, the Law having been finally re-promulgated via the medium of two new slabs of granite—identical in symbol with the twin slabs of the Antechamber's supremely Messianic Granite Leaf—the people are finally led by the Ark of the Covenant, which they themselves bear, towards the Promised Land—the ark, with its heavy Law-Tablets and protective tabernacle, being a kind of 'entry

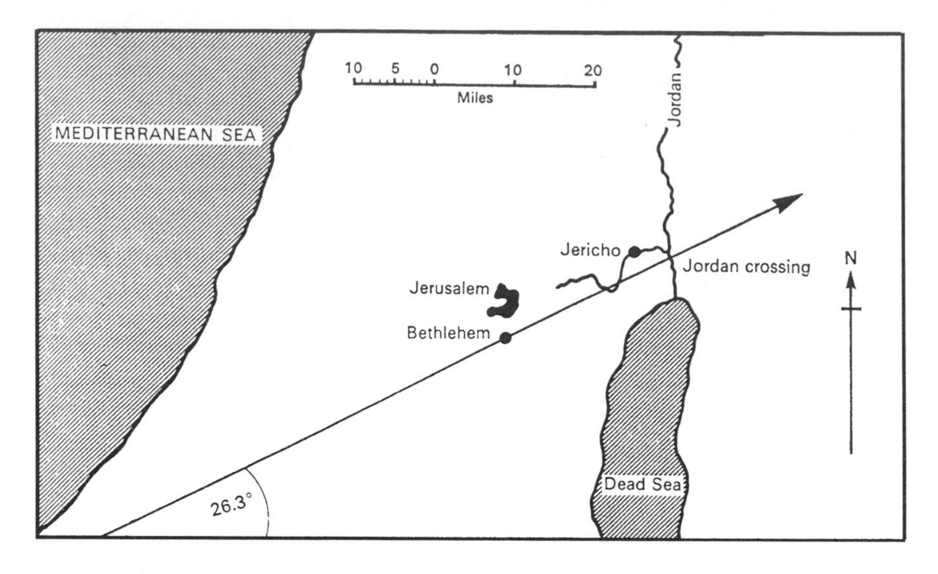

The Great Pyramid's Bethlehem-line and the crossing of Jordan.

permit' to the Final Age, and signifying the acceptance by its bearers of the stern conditions laid down for final admittance into the spirit-planes. And as if in confirmation of the point, the ancient text records that the sacred tabernacle containing the ark was set up forty-two (6 × 7) times during the journey through the wilderness—thus identifying that journey unmistakably with the preparation of spiritual perfection in man.

Eventually the Promised Land is sighted, but at this point, with Aaron the mouthpiece already dead (and man thus symbolically denied for a time his ready source of external enlightenment), Moses the Messianic leader himself retires from the scene—his alleged age of 120 apparently announcing, in symbol, the advent of the human millennium (10 × 12). And so, if we read the symbols aright, it seems that each individual man must still make that final escape alone and unaided.[14] This he must do by 'crossing Jordan'—the last river of death[15]—and the crossing can be achieved only by observing the strict conditions represented by the ark which, in the Exodus story, alone holds back the all-engulfing waters while the precious remnant passes through. It is each man's self-borne ark of enlightenment, in other words. which must see him across that final river, and not the vicarious enlightenment (Aaron) of some remote

[14] Compare Gospel of Thomas 75: 'Many are standing at the door, but solitary are the ones who will enter the bridal chamber.'

[15] Compare Job 33:18 and 36:12, which both describe death in terms of a river.

saviour-figure. This time there is no Moses, no miracle-working superman to act on the people's behalf—and yet there is one whose example and willingness to follow the divine commands gives heart to the struggling masses, and his name, significantly perhaps, is Jesus (Joshua).

And at this point it is worth noting that the Great Pyramid's Bethlehem-line (see opposite) not only crosses the present Suez Canal just south of Lake Timsah—the likely site of the crossing of the Reed Sea; after passing through Bethlehem itself, it also cuts the river Jordan just opposite Jericho, and thus at the reported site of the Israelites' crossing of that river under Joshua. These facts are remarkable enough in themselves, but there is something even more remarkable about them: the line in question (derived from the Pyramid's passage-angle, which specifically signifies man's spiritual destiny) leads symbolically from the Pyramid itself, via the key event of the Mosaic Exodus, to the new dispensation of Bethlehem, and thence onward to the entry into the Promised Land under 'Jesus'. In other words it seems, in symbol, to sum up (as Paul seems to have realised at 1 Corinthians 10:11) the whole Messianic rescue-plan—from the pyramidal symbol of the earth itself and its hidden potentialities for the future (indeed, of the whole Plan for the spiritual evolution and harvest of planet earth), via the Old Covenant to the New, and thence finally, through the Messianic initiative, to man's ultimate re-entry into the spirit-planes.[16] And if the Reed Sea crossing, undertaken at night, symbolises the preparatory death and eventual rebirth of the chosen ones at the onset of the Final Age, then the Jordan crossing, undertaken in broad daylight, would seem to suggest that that final spiritual transmutation will somehow take place not at death, but in the midst of life itself.

As if to confirm our general interpretation, the biblical account states specifically that Joshua's crossing of the final river took place *at the time of the harvest* (Josh. 2:15), and that he caused twelve stones to be shifted from the bed of the Jordan and to be set up in the first camp in the Promised Land. Twelve other stones, we are told, were also set up on the river bed itself, where subsequently they would once more be covered by the returning waters. The text could scarcely make it plainer that the events here described refer symbolically to the final harvest which will result in mankind's gaining his

[16] If John Michell is right in his thesis (*The View over Atlantis*) that the whole world is criss-crossed by what he calls 'ley-lines' which represent paths of some hidden power, then this line must surely be the ley-line *par excellence*!

first foothold in the Promised Land which he had forsaken so long before: the setting up of the twelve stones, in other words, we may interpret as a code-statement for the final resurrection of man. The fact that Joshua also sets up twelve other stones *within the river* suggests, meanwhile, that at this same time there will nevertheless be other men who will not make the grade, and who will therefore continue to be subjected to mortality until the advent of yet another Messianic Age (see page 146).

Finally, the very name of Joshua's first encampment to the west of Jordan—the site of the alleged twelve-stone memorial—adds even more weight to this symbolic interpretation. For it is reported at Joshua 4:19 that the Israelites made their first camp at a place called Gilgal, which is interpreted at 5:9 to mean 'rolling stones'. But this very term is used to this day by the eastern Jewish Hasidic sects to refer to the eternal going round and round of reincarnation (cf. Jiri Langer's *Nine Gates*). In other words the twelve rolling stones that are rescued from the last river of death and set up by Joshua in the Promised Land (where of course they will roll no more) are almost certainly meant to be symbolic—as we suspected above—of the elect who are eventually to succeed in escaping from the world of rebirths and redeaths into that of immortality. Hence, we may further suspect, the symbolic baptismal rite of the much later John *in Jordan, within ten miles of this very spot*, with its message of death and new birth, leading to an eventual baptism of fire. (See Matthew 3:11.)

And thus it is that the city of Jericho—probably a symbol of the old World Order at the inception of the Final Age—is at last stormed and destroyed by the chosen people, though not before two spies (symbolic, perhaps, of prophetic figures) have reconnoitred it and returned to tell the tale. They owe their survival to the assistance of a prostitute called Rahab, who is promised indemnity for her willing cooperation with the Israelite plan. Even the concept of Messianic forgiveness, then—the notion that the meanest worker for the Messianic cause is accorded some kind of special protection during the birth-pangs of the New Age, given a sufficiently wholehearted self-commitment—even this supposedly New Testament idea is present already in the Exodus account.

The walls of Jericho finally fall to the Israelites when—led by seven priests bearing seven ram's-horns and followed by the Ark—they have marched round the city on seven successive days, concluding with seven circuits on the seventh. The text could scarcely make it clearer that the city's walls fall to the invaders as a specific result of their utter spiritual perfection.

With the destruction of the city, Joshua now goes so far as to lay a curse on it. Never again, it seems, may the Old Order be rebuilt. And even now the epic task is not finished, for the chosen people still have to defeat a number of further kings. A number of obstacles, in short, still have to be overcome before the re-occupation of the promised homeland can be completed. Man, we may perhaps interpret, has yet further stages of initiation to undergo, even once he has gained what we have elsewhere described as the plane of initial reception. The same idea appears to be reflected in the symbolism of the Great Pyramid's Construction Chambers (see pages 129–34) and in Jesus' mysterious statement, 'In my Father's house are many mansions' (John 14:2).

In short, then, the story of Moses and the great Israelite Exodus from Egypt seems to have been planned from end to end in such a way as to foreshadow almost the whole of the Messianic Plan, as depicted in the Great Pyramid and corroborated by the Judæo-Christian scriptures and other sources of prophetic gnosis. Only the reader can decide whether the whole series of happenings was accidental, consciously acted out, divinely inspired, or merely edited after the event. But that this extraordinary episode of scripture does constitute an unmistakable 'memory of the future' is something that can no longer be seriously open to doubt.

Who was 'Jehovah'?

We have already noted (pages 273–85) that the symbolic events of the Israelite Exodus from Egypt correspond with remarkable exactness to the Messianic Plan enshrined in the Great Pyramid of Giza. We have also noted that the events of the Exodus seem to have been planned with precisely this aim in view, as Paul himself was later to suggest (see page 275). Furthermore, in view of Stephen's statement at Acts 6:22 that 'Moses was trained in all the wisdom of the Egyptians', it seems reasonable to deduce that Moses' planning and execution of the Exodus was directly based on his knowledge of the Great Pyramid's prophecy. The Pyramid of Khufu, in other words, may have provided the direct motivation and source of knowledge on which the whole Israelite Exodus was based.

On the other hand, the biblical account of Exodus states quite firmly that Moses was personally ordered *by his God* to initiate the plan. At this, Moses asks the rather odd question: 'If I go to the Israelites and tell them that the God of their forefathers has sent me to them, and they ask me his name, what shall I say?' The celebrated and awesome reply, recorded at Exodus 3:14, is: 'I am, that is who I am . . . You must tell the Israelites this, that it is Jehovah the God of their forefathers, the God of Abraham, the God of Jacob, who has sent you to them.' The actual spelling of the sacred name, which even today no faithful Jew may speak, transliterates as 'YHWH'—but 'Jehovah' is the form in which the name 'I am' is generally read.

At Exodus 6:2–8 the matter is elucidated further: 'I am the Lord,' Moses is told. 'I appeared to Abraham, Isaac and Jacob as God Almighty. But I did not let myself be known to them by my name Jehovah.' And there follows a further summary of the plan to rescue the Israelites from Egypt—or indeed mankind from his physical slavery—and bring them to their Promised Land, a plan which the deity proposes to accomplish 'with arm outstretched and with mighty acts of judgement'.

Now unless we subscribe to the anthropomorphic nonsense that the deity actually treasures a special 'taboo' group of sounds as its own sacred name,[17] the passage at Exodus 6:2–8 must be regarded as meaning that 'Jehovah' is a particular and newly revealed manifestation of the Divine Will, which is accessible to Moses, but was not accessible either to Abraham, Isaac or Jacob. Interestingly enough, this description also seems to fit the Great Pyramid's revelations themselves very aptly. The subsequent promise, made 'with uplifted hand', to redeem Israel and lead it back to its Promised Land 'with arm outstretched' would serve as a magnificently appropriate allegory for the redemption of mankind *via the geographical application of the Bethlehem-line* (see maps, pages 19, 276, 282).

This apparent link between the Jehovah-notion and the Great Pyramid naturally prompts one to look more closely into the way in which the name Jehovah is used in the Old Testament. Of the many instances of the word's use in the Hebrew scriptures, the compilers of the Authorised Version seem to have felt that the text calls for the specific emphasis and reproduction in English of only seven (eight, if we include Ex. 3:15), and in these cases the name is used in only two quite clearly delineated senses.

First of all, it is used at Exodus 6:3, Psalm 83:18, Isaiah 12:2 and Isaiah 26:4 in a quite unambiguously Messianic sense. In Psalm 83, Jehovah is 'God the avenger and rescuer', and in Isaiah 12 the

[17] As Simon Roof has pointed out, with unassailable linguistic soundness, 'A name is useful only to distinguish one entity from another; therefore what need is there for a name for the One which contains all which is, apart from which there is no other?' (*Journeys on the Razor-Edged Path*).

'deliverer'. In fact the latter reference occurs in the highly Messianic passage starting at 11:12 in which it is prophesied that the Lord 'will raise a signal to the nations', and that Judah will be once more brought together 'from the four corners of the earth'—both notions being highly reminiscent of the Messianic message of the Pyramid itself. The text goes on to assure the chosen people that they will 'draw water with joy from the springs of deliverance'—an expression having possible connections with the Pyramid's Well of Life—and to proclaim that 'the Holy One of Israel', like the Pyramid itself, 'is among you in majesty'.

Meanwhile the reference to Jehovah at Isaiah 26 is, as we have already seen, quite astonishingly suggestive of the Great Pyramid and its symbolism. 'We have a strong city', exults the text, 'whose walls and ramparts are our deliverance. Open the gates', it goes on, in a striking allusion to the Osirian ritual of the Chamber of the Open Tomb, 'to let a righteous nation in, a nation that keeps faith.' And at this point the apparent link between Jehovah and the Great Pyramid is proclaimed almost explicitly: Jehovah, states the text, 'is an everlasting rock . . . the path of the righteous is level, and thou markest out the right way for the upright.' And as if the clear reference to the Pyramid's passage symbolism is not enough, the poem then launches into the famous passage which leads to the proclamation of the rebirth of the faithful during the Final Age—'But thy dead live, their bodies will rise again. They that sleep in the earth will awake and shout for joy; for thy dew is a dew of sparkling light, and the earth will bring those long dead to birth again.'

But at this point we need to consider the three compound uses of the name Jehovah in the Old Testament. They occur in the books of Genesis, Exodus and Judges. At Genesis 22:14 Abraham, after the sparing of Isaac in the story of the burning bush, names the spot *Jehovah-jireh* (translated as 'The Lord will provide')—and the text adds the commentary that 'to this day the saying is: "In the mountain of the Lord it was provided." ' The implied link between the name and the notion of a mountain is further strengthened at Exodus 17:15, where Moses, after the battle with the Amalekites, 'built an altar and named it *Jehovah-nissi*.' Similarly, at Judges 6:24, Gideon responds to the divine command to free his people from the Midianites, by building an altar and naming it *Jehovah-shalom* ('The Lord is peace'). Research into the nature of the Israelite altars of witness, which both of these clearly were, shows (see Cruden's *Concordance*, page 636, under 'stone') that they normally took the form of great heaps of stones—a description likewise clearly applicable to any 'stone mountain' or pyramid.

On the basis of these particular references, then, there seems to be some evidence at least of a possible link between the name Jehovah and the Messianic Plan as symbolised above all in the Great Pyramid. 'Jehovah' seems to represent the specifically Messianic and commemorative aspect, or *persona*, of the deity, and it is thus a possibility worth considering that Moses may have seen, in the Pyramid, Jehovah's principal concrete manifestation on earth. The Pyramid of Khufu, in other words, might equally well have been seen by Moses as the Mountain of Jehovah.

And at this point an intriguing linguistic possibility springs to mind. As we have already seen, the true vowels of the sacred name are unknown—we merely have the consonants YHWH, of which Jehovah is the most widely accepted reading. But the same conditions apply to the name Khufu: here again the vowels are absent, and the transliteration of the name gives us only HWFW.

Now it is well known—the Hebrew text is quite explicit on the point—that the name YHWH derives directly from the Hebrew verb *hava(h)* (I am). And from HWFW (or Khufu) to *hava(h)* is linguistically an extremely short step—scarcely greater than that between the *contemporaneous and interchangeable* northern and southern English spoken forms of the word 'brother'. Thus it is entirely possible linguistically that the name of the Divine 'I am' could have arisen in the first instance on the basis of the Egyptian name HWFW, which is historically the older name of the two. *In YHWH, in short, we may actually have a Hebrew version of the Egyptian name HWFW.*

But if so, how could this astonishing link have come about? Perhaps the explanation lies in the well-known Hebrew tendency to attach meanings to names which they could not otherwise explain, via what (it must be admitted) is an often highly dubious process of word-association. Their early literature positively bristles with instances of this process, from the association of the name Adam with the 'dust of the ground' at Genesis 2:7 (a play on the Hebrew words *adam* and *adamah*); by way of the naming of Seth at Genesis 4:25 ('God has granted [= *Seth*] me another son'), Noah's blessing on Japheth at 9:27 ('May God extend [= *japht*] Japheth's bounds') and the naming of Moses (Mosheh) at Exodus 2:10 ('because, she said, I drew [from the verb *mashah*] him out of the water'); right through to the later associations between the Hebrew word *netzer* ('stem'), the Nazirites, the Nazarenes (*Notsrim*) and the (possibly unhistorical) town of Nazareth. Meanwhile we should note that in the case of Seth, the explanation of the name in terms of a 'replacement' for the dead Abel (one of the less convincing explanations, incidentally) may well have been necessitated by the fact that the word was (I conjecture) a foreign one—namely *Set*, the Egyptian name for the dark aspect (or 'replacement') of Horus: in other words, in Seth son of Ishshah (Eve) we may conceivably have a Hebrew version of the Egyptian Set, son of Isis. Similarly, Moses himself, who was apparently named after Tuthmosis (the pharaoh of the time of his birth), has his foreign name justified in terms of the Hebrew verb 'to draw forth'. Both names, in other words, are justified in the official scriptures in specifically Hebrew terms.

Similarly, then, it is quite possible that the Israelites in Egypt, having heard that the glorious king within the golden mountain, the legendary occupant of the Great Pyramid, was named HWFW in Egyptian, may have noticed a similarity between that name and their own current expression for 'I am' (whatever the spoken form then corresponding to that verb or to YHWH may have been).[18] Thereafter they would naturally have used the translated form of the name and referred to him in those same Hebrew terms, and it needed only the advent of a Moses, his initiation into the Pyramid's mysteries, and his consequent conviction that its message enshrined the divine will for his own people and for mankind as a whole, to produce the concept of a *divine* 'I am'. Except that in Moses' case the whole concept of the (non-existent) 'god' within the earth-symbolising Pyramid was expanded to refer to the (equally invisible) Lord of the physical Earth itself—a greater 'I am' whose immediate purpose was to lead his servant nation Israel to its Promised Land. Meanwhile the notion of 'essential being', which the term suggests, was in any case a peculiarly felicitous way of describing the deity.

Certainly the fact that the Jewish Jehovah is above all the God of the Exodus tends to support the association we have suggested. The name appears to have been in free use only during the few weeks between the beginning of the Exodus and the Israelites' arrival at Mount Horeb. For on arrival at the sacred mountain Moses promptly imposes a strict taboo on the name he had himself revealed only a short time before.

The reasons for this interdict are far from clear. But since one does not normally ban things unless they have already started to occur,

[18] We have little or no evidence on the state of Hebrew and Aramaic at the time of the Exodus. But we do know that even at the time of the Babylonian captivity over eight hundred years later, when the present version of the Hebrew text was being compiled, the linguistic correlations mentioned above still held good.

it seems probable that one (or perhaps both) of two things had come about. First of all, it may be that the Israelites, through association of the phrase 'I am' with the Supreme Being, started to apply the term *to themselves* (either individually or collectively) and to regard the Israelitish nation as God incarnate or, to use the customary orientalism, the Son of God—a notion which we certainly know to have been widespread among the Jews at a later date, and even in our own day. To Moses such a use of the term Jehovah would presumably have seemed an out-and-out blasphemy.

Alternatively, it may be that the Israelites began to speak of the Jehovah of the Pyramid as though this entity were in some way a separate 'person' from the One God, the concept which Moses was above all anxious to promote. This tendency, too, would therefore have been anathema to Moses, who may not have realised until too late that the common man is prone to identify appearance with reality, and to confuse the manifestations of a power with that power itself.

Either development would help to explain why the insistence on the oneness of God, with which Moses subsequently starts his celebrated Ten Commandments, is followed immediately by the banning of the worshipping of graven images, and of the misuse (indeed, for all practical purposes, the mentioning at all) of the name Jehovah—as though that, too, had become a kind of graven image.

To sum up so far, then, the fact that Moses seems to have received his instructions from the Great Pyramid, while the Bible reports him as receiving them from Jehovah; the fact that the word YHWH may conceivably have derived indirectly from the name HWFW; and the fact that the currency of the free use of the name Jehovah appears to be contemporaneous with that part of the Exodus which lay nearest to the Great Pyramid—all these suggest quite strongly that the similarities between the rôle of Jehovah and that of Khufu's Pyramid may indeed be no accident.

Nevertheless there is one aspect of our argument thus far which seems to merit further special investigation of its own. I refer to the passage at Exodus 6:2–8 in which Jehovah recalls his promise, made 'with uplifted hand', to lead Israel to the Promised Land 'with arm outstretched'. We have already commented on the aptness of this imagery to the geographical application of the Bethlehem-line itself, but the *double* insistence in this passage on the notion of the outstretched arm seems so extraordinary—particularly in respect of a supposedly unseen God—that the text as a whole seems to invite further research into the features of the Pyramid in order to see

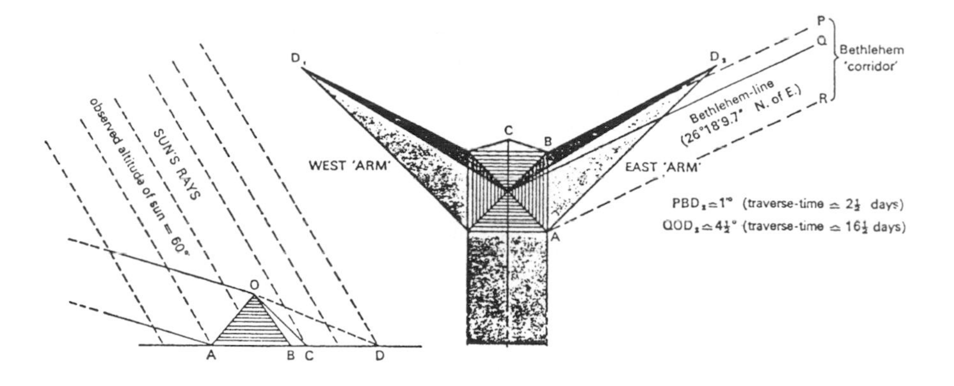

Elevation and Plan of the Great Pyramid's noon-reflections at the Spring Equinox, with Bethlehem-line and Bethlehem 'corridor' superimposed (capstone assumed for diagrammatic simplicity).

whether there may be some even more specific pyramidal significance in the use of these expressions.

And immediately one is struck by the fact that the original Pyramid *did have 'arms'*—the two arms of its solar noon-reflections (see diagrams page 337), whose outlines and peculiarities were thoroughly familiar to the ancient Egyptian priesthood, conceivably being used by them to regulate the seasons (see Davidson and Aldersmith's research into this point). Furthermore, the right arm—which, in connection with the deity, is always the one described as performing mighty deeds—lay in close proximity to the Bethlehem-line already described. Now during the winter (see diagram page 337) the axis of the right arm (i.e. the 'apex ridge' of the reflection) lay well north of the true Bethlehem-line, whilst at midsummer it lay well to the south. At some date during the spring, then, the right arm of the noon-reflection must start to align itself with the Bethlehem-line. The first part of the arm to align itself in this way would be the northern edge of the arm (see plan above). This would be followed by the coming into alignment of the apex ridge of the arm with what could likewise be described as the apex ridge of the Bethlehem corridor (see diagram)—at which point the arm would be entirely interior to that corridor, and thus at the culmination of its alignment. Eventually, as the eastern arm moved progressively southwards and the arm as a whole once again moved out of alignment, the south edge of the arm would in turn take up the Bethlehem alignment—after which the alignment period would come to an end.

It is thus quite possible that this conjunction of the right arm of the noon-reflection with the line of the Bethlehem-angle could have been regarded by Moses as of some special significance with regard to the timing of the Israelite Exodus—and if this could be conclusively demonstrated, then there would be very strong grounds indeed for deducing that the 'arm outstretched' and 'uplifted hand' of Exodus 6:2–8 are direct references to the Great Pyramid and its reflections, and consequently that 'Jehovah' and the message of Khufu's Pyramid were identical.

Reference to the data supplied by Davidson and Aldersmith shows that the seasonal reflection configuration which most nearly provides an alignment of the right arm with the Bethlehem corridor is that of the spring or autumn equinox (see diagram above)—and of the two the spring equinox would seem to be the more appropriate to the matter in hand, in view of the spring's universal message of life and hope, as well as the possible difficulties of a winter Exodus.

At the spring equinox, then (notionally, midnight on 20th/21st March), the theoretical noon-bearing of the north edge of the arm can be calculated as lying marginally north of 063° (True), or almost exactly one degree north of the Bethlehem alignment. As the same north edge would later bear 097° (True) at the summer solstice (20th/21st June), we can calculate that, since the bearing of the north edge would thus have increased by some 35° in 92 days, the average number of days required for a traverse of one degree would be 2.6. On this basis, the beginning of the theoretically favourable period for an Exodus controlled by this phenomenon (i.e. the first alignment of the north edge of the arm with the Bethlehem-line) would be fixed at 22nd/23rd March.

From this moment onwards—if we have read the symbolism correctly—the favourable signs for the Exodus would increase until they reached their culmination at the moment when the right arm's apex ridge was in alignment with the apex ridge of the Bethlehem corridor. Now the theoretical noon-bearing of the apex ridge on 20th/21st March was 30°43′ north of true east. while at the summer solstice it would bear only 6°9′ north of true east. It would thus traverse the whole angle at an average of 3·7 days per degree, so giving a time from spring equinox to Bethlehem alignment of $16\frac{1}{2}$ days. Some time on 5th/6th April would thus be fixed as the culmination of the favourable period for the Exodus. if our theory is a sound one—after which the favourable signs would start to decline as the arm started to swing away southwards.

Thus, if Moses had indeed interpreted the signs in the way we have suggested, he would doubtless have come to the conclusion that the most propitious time for the planned Exodus would be these same fourteen days between 22nd/23rd March and 5th/6th April. Moreover the most obvious date during this period for taking the initial action would probably appear to him to lie exactly midway between the two—i.e. sometime on 29th/30th March. This would seem to assure him of quite favourable conditions for initiating the plan, and would leave a further week of increasing propitiousness in which to reach the Egyptian frontier in the area of the Red Sea, in order that the final and most critical part of the escape from Egypt might be made at the most favourable possible moment. Like the prudent mariner planning to put out across a dangerous bar, his best plan was to leave not at the full flood, in other words, but on the rising tide.[19]

Reference to the map on page 276 will show that the distance in question, measured to the specific crossing-point indicated by the Pyramid's Bethlehem-line, is no more than eighty or so miles, giving a required average, in this case, of some eleven or twelve miles per day across relatively flat country. This is known to be within the speed at which cattle can be driven for long distances, and well within the capability of a highly motivated army of refugees undertaking a forced march by day and night, despite their colossal numbers.

Thus, if our interpretation of the biblical account in terms of the arms of the Pyramid's reflections is valid, and if we have succeeded in reading correctly Moses' probable assessment of things, we should expect to find evidence that the Israelites did indeed commence their journey some time during 29th/30th March, and completed their fording of the Red Sea by 5th or 6th April of the year in question.

Now the Pyramid's own chronograph appears to fix the Exodus

[19] These calculations are based on *average* traverse-times from spring equinox to summer solstice. The speed of change in the sun's declination (and thus in the Pyramid's noon reflection angles) is not constant, however, being maximum at the equinoxes and zero at the solstices. Thus, taking instead the actual *daily* positions of the Pyramid's right arm, the north edge would have started to align with the Bethlehem corridor on 24th March, while the apex ridge would already have reached full alignment on 1st April (notionally, shortly after midnight). On a strict interpretation of the astronomical ephemeris, then, the optimum time for departure would be reduced to a 'window' of only nine days—yet 30th March is still comfortably inside that window. Meanwhile it is interesting to note that the coincidence of the Bethlehem-angle with the right arm of the Pyramid's cross-shaped reflection falls on the very day which was later to mark (in Pyramidal terms at least) the death *on a cross* of the man *born in Bethlehem*.

(see pages 56–7) at a point 688·0245 years after the spring equinox of 2141 B.C.—or in the year 1453 B.C., *sometime during the morning of 30th March.* An early morning departure is indeed described by the biblical account, and Rutherford has shown conclusively and at some length (in Vol. II of his *Pyramidology*) (a) that the year 1453 B.C. is an acceptable dating in terms of the ancient chronologies and surviving documentary evidence, and also (b) that the dating of 30th March of that year accords exactly with the astronomical requirements relating to the Jewish Passover, and thus to the eve of the Exodus—since the first full moon after the spring equinox of 1453 B.C. fell on 29th March. Meanwhile, from the purely practical point of view, it is clear that a full moon would be of major assistance in effecting a night-time Exodus followed by a day-and-night route march. As for the date of the Red Sea crossing itself, we have no specific documentary evidence, but given 30th March as the date of departure, 5th/6th April certainly seems a reasonable date for this culminating event of the initial flight from Egypt, while the Exodus account places the completion of the crossing at around daybreak.

But it is at this point above all that one becomes struck by the plan's extraordinary ingenuity. The route, as we have seen, already appears to have been defined by the Great Pyramid's applied passage-angle. A glance at the map on page 276 will show that, to an observer spying out the land from the Pyramid, the Wadi Tumilat immediately appears to be specifically marked out as of some special significance, since the entrance to this particular 'mouth' of the Nile is the first geographical feature of note to occur along the path marked out by the Bethlehem-line. That a crossing of the apparently impenetrable barrier of the *Yam Suph* (Hebrew for Sea of Reeds, normally translated in the Exodus accounts as Red Sea) is next envisaged, seems to have been symbolised by the Pyramid's own 'Hidden Lintel'—the Egyptian name for the removable limestone block which originally concealed the entrance to the Ascending Passage (see page 55). The Sea of Reeds, in other words, was to provide the 'hidden doorway'—the unsuspected exit—which was to lead to the Israelites' final escape. That the next blockage of the Ascending Passage is the immovable Granite Plug is symbolic of the next, much more serious, obstacle to confront the Israelites—their own unwillingness to accept the Divine Law in all its fullness, as promulgated at Mount Horeb.

Meanwhile the ingenious strategy of the Exodus—apparently foreseen by the Pyramid's designer—starts to become clear. The escape route must first of all avoid the normal trade route exits from

Egypt—and thus the armed forts guarding them—and the unexpected fording of the Sea of Reeds would admirably satisfy this requirement. Moreover lunar, tidal and meteorological conditions would need to be carefully selected to ensure that subsequent Egyptian pursuers would be unable to follow the same route, and would instead be forced to make a fifty-mile détour around what is now Lake Timsah (see page 276), finally losing their quarry in the process.

The selection of the period from 29th/30th March to 5th/6th April for the plan's inception fits these requirements with amazing perfection. The departure on the night of the full moon following the spring equinox first of all meant—as we have seen—that there would be sufficient light for a day-and-night route march. Assuming that the Israelites arrived at Pihahiroth, on the Sea of Reeds, on 5th April, their crossing of the following night would then have occurred seven days after the night of the full moon.

Now the period of the vernal equinox is characterised by a variety of tidal and meteorological conditions. For a start, it is a matter of statistical observation that (in the words of Reed's *Nautical Almanac*) 'the greatest spring tides will occur after a Full or New Moon, which is in Perigee near the Equinoxes'—the actual time-lag between the full or new moon and the greatest tides being some two or three days. Now at 'springs' the high tides are exceptionally high and the low tides exceptionally low—so that an Israelite low-tide crossing of the Sea of Reeds, some three or four days after the greatest spring tides, would probably have occurred on one of the last occasions when the tide was low enough to permit it; thereafter the approaching 'neaps' (with their much smaller range) would have resulted in a failure of the sea-bed to uncover at all at low tide for at least a week—thus cutting off all effective pursuit.

It may of course be objected that the Red Sea *has* no tides to speak of—but it should be remembered that the Pyramid's passage-angle defines (and the text of Exodus describes) a crossing *not* of the Red Sea but of the *Yam Suph*—the Sea of Reeds, or the northern extremity of the then Gulf of Suez, which seems to have extended at the time as far as the biblical Succoth (see page 276). In fact, both the Gulf of Suez and the Gulf of Aqaba do experience quite marked tides as a result of the considerable narrowing and shallowing of the Red Sea on either side of the Sinai peninsula[20] and it is clear that the effect

[20] The tidal rise at Suez is put at some 6ft. by the Admiralty's *Red Sea and Gulf of Aden Pilot*.

of these tides would have been even more pronounced in the even more restricted narrows of the original Sea of Reeds as depicted—possibly even producing some kind of tidal bore in the narrow strait opposite Pihahiroth, especially at the time of the largest 'springs'.

Meanwhile, the prevailing wind in the area tends to be northerly or north-westerly during the winter, while the more varied summer pattern includes occasional experience of the north-east trades. The time of the change-over tends closely to follow the equinoxes—as indeed do most seasonal world wind-pattern changes. Moreover, because of this change of weather patterns, the period of the equinoxes tends to be characterised by abnormally unsettled weather conditions and consequently by stronger winds than usual. The biblical account suggests that the Exodus did indeed occur at such a time of boisterous meteorological conditions: only a matter of days before the Exodus, for example, a large swarm of locusts had been swept in by a strong east wind, only to be blown away again shortly afterwards by a *westerly gale* (see Exodus 10:12–20). Thus it seems more than likely that the date laid down for the crossing of the Sea of Reeds was planned deliberately to coincide with the beginning of the summer wind pattern, with its periodic north-east winds.

The effect of strong winds on shallow waters is, of course, to 'pile them up' in the direction of the air movement: the level of the by no means shallow Red Sea itself, for example, can vary at least two feet at either end according to the wind direction. On the other hand, a strong north-easter would have had little effect on the level of the Red Sea: what it *would* have affected was the level of water in the Wadi Tumilat. This in fact was the original 'Suez Canal', and had once been navigable to ships passing from the Nile to the Red Sea, at a time before the local earth's surface had risen to its present level. By the time of the Exodus, however, the channel had become partly silted up and choked with reeds, with the result that it had to be newly excavated under Seti I and his successors. Since the ancient Egyptians did not go in for deep-keel boats, it seems likely, therefore, that the mean depth of water in the Wadi Tumilat could not have been much more than six feet, and may well have been much less—as also may the depth of the Reed Sea opposite Pihahiroth. (It should also be remembered that the depth of water in the Wadi would have been at its lowest in the spring, since the Nile inundation of Egypt does not commence until June or July.)

Thus we have a period of declining spring tides and also conceivably of strong north-easterly winds—a combination which would permit the waters of the shallower parts of the Sea of Reeds to drain away southwards at low tide, while preventing the in any case low waters of the Wadi Tumilat from flowing eastwards to replace them. The conditions for a crossing of the bed of the Sea of Reeds were thus ideal.

But that is not all. The plan, as interpreted by Moses, clearly involved a surprise crossing of the sea-bed *at night*—a daring operation involving considerable risk of life and limb unless certain critical conditions were fulfilled. For a start, the sea-bed must be *dry*: even a foot of water would have made it impossible for the army of Israelite refugees to pick their way across the shallowest parts of some two miles of largely invisible sea-bed without straying into deeper water—thus producing mass panic and an inevitable watery catastrophe. Secondly, there must be some form of navigational beacon towards which they could direct their path, for a featureless sea-bed at night is an obvious recipe for losing one's bearings. (Meanwhile there would be a tendency for night-time surface cooling to produce a katabatic wind off the Sinai massif, thus tending to add an even more easterly point to the wind).

Yet it is clear that even these conditions were foreseen by the plan. For a date a week after full moon tends to produce a moonrise during the early part of the night (a feature which can be checked for the date and place in question by astronomical calculation), while there is a tendency (again quoting the words of Reed's) 'for low water to occur at Moonrise'. Thus, by commencing their crossing during the early part of the night, the Israelites could use *the rising moon* as their leading mark. The whole operation was apparently complete by around 3 a.m., by which time, according to the biblical account, the Egyptians were already in pursuit, having made the fatal mistake of sending out their chariots—already notoriously unstable vehicles—across the soft sea-bed. Inevitably they became bogged down, and before they could extricate themselves the tide started to return—quickly, it seems, and with some force, as is normal in such flat, shallow areas—and the pursuers, unable to run fast enough to escape what may well have been nothing less than a tidal bore (i.e. a veritable 'wall of water') were swept northwards to their deaths in what is now Lake Timsah. And by daybreak, with the moon now past its zenith, the tide was once more fully in and the Israelites' escape route finally sealed behind them.

Thus one can see that the 'divine' choice of the night of the spring full moon for the 'Passover' and the inception of the escape plan has little or nothing to do with mere ritual niceties: it is evidence of the most brilliant military planning, and of painstaking reconnaissance apparently carried out by Moses himself (probably during his many years as a 'shepherd' in the Sinai peninsula area), under the 'remote guidance' of the Pyramid's designer. And meanwhile both the timing of the plan and the controlling geometry of the right arm of the Pyramid's noon-reflections (let it be observed) are intimately and inextricably bound up with the known, *and therefore predictable*, motions of the sun and moon.

Thus we appear to have demonstrated beyond all reasonable doubt that the Jehovah of Exodus 6:2–8—the controlling genius of the whole operation—may justly be identified with the 'voice of the Pyramid', while the ancient titles HWFW and YHWH were in all probability no more than separate, if related, ciphers for a single, underlying reality. 'Jehovah' and 'Khufu', in short, were one and the same.

The Legend of the Three Kings

The age-old legend of the Three Kings bringing their symbolic gifts of gold, frankincense and myrrh to the new-born Messiah, and led by his star to Bethlehem, seems to represent nothing less than the survival, in anecdotal form, of an ancient tradition specifically linking the Pyramid with the birth of Jesus of Nazareth. As we have seen, the original Great Pyramid, formerly known as The Light, cast a *star*-shaped reflection on the desert around it at noon on the summer solstice; it and its famous neighbours, the Second and Third Pyramids of Giza, have always been supposed to contain between them the remains of three ancient *kings;* and the Great Pyramid's passage-angle has a direct geographical association with the town of Bethlehem—and with the symbolic events of the Israelite Exodus from Egypt. It seems likely, therefore, in view of these combined facts, that the three main Giza pyramids must be the originals of the legendary Three Kings who are led by the Messianic star (the Great Pyramid) to Bethlehem. The biblical legend and the surviving stones make a perfect fit.

In Matthew's account (chapter 2) the Three Kings have become 'astrologers'. As such, they not only know that a Messiah is to be born, but also have a guiding star, which leads them to the very site of the birth (the 'rising of the star' mentioned by Matthew appears to be directly connected with the laying-off of the Pyramid's passage-angle *towards the sunrise*—a phenomenon which has long

Messianic associations). They thus know, or at least are able to discover, both when and where the birth is to take place. But, even more than this, they also know what his future holds—for their strange, symbolic gifts seem to represent nothing less than the presenting of the infant with his Messianic destiny. His kingly and priestly rôles are clearly indicated by the gold and frankincense, while the myrrh directly foreshadows his death and burial: none the less it should be remembered that the funerary myrrh was used specifically *as an embalming oil* (see John 19:39–40), and that the purpose of embalming was, among the Jews as among the Egyptians, the preservation of the body *so that it might in due course be 'resurrected', or rather reanimated.* In other words, even the Messiah's eventual resurrection is foreshadowed in the 'royal gifts'.[21]

Like the Great Pyramid, then, the Three Kings bear with them the knowledge of the time and place of the Messiah's birth and the foreknowledge of his destiny, and if we are justified in identifying them with the three major pyramids of Giza, then it may be that the Second and Third Pyramids also enshrine some kind of didactic or prophetic message. Meanwhile, Matthew's use of the term 'astrologers' is a natural one, in view of the travellers' apparently celestial motivation and ability to foretell events. This is a gift which the author would naturally associate primarily with the practices of the Eastern magicians known as Chaldeans (or Magi), and in particular with their science of astrology. Moreover this notion may not be so wide of the mark as it may seem if, as suggested on page 331, some of the traditional astrological notions *actually derive from* the foreknowledge of the Pyramid's architect and its symbolic codification.

It is a distinct possibility, then, that the presentation by the Three Kings of their symbolic gifts to the infant Messiah represents the detailed forecast of the Messianic rôle enshrined in the Pyramids of Giza—and notably in the Great Pyramid itself. But if the tradition in question did so originate, then we would expect to find some evidence that Jesus visited Egypt at an early date in order to learn at first hand about the vital rôle there assigned to him. Since the

[21] One is reminded at this point that the ancient Egyptian characteristic sign of the verb *(r)di* 'give' was △—a sign oddly reminiscent of the design of the Great Pyramid, with its fifth-scale inset triangle apparently portraying the Messianic 'kernel' of the full design. Egyptologists, however, will point out that the sign for pyramid or tomb was △, and prefer the suggestion that the former sign originally signified some kind of ritual 'cake'. But in this case one is bound to point out that one school of thought at least insists that the word pyramid *itself* derives from a Greek word meaning 'wheaten cake'.

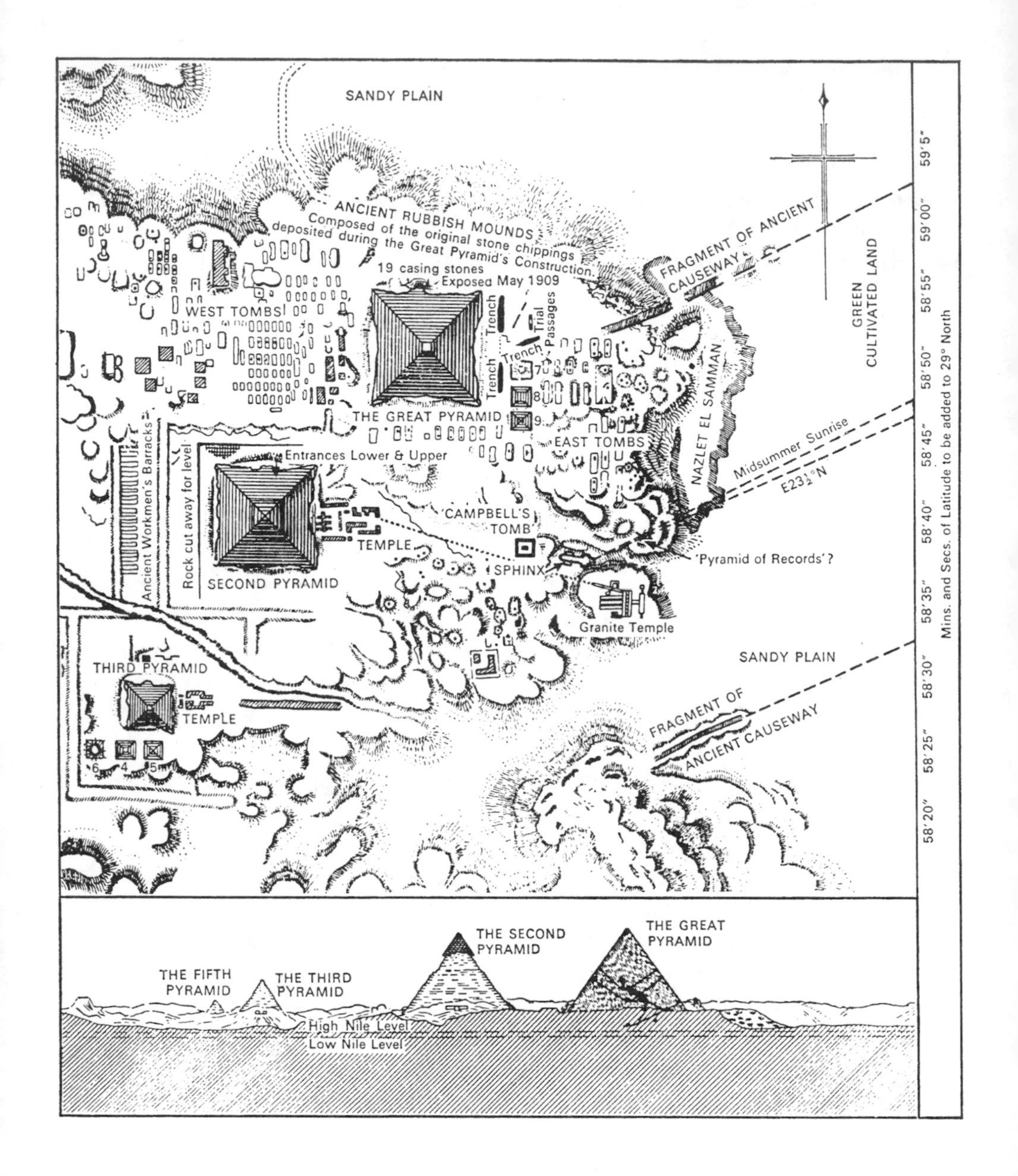

The Giza Plateau, showing the three major pyramids—the probable Three Kings of the Nativity traditions (based on original map from Davidson & Aldersmith). The Sphinx, which faces due east, is here shown slightly out of alignment. Campbell's Tomb, immediately to the west of it, closely fits Herodotus's description of the tomb of the pharaoh Khufu. The dotted line drawn on a bearing of E. $23\frac{1}{2}°$ N. from the Sphinx's paws would appear to be that which (according to Cayce) denotes the site of the yet-to-be-discovered Pyramid of Records (see page 236). It also appears to mark the centreline of an ancient Avenue of the Summer Sunrise.

'mountain' in question was unlikely to come to 'Mahomet', Mahomet must necessarily visit the mountain, if the Kings were to present their ancient gifts to him. *And such a visit is described in the course of this same, strange story*. For we find, in Matthew's version, that as a direct result of the visit by the 'astrologers', the young child is taken post-haste to Egypt. Admittedly Matthew attributes this to an impending massacre of the innocents by Herod, but, on the other hand, it seems virtually certain that the event in question was not among that king's historical crimes, and that this particular aspect of the story has been borrowed by the gospel's author from the Old Testament account of the escape of the infant Moses from the pharaonic massacre of the firstborn (see page 273). At all events, for one reason or another, the child was apparently packed off to Egypt in accordance with at least one biblical prophecy (see 2:15), and the various traditions of his training and studies in that country are thus not entirely without scriptural foundation.

Meanwhile, the author of Luke's account clearly knows nothing of the legendary Kings. Instead he has the child visited by a group of shepherds (chapter 2) who have left their sheep (which they had been watching in the fields) on account of a vision announcing the child's birth. Now, apart from remarking that this account portrays the shepherds as astonishingly irresponsible, and that it would tend to place the birth in the summer[22] (the sheep are never left out in the fields around Bethlehem during the winter), it is interesting to note that there *is* a possible, though tenuous, historical connection between the legendary Kings and the Lucan shepherds. For Egypt had, in the remote past, been ruled for some 150 years by a dynasty of kings known as the Hyksos, or 'shepherd-kings'. They were a long-execrated race of conquerors, apparently of Semitic origin, who had swept in out of Asia (and thus *from the east*), had apparently subdued Egypt without using military force, and were finally driven out by the Egyptians around 1555 B.C. Now the Israelites were *themselves* living in Egypt during this same period, while the Israelite Exodus under Moses (who himself had worked for some time as a shepherd) seems to have taken place within a century or so of the final expulsion of the Hyksos.

One suspects, in other words, that there is more than meets the eye in the apparently unrelated sojourns of the Israelites and of the shepherd-kings in Egypt. Both sets of alien 'invaders' were Semitic; both entered the country peacefully, and from the east; both

[22] The Pyramid's chronograph places it in September.

achieved great power in Egypt; both grew to be hated by the Egyptians; and both either left or were expelled eastwards with precipitate suddenness. Moreover, the periods of their occupation of Egypt appear to overlap, insofar as either the biblical or the ancient Egyptian datings can be trusted. Admittedly the biblical account describes the Israelites as oppressed slaves who fled of their own accord, while the Egyptian records describe the Hyksos as tyrannical masters who were expelled; but then the production of diametrically opposite interpretations of a single set of events is not a propagandist phenomenon limited to the Mosaic era alone.

However, the whole web of historical Israeli-Egyptian interrelationships is a tangled one, and whether or not there may be some single, obscure Messianic connection between the Mosaic Exodus, the mysterious Hyksos, the Pyramids, the Three Kings of Matthew's account and the shepherds of Luke's, only time and future research are likely to reveal.[23] What is certain, however, is that there is far more to the various episodes in question than most readers of the biblical accounts ever suspect.

Meanwhile the identification of the Three Kings with the three major pyramids of Giza is supported by some interesting evidence:

(i) The passage-angle of all three pyramids is similar (diagrams page 4) and all three therefore appear to share in the significance of the Bethlehem-angle.

(ii) Tradition has it that one of the Three Kings was black-skinned. Similarly, unlike the other two pyramids, the Third Pyramid was originally sheathed to at least a quarter of its height with casing-stones of red granite—*which, once weathered, appears black or dull purple.*

(iii) The names attributed to the Three Kings by the Armenian Gospel of the Infancy are Balthazar (from Arabia), Gaspar (from India), and Melkon (from Persia). The names Gaspar and Melkon appear similar enough to Khafra and Menkura (the reputed builders and occupants of the Second and Third Pyramids respectively) to suggest a possible link. The name Balthazar, however, has little obvious connection with the Great Pyramid. It may perhaps be of

[23] The true explanation of the shepherd incident may of course be a much simpler one. The tradition that Miriam and Joseph were engaged in some sort of journey at the time of Jesus' birth is a strong one, and the author of the Protevangelium actually describes the birth as occurring in open country during the course of it. Joseph leaves his wife in the shelter of a nearby cave—a feature also hallowed by long tradition, albeit possibly derived from the nativity-cave of Mithras—and goes to seek help. And it is natural enough to assume that the first help that arrived took the form of a group of local shepherds.

some significance, though, that the word appears to derive from the Hebrew Belteshazzar, meaning 'the lord's leader'. Moreover, this name is also found in the form Belshazzar, a king whose name later Jews could conceivably have associated with the celebrated Great Ziggurat, or step-pyramid, of Babylon, with which the Great Pyramid could have become identified in the racial memory.

(iv) The geographical origins variously attributed by the same gospel to the Three Kings offer an intriguing hint that the knowledge enshrined in the pyramids may represent a distillation of the ancient wisdom gleaned by the Egyptian initiates from as far afield as India, Persia, and 'Arabia'. It may be that, as we postulated earlier, the Messianic Plan attested to by the Great Pyramid is truly *a world-plan*, embracing the essentials not merely of Christianity, Judaism and the Egyptian teachings, but also of the Hindu, Babylonian and Zoroastrian traditions. Alternatively we may see here an allegorical reference to the tradition that Jesus himself delved at first hand into the wisdom of the lands in question.

(v) According to the same gospel, the Three Kings bore with them 'the testament bequeathed by Adam to Seth'. It is difficult to imagine why they would do this, other than because the testament, like the legendary gifts, enshrined a prophecy of things to come—constituting, in fact, the fore-ordained destiny of the new-born Messiah. But since this notion appears to correspond to the message of the Great Pyramid itself, we may once again suspect that the reference is in fact to the pyramids of Giza. Thus we may conclude that the testament in question represents nothing less than the Knowledge handed on by the primeval 'sons of the gods' (Adam) to the ancient Egyptians, as personified by Set (the 'dark aspect' of Horus), and incorporated by them into the Great Pyramid's design—an interesting footnote to our speculations about the destruction of an earlier, advanced civilisation by a primeval cataclysm cognate with the biblical Flood.

Ancient Biblical and Messianic Parallels in Central America

The ancient Mayan and Aztec religions were so full of direct biblical parallels as to convince the earliest Christian missionaries that they had come across some kind of bizarre caricature of Christianity, designed by the Devil himself to undermine their efforts in advance—an unnerving experience which they shared with earlier Christian missionaries to Tibet.

For a start, the Mayan Year One was 3113 B.C., when (on August 12th, to be exact) 'a white, bearded man "descended from the sun" landed on the Gulf of Mexico with a stimulating party of scholars, astronomers, architects, priests and musicians' (the words are Thor Heyerdahl's). This date is of course remarkably close to Year One of the ancient Egyptian Age of Horus—probably 3141 B.C.—which came immediately after an age of destruction apparently cognate with the biblical Flood.

In almost countless respects, the ancient Central American traditions parallel to an uncanny degree those of ancient Egypt. Like the Egyptians, the Maya and their remote precursors organised their society as a theocracy governed by religious beliefs and traditions of considerable sophistication. Like the Egyptians they set out those beliefs in a hieroglyphic script of great complexity; like the Egyptians, they devoted an enormous part of their wealth to the construction and decoration of massive temple-architecture, characterised above all by the shape of the sun-symbolising pyramid, and designed throughout on the basis of symbolic mathematics; like the Egyptians, they seem to have been obsessed by number and numerology, with particular reference to dates and astronomical events.[24] And, like the Egyptians again, they built reed-boats, made clothes of cotton, embalmed their dead, told legends of creations, floods and lordly founders of civilisation, taught that history is governed by astronomical cycles, and pronounced sage, symbolic discourses upon the evolutionary destiny of the human soul. Such are the mutual resemblances, indeed, as to convince many a scholar that the Central American civilisations owed their very origin to the efforts of Egyptian colonists—a fascinating, if simplistic, proposition to which Heyerdahl's celebrated 'Ra' expeditions were of course both dedicated.

In the religious sphere in particular, Central American tradition (like that of Egypt, Babylon, Greece, Syria, Persia, India and other parts of the ancient world) was already 'counterfeiting' that of Christianity in many respects long before the birth of Jesus of Nazareth (as J. G. Frazer, of *Golden Bough* fame, was not slow to realise).

Thus, among the 'oddities', we find:

[24] In particular they had calculated to a fine degree of accuracy the lengths both of the Solar Tropical *and of the Venusian* years (compare note 13, page 20), while their calendar revolved around a 'week' of thirteen days, each of which had its own symbolic significance (compare the Great Pyramid's arithmetical code, and see page 307).

(i) The cross, or quincunx, had long been a well-known symbol, indicating man's existence at the intersection of the (horizontal) physical plane and the (vertical) plane of eternity.

(ii) Rebirth was a familiar notion, and was associated, as in Egypt, with the direction 'east'.

(iii) The Nahua cosmology included three heavens or paradises—the lowest (Tlalocan) was the land of water and mist, from which the soul returned and reincarnated after some four years; the next (Tlillan-Tlapallan) was a *nirvana*-like heaven of non-attachment to the body, attained only by initiates; and the highest (Tonatiuhican) was the Home of the Sun, the abode of eternal happiness, reached only by the fully enlightened. At the lower end of the scale, there was also a kind of limbo known as Mictlan, where the souls of the condemned would endure a rather colourless, though quite painless, eternity.

(iv) Creation was described by the Maya in decidedly biblical terms: 'Where there was neither heaven nor earth sounded the first word of God. And He unloosed Himself from His stone, and declared His divinity. And all the vastness of eternity shuddered. And His word was a measure of grace, and He broke and pierced the backbone of the mountains. Who was born there? Who? Father,

Yiacatecuhtli, Aztec Pre-Conquest Lord of the Vanguard. His name presumably signifies 'Lord who leads the way', and he was the god of the wandering vendors and missionaries known as the *pochtecas*.

Thou knowest: He who was tender in Heaven came into Being' (from the *Chilam Balam of Chumayel*, quoted in *Mythology of the Americas*, by Burland, Nicholson and Osborne).

(v) The ages were seen by the Maya, as by Buddhists, as having a cyclic form, destined eventually to reach its appointed completion: 'All moons, all years, all days, all winds, reach their completion and pass away. So does all blood reach its place of quiet, as it reaches its power and its throne. Measured was the time in which they could praise the splendour of the Trinity. Measured was the time in which they could know the Sun's benevolence. Measured was the time in which the grid of the stars would look down upon them; and through it, keeping watch over their safety, the gods trapped within the stars would contemplate them.' (op. cit.)

(vi) Both baptism and eucharist (the symbolic eating of the god's body and blood) were commonplace notions long before the known arrival of European culture in Central and South America.

(vii) Man's spiritual pilgrimage was represented by the twenty Mayan days, or 'footprints of the God', reminiscent of the symbolic 'creation week' of Genesis:

'There are twenty steps up and down the ladder, beginning with Imix, from Im, the womb. The first day starts the child off on his journey through life. On the second, Ik, spirit or breath, is bestowed upon him when he is still within the womb. On the third, Akbal, he is born of water. On the fourth, Kan, he begins to know evil; and on the fifth, Chicchan, he gathers together all the experience of his life. On the sixth, Cimi, he dies. On the seventh, Man-Ik [from Manzal-Ik, "Pass through the spirit"] he overcomes death.

'Now he must plunge into the lower regions; he must struggle to overcome the material state. This is the eighth day, Lamat, the sign of Venus. On the ninth day, Muluc, he reaps the reward of his effort; and on the tenth, Oc, he enters fully into the uttermost depths of matter in order, on the eleventh day, Chuen, to burn without flame. [In other words he suffers the greatest possible agony.] On the twelfth day, Eb, he begins to climb the ladder, a long process which continues through the thirteenth day, Ben [which represents growing maize], until on the fourteenth, Ix [the Jaguar god], he is washed entirely clean. This allows him on the fifteenth day, Men, to become perfect, but still he does not possess the full light of consciousness, which comes to him on the sixteenth day, Cib. On the seventeenth day, C'haban, he shakes off the last traces of ash clinging to him from the material world. ["Ash" is the word specifically used, and it suggests that he has been purged by fire.] On the eighteenth day,

Day-number	*Day-name*	*Symbolism*	*Number in 'soul-cycle'*	*Possible Pyramidal interpretation*
8	Lamat	The morning star . . .	1	The Divine . . .
9	Muluc	. . . reaps its reward . . .	2	. . . produces . . .
10	Oc }	and enters the depths }	3	. . . the utterly
11	Chuen	of matter to suffer agony.[25]	4	. . . physical.
12	Eb	The soul . . .	5	The initiate . . .
13	Ben	. . . starts to climb the ladder again . . .	6	. . . is prepared . . .
14	Ix	. . . and is purified.	7	. . . for spiritual perfection.
15	Men	Having become perfect . . .	8	Rebirth/ resurrection. . .
16	Cib	. . . it next gains the light of full consciousness.	9	. . . leads to the utter perfection . . .
17	C'haban	Thus it survives the ordeal and attains freedom from the material world.	10	. . . of the Millennium.
18	Edznab	Now finally perfected . . .	11	With the achievement . . .
19	Cauac	. . . it becomes godlike . . .	12	. . . of (true) manhood . . .
20	Ahau	. . . and achieves union once more with the Divine.	13	. . . man once more becomes Pure Soul.

Edznab, he is made perfect. By the nineteenth day, Cauac, his divine nature is manifest. On the twentieth and last day, Ahau [god], he becomes one with the divinity.' (op. cit.)

Of the above, the first seven days clearly refer to the cycle of a single human life. The remaining thirteen seem to represent a kind of 'soul-ladder'—*their symbolism remarkably close to that of the numbers 1 to 13 in the Great Pyramid's number-code.* For the pyramidal series 1 to 13 could likewise be 'read' as a kind of soul-ladder symbolic of the process of rebirth leading to eventual union with the Divine. Consider, for example, the table above, which sets out the two symbolisms side by side for comparison.

(viii) One Mayan tale tells of a god called, surprisingly, Votan, 'who declared himself to be a serpent' (i.e., presumably a possessor of Knowledge). From some unknown origin he was ordered by the gods to go to America to found a culture. So he departed from his home, called Valum Chivim and unidentified, and by way of the 'dwelling of the thirteen snakes' he arrived at Valum Votan. From there he travelled up the Usumacinta river and founded Palenque. Afterwards he made several visits to his native home, on one of

[25] Compare the ancient legend of Lucifer who, like the soul of man, was 'cast down from heaven' and is symbolised by the Morning Star (Isa. 14:12).

which he came upon a tower which was originally planned to reach the heavens but which was destroyed because of a 'confusion of tongues' among its architects. Votan was, however, allowed to use a subterranean passage in order to reach 'the rock of heaven' (op. cit.).[26]

The foregoing seems like an odd mixture of Norse legend, Jewish mythology and Pyramidal symbolism. It is worth noting particularly that the Norse Wodan, or Odin, was worshipped *especially by noble families*—who claimed direct descent from him—and was held to be the patron of culture, inventor of the runes, god of wisdom, poetry, magic and prophecy. He roamed the world, staff in hand, disguised as a one-eyed old man, and his eternal home was called Valhalla (compare Valum Chivim). And it is perhaps of particular significance that the name Wodan/Odin—he of the 'winged hat'—is traditionally associated (partly because of this alleged sartorial distinction) with Mercury the winged messenger, and thus also with the Greek Hermes. Hence, of course, the naming of the fourth day of the week after Wodan/Odin in English (Wednesday), but after Mercury in French (*mercredi*). Hermes, meanwhile, is apparently to be identified with the Egyptian Thoth or Tehuti, who, like Odin and the Mayan Votan, was held by his devotees to have been one of the founders of true civilisation. When one adds to this remarkable chain of circumstances the fact that the Egyptian *Tehuti* appears to be the same word as the Aztec *tecuhtli* (meaning 'grandfather' or 'lord', as in Yiacatecuhtli on page 305), it starts to become apparent that all these various legends of ancient founding-fathers of civilisation may indeed derive from a single common source.

(ix) The central figure of Central American myth was the semi-historical, semi-mythical Quetzalcoatl, or Kukulcan, the 'plumed serpent'. He was born (after a heavenly 'annunciation') of the virgin 'queen of heaven', and became a great lawgiver, civiliser and inventor of the calendar. He was an innovator in arts and crafts, a preacher

[26] The biblical story of the Tower of Babel, apparently referred to here, may itself represent a memory of the Great Pyramid, transferred to the Babylonian setting—where the Great Ziggurat offered itself as an obvious substitute-symbol to the captive Israelites. For the non-completion of the Pyramid, and the consequent 'separation' of its four faces, does seem to be symbolic of human division (see p. 198), while its eventual completion would seem, as the story suggests, to symbolise the attainment of 'heaven'. Meanwhile Votan's use of a subterranean passage to reach the rock of heaven would certainly square with the design of the original Pyramid, in which access to the granite house of the King's Chamber Complex—specifically symbolic, in terms of the reconstructed code, of the 'rock of heaven'—could be gained only via the subterranean entrance to the Well-Shaft.

of love and compassion, and was held to have been responsible for the introduction of maize as an item of diet.[27] He was 'tall, robust, broad of brow, with large eyes and a fair beard . . . He wore a conical ocelot-skin cap . . . His jacket was of cotton, and on his feet he wore anklets, rattles and foam sandals' (op. cit.).

Rather like the Hindu Krishna, Quetzalcoatl is a figure of paradoxes. 'In so far as he was man, he was of the breed of heroes from whom myths are created. In so far as he is symbol, he presides like the wind over all space. He is the soul taking wing to heaven, and he is matter descending to earth as the crawling snake; he is virtue rising, and he is the blind force pulling man down; he is waking and dream, angel and demon . . . He represents daylight and also, when he journeys to the underworld, night. He is love with its transmuting power, and carnal desire that wears chains . . . In Quetzalcoatl's hand is a staff sprouting life, and he also carries the spear of the morning star' (op. cit.). The staff sprouting life seems suspiciously reminiscent of the Egyptian *ankh*.

In other respects he resembles the Egyptian Horus, one of his first tasks being to recover the dead body of his father (the sun) from where enemies have buried it in the sand; again, his 'twin' is the dog Xolotl, just as Horus' is the dog Anubis.

Having succumbed, Adam-like, to human temptation, Quetzalcoatl (it is said in one account) dismissed all his followers and ordered 'that a stone casket should be built in which he was to lie four days and nights in strict penance. This done, the pilgrims marched to the seashore; and there Quetzalcoatl dressed himself in his feathered robes and his turquoise mask. Building a funeral pyre he threw himself upon it and was consumed in flames. The ashes rose into the sky as a flock of birds, bearing his heart which was to become the planet Venus' (op. cit.) — i.e. the morning star.

In this version of the story, Quetzalcoatl is also associated with the sun, which daily 'dies' and is 'resurrected'. Similarly the Egyptian Osiris was daily reincarnated as Re or Ra, the sun, 'at the fifth

[27] The same functions seem to be attributed almost universally to the ancient 'gods'. The Sumerian gods, for example, were traditionally responsible for the development of writing, the making of metals and the cultivation of barley. The Egyptian Osiris and Thoth between them exercised a like function, even giving directions for the making of a *fermented drink* from barley, as well as bequeathing to mankind the celebrated Hermetic writings. In short, there seems to be good reason for linking the appearance of the 'gods' with the foundation of the first agricultural communities and with the end of the cultural 'stone age'—a topic on which von Däniken has much to say.

hour, when the god's boat slid over *a pyramidal shape* which protected the divine egg from which the sun emerged' (from *Tutankhamen*, by C. Desroches-Noblecourt).[28] It was thus through association with Osiris that the dead pharaohs expected to be resurrected to life in a new, spiritual body, just as most Christians expect to be similarly 'resurrected' by association with their own Divine Saviour's resurrection.

Quetzalcoatl's disciples, or *pochtecas*, spread his teachings through South America, their mission described in words reminiscent of Jesus' to his disciples in Matthew 10: 'You are to wander, entering and departing from strange villages . . . Perhaps you will achieve nothing anywhere. It may be that your merchandise and your items of trade find no favour in any place . . . Do not turn back, keep a firm step . . . Something you will achieve . . . Something the Lord of the Universe will assign to you . . .' (*Chilam Balam*, op. cit.) The *pochtecas* became powerful figures. They were 'itinerant vendors who formed a guild or brotherhood with the material purpose of trade but with a central set of ethical principles that were more important to them than money-making.' Indeed, 'If they accumulated too much wealth, they organised religious banquets and quickly got rid of it.' (op. cit.)

In another version of the Quetzalcoatl story, the god-king successfully overcomes a series of sensual and sexual temptations, and when he is eventually forsaken by most of his followers, he and his last remaining companions set out on a last pilgrimage to Tlapallan 'in order to learn.' After building a bridge to cross a river he is waylaid by devils, who refuse to let him pass until he has surrendered all his skills, his knowledge and his jewels—which he duly does. The story, clearly symbolic of departure from the physical world for the spiritual planes, continues as he climbs two symbolic volcanoes and builds a house in Mictlan, the Land of the Dead ('he descended into hell', in other words). And finally he sets off to an unknown destination on 'a raft of serpents' (an underworld notion likewise frequently met with in Egyptian tomb-paintings), whence he is expected to return at some date in the future. According to one set of calculations, his return was expected in the very year which in fact took the bearded Cortez to America, with the calamitous

[28] The presence, in Tutankhamun's tomb, of a 'sacred goose' perhaps suggests a link with the ancient notion of 'the goose that laid the golden egg', while the egg itself is the almost universal symbol of the ultimate origin. (Von Däniken even sees it as a memory of the 'spaceship' from which the primeval, civilising 'gods' originally emerged.)

results for Central American civilisation which are now a matter of history.

(x) Very much after the style of the biblical Israelites, the Aztecs, during part of their history starting in about A.D. 1160, underwent a period of wandering in search of their Promised Land—Anahuac, the 'place in the midst of the circle', the source and inspiration of their being. In this, they were led by two warrior-brothers (reminiscent of the biblical Moses and Aaron) named Gagavitz and Zactecauh.

On one occasion these two, on reaching the sea, fell asleep and were overcome by enemies. But then they produced a red staff which they had brought from the holy land of Tulan, and drove it into the sea, whereupon the waters parted and the pilgrims walked across to the further shore. The parallel with the Israelites' Red Sea crossing is extraordinary.

Elsewhere one of the brothers is credited with the ability to bring down flying birds at a distance using a blowpipe but no arrow—exactly the sort of description which one would expect a primitive people to give to a modern gun. At a later stage, after the disappearance of Zactecauh following an accident, the tribe arrived for the second time at a white volcano, which Gagavitz then ascended with one Zakitzunun. Gagavitz then spent a lonely vigil on the mountain, to the increasing dismay of his followers, only to reappear from inside the mountain so that the warriors cried out: 'Truly his power, his knowledge, his glory and majesty are terrible. He died, yet he has returned . . . When the heart of the mountain is opened the fire separates from the stone . . .' And at this point we seem to have a direct echo of the so-called Emerald Tablet of Hermes Trismegistus: 'Throw it upon earth, and earth will separate from fire. The impalpable separated from the palpable. Through wisdom it rises slowly from the world to heaven . . .' (Tr. Idries Shah).

Throughout one is reminded, of course, of the Jewish Exodus, led by the brothers Moses and Aaron, even to the magic staff, the crossing of the sea dry-shod and Moses' ascent of the Holy Mountain (indeed, the Edgar Cayce readings seem to suggest a *direct Hebrew link*—remarkable as that idea may seem). Moreover, the fact that the Aztecs' mountain was white reminds one strongly of the Great Pyramid of Giza, as it was left by the builders. And remembering that the word pyramid may well be connected with the Greek root *pyr-* meaning fire, one may be forgiven for suspecting a link between the Pyramid, with its eventual shining, sunlike capstone, and the fire-crowned volcano of the Aztecs' holy mountain, from which the Messianic 'Gagavitz' is destined one day to emerge in glory.

(xi) Pierre Honoré, in his *In Quest of the White God*, claims to discover equally striking links between the ancient Central and South American civilisations and those of the Middle East. In a detailed exposé he traces the chief characteristics of Aztec civilisation back through Toltecs and Maya to the Olmecs and the La Venta civilisation of around the beginning of our own era. The Incas' roots he similarly traces back through Tiahuanaco and the Chimu to the Chavin civilisation of around 700 B.C..

Throughout, Honoré discerns constant artistic and cultural links not only with China and Indo-China, but also with the ancient Middle East. His most startling revelation, however, is to the effect that fifteen of the ancient Mayan glyphs are (as he illustrates) almost identical to the Cretan script known as Linear A. He therefore deduces (not unreasonably) that the Central and Southern American peoples or their predecessors have come at various times in the past under the influence of Chinese and Indo-Chinese, and more especially of the ancient Minoan and/or Phoenician cultures.

While on the subject of the Mayan tongue, an arresting footnote is provided by the Guatemalan monk Antonio Batres Jaurequi. In his *History of Central America* he makes the challenging claim that the last words of Jesus of Nazareth as reported by Matthew ('Eli, Eli, lama sabachthani') are in fact the Mayan for 'I faint, I faint, and my face is hid in darkness' ('Hele, hele, lamah sabac ta ni'). Churchward, for his part, claims a similar link with the ancient 'language of Mu', which he believes was known as a kind of sacred *lingua franca* to initiates of the ancient mysteries. At all events, Jaurequi's claim would seem to merit the attention of Mayan specialists.

The Atlantean Tradition

At this point a number of linguistic aspects of the Central American traditions, together with the obvious ritual and symbolic similarities between Central American, Egyptian and Hebrew religious ideas, lead us almost inevitably to consider the possibility that all of these phenomena may have had some common origin. Could, for example, the legend of the lost Atlantis (as outlined by Plato in his *Timaeus* and *Critias*) somehow provide a possible explanation and common origin for the whole complex of knowledge which we are presently considering? Plato, it should be remembered, claimed that the knowledge retailed by him in these extracts was obtained from an

Egyptian priest at Saïs—where the Saïte recension of the *Egyptian Book of the Dead*, complete with its apparent identifications of the passages and chambers of the Great Pyramid, also first saw the light of day. Moreover, it is in these same works that Plato quotes the priest concerned as expounding a cyclic view of history involving periodic world cataclysms (see page 14, note 8)—a view apparently also shared by the Great Pyramid's designer. Could the original 'sons of the gods' and their 'secret knowledge' (compare the Mayan caste of Almenhenob, or 'sons of the true men') have Atlantean connections?

The proposition may at first sight seem far-fetched. Certainly the subject of Atlantis has attracted more than its fair share of documentary and archaeological debate down the years. What is often overlooked, however, is the *linguistic* evidence on the question, of which I shall now list what appear to me to be the salient features, together with a few speculations of my own:

(i) In the name Quetzalcoatl, the last syllable—*atl*—is the Nahua word for water. A related word, *atlatl*, means arrow.

(ii) The god Atlaua is thus known as Master of the Waters—a name oddly reminiscent of the function of the early pharaohs. He is recorded as singing: 'I leave my sandals behind. I leave my sandals and helmet . . . I cast off my arrows, even my reed arrows. I boast that they cannot break. Clad as a priest, I take the arrow in my hand. Even now I shall rise and come forth like the quetzal bird.' (op. cit.)

(iii) If the name Atlaua had reached the ears of Plato in connection with his account of Atlantis, he would doubtless have transcribed it into Greek as Atlas—and he specifically admits in that account to having hellenised the names involved. Atlas is named by Plato as the first king of Atlantis, a land of intense irrigation.

(iv) The consonantal combination *tl* is exceedingly rare in most European languages, including Greek, as part of a root-morpheme (and especially initially in a stressed syllable), but among the rare exceptions to this are the words Atlas, Atlantic and Atlantis. In Nahua and related Central American languages, however, the *tl* combination is exceedingly frequent—perplexingly so for European tongues. This may suggest that the source of the word Atlantis may have been an area in close linguistic contact with Central and South America.

(v) Indeed, the only locality which would completely fit Plato's description of Atlantis without a great deal of 'editing' would be an island (now submerged) some 300 miles long and 200 wide, partly encircled by mountains (especially on the north) and in a geologically unstable part of the North Atlantic now marked by a large area of shallow muddy water. Other islands in the vicinity provided 'stepping stones', as it were, to the 'real continent' which lay beyond.

Only the West Indies or Caribbean area can completely fulfil all these conditions, being close to the muddy swamplands of Florida and the Mississippi delta area. Research in the area, prompted by Edgar Cayce's predictions, has now revealed enormous submerged stone structures off Bimini, near Florida, one of which is in the shape of a gigantic *arrow* (cf. *atlatl*).

(vi) The Votan/Wodan parallel quoted earlier suggests that the Germanic and Central American stories could conceivably have had a common origin—in a lordly 'wanderer' who visited, either in person or in legend, both sides of the Atlantic. One wonders, in this case, about the precise locality of the original Valhalla and Valum Chivim.

(vii) Plato's account of the vast irrigation schemes of the doomed Atlantis once again reminds one of Atlaua, the Master of the Waters, and of the fact that both the Egyptian and the Olmec civilisation of Central America were founded originally in vast swamp-areas where irrigation was therefore of first importance. Moreover the Aztecs' city of Tenohtitlan (the present Mexico City) is known to have been built on a lake-island surrounded by concentric canals, specifically because this arrangement reflected the original topography of Aztlan, the homeland of their ancient traditions. *Yet this same arrangement is found in Plato's description of the Atlanteans' capital city.*

Meanwhile, the Aztecs' non-use of the wheel (although it was known to them, as surviving wheeled toys suggest) may thus be connected with the availability of water transport. On the other hand it could be regarded as evidence that their ancestors may have had access to means of transport superior to the wheeled surface-vehicle (*something* must presumably have used the magnificent roads they constructed). The possibilities are intriguing, to say the least.

(viii) A number of Mexican place-names (e.g., Mazatlán, Miahuatlán) actually contain the word *Atlán*, while the Aztecs themselves, as we saw above, always insisted that they had come from an island home called Aztlan. Both words could conceivably represent a direct survival of the name Atlantis. Meanwhile the meaning of the word *Maya* itself would appear to be quite consistent with the possibility that the Maya's remote ancestors may have been a small group of survivors of the supposed Atlantean cataclysm, as Edgar

Cayce seems to have hinted (see page 269). For in Mayan the word *Maya* means 'not many'—in other words, 'the few'.

(ix) Aztlan, the 'land of the sun' from which the primeval ancestors of the Nahua and Maya are said to have come, is also variously referred to as Tula, Tollan, or Tonalan (various Mexican towns and villages are so named to this day). These names are all oddly reminiscent of the legendary North Atlantic island of Thule, first mentioned by Pytheas the Greek in the fourth century B.C., and later taken up by Virgil, who referred to it as Ultima Thule (why, one asks oneself, 'ultima', if not to indicate that the island represented the last piece of Thule remaining?). Meanwhile, bearing in mind the constant interconnection between *atl* (water) and *atlatl* (arrow), it is an even odder fact that known Old High German contained only one remotely similar word—namely *tulli*, which meant, of all things, 'arrow-head'.

(x) Sticking for a moment, out of sheer curiosity, to the ancient Germanic context, it is interesting to note that the nearest Germanic word to *atl* is *adel* (formerly *adal*), meaning noble, while the word *Adler* (originally meaning noble one) is the word for eagle—which, after the quetzal, was the most sacred and noble bird in the Nahua/Maya pantheon. (*Atzel*—a word somewhat reminiscent of quetzal—was formerly the German word for magpie.)

(xi) To sum up, then, if a lost island called, perhaps, Atalán or Aztlán or Atalánti had produced an advanced civilisation whose ideas were subsequently taken by missionaries and refugees both to Europe and to Central America, then it is by no means unlikely linguistically that the words Atlantis, Tollan (from a presumed shortened form such as Talan), Tula and Thule could all have derived from its name.

Moreover, its associated forms *atl*, *atlatl* and *Atlaua*—linked semantically with both water and arrows, as well as possibly with a race of lordly invaders—could conceivably have had direct etymological connections with such European words as *tulli*, *Atlas*, *adel* (= noble) and its derivative *Adler*. And the fact that the Norse Odin-legend was linked specifically with families of noble origin could also be significant in this context.

Indeed, to take the above free speculation to its extreme point, it is not difficult to envisage the possibility that if a Germanic tribe of nomads had come, at some remote point in the past, under the influence of strangers bearing advanced techniques and referring to their western homeland as Atlánti, they might have adopted the word themselves for referring to the (for them) relatively novel notion of 'fixed homeland', or 'land in the west'. In this case the known Germanic tendency to associate the stressed syllable of a word (here *-tlan*) with its first one (as a result of which, for example, the Greek *episkopos* became the English 'bishop') would almost certainly have produced the form *lant*. And this indeed is known to have been the Old High German form of the modern English word 'land', which in its French form (*Landes*) even retains still its feeling of 'westernness'—being the name of a region in the far west of France, bordering on the Atlantic ocean. In fact, the region in question stretches right up to the Pyrenean foothills—an area specifically described by Cayce *as having been settled by Atlantean refugees*. Meanwhile the corresponding Celtic treatment of the same word produced a word similar to the Welsh word *llan*—a word today means 'church', but originally meant 'enclosure'; and this is of course a concept characteristic of the original founding of any fixed settlement on virgin land. The proposed derivation may seem improbable, and yet no other convincing source has so far been suggested for the two modern words in question.

(xii) *Some* of the apparent linguistic similarities listed above—to which may be added such oddities as the apparent similarity between the words Noah and Nahua, between Sidon (the name of Noah's great-grandson in the biblical account) and Poseidon (founder of Atlantis in Plato's account), or between the formerly Nile-dwelling Ashanti and our postulated *Atlánti*—are no doubt merely coincidental. On the other hand it seems improbable that all of them are. Adding this evidence to the well-known similarities between the ancient South American and Egyptian cultures (not forgetting the considerable influence of the latter on Jewish thought), one suspects that, as usual, it is less likely that the one effect gave rise to the other than that both arose from a common source. The ancient Egyptian tradition that the Egyptian founding fathers came from somewhere in the west, and the occult tradition that Osiris himself was, like Thoth, a historical figure—nothing less, in fact than a former Atlantean priest-reformer, crowned king and subsequently deified—merely add fuel to an already smouldering fire of probability. And a 'lost' Atlantis, producer of a highly advanced culture and technology, but long since vanished almost without trace, would seem to be as good a hypothetical common source (even ignoring Cayce's remarkable claims) as any other so far suggested.

* * *

It must be pointed out, however, that the conclusions suggested by the foregoing linguistic evidence in no way accord with the popular theory—often referred to as though it were established fact—that Plato's description of Atlantis was really based on Crete and the Minoan island of Thera or Santorini in the eastern Mediterranean, and on the cataclysm which apparently overwhelmed both in around 1600 B.C. The idea is admittedly an attractive one, originally put forward by A. G. Galanopoulos (of the Athens Seismological Institute) in his splendidly illustrated book on the subject, *Atlantis: The Truth behind the Legend*. None the less, if it is correct, then Plato was completely wrong about the site, the date and the size of Atlantis—which presumably can only mean that he was writing about a different civilisation entirely.

Indeed, on a careful reading of Galanopoulos's book, it becomes obvious that his conclusions actually invalidate his own reasoning. If, as Galanopoulos contends (op. cit., page 37), Plato's figures for the overall dimensions of Atlantis need to be multiplied by ten, and the real Atlantis was therefore too big to be identified with any known sunken landmass in the Atlantic, then clearly it is even less reasonable to place it (as Galanopoulos does) in the eastern Mediterranean. If, on the other hand, Plato's figures are ten times *too big*—as Galanopoulos later suggests (op. cit., page 133)—then the latter's initial argument is immediately invalidated, and we are back to square one. Meanwhile he seems to imply, *en passant* (op. cit., page 135), that the ancients were incapable of distinguishing between floating pumice and shallow mud, and also implies (op. cit., page 30) that Plato's typically 'bronze age' description proves that the civilisation described was a bronze age one—apparently oblivious of the fact that a bronze age description of *any* civilisation (including our own) must necessarily be expressed in bronze age terms.

Finally, Galanopoulos falls into several further age-old traps. He assumes (op. cit., page 21) that the term 'bronze age' means a period of historical time, rather than a stage of technical development (there are, for example, several 'stone age' societies still in existence today); he is prepared to dismiss Plato's datings because they do not accord with such dubious theories (viz. that all 'bronze age' societies flourished between certain fixed dates); he assumes, on the basis of current theory, that the ancients (in this case Plato) were too 'primitive' to be capable of meaning what they so clearly said; and he misrepresents through paraphrase the clear and categorical statements of the writer under review. Plato *does not*, for example, say anywhere in his *Timaeus* and *Critias* that Atlantis disappeared in a day and a night (compare Plato's actual text as translated on page 179 of the book), *cannot* have used the term 'the Pillars of Heracles' (op. cit., page 97) to refer to any part of the Peloponnese (since, in his text, Plato places Libya, Egypt and Tuscany *within* them), and *does not* (as is claimed on page 38 of the book) describe Atlantis in terms of two separate islands. Nor does Plato anywhere suggest, as Galanopoulos assumes, that the Athenian army which allegedly defeated the Atlanteans was necessarily a Greek-speaking one. Galanopoulos's account, in short, is exasperatingly misleading in almost every respect.

In any case, however, his theories seem to be in the process of being overtaken by actual research in the area of Bimini in the Bahamas. As reported in *The Observer* of 17th December 1971, and as described at greater length in Charles Berlitz's *Mysteries from Forgotten Worlds*, this research is being carried out by Dr. J. Manson Valentine, of the Miami Museum of Science, in cooperation with the Marine Archaeology Research Society. The latter's work dates from 1966, while Dr. Valentine's discovery, just north-west of Bimini, of a great double-causeway of huge blocks of 'foreign' limestone weighing up to 40 tons, lying in 16 feet of water, dates from 1968. The remains give the impression of forming part of a vast harbour of Phoenician type.

Curiously enough, 1968 was the year for which Edgar Cayce foretold the reappearance of parts of Atlantis in the area in question, during a trance-session in 1933. The various finds are currently under investigation by the School of Marine and Atmospheric Sciences of the University of Miami, in cooperation with the National Geographic Society.

Meanwhile the celebrated Commandant Jacques Cousteau's underwater explorations in the area of the Bahama reefs have revealed a series of remarkable submarine caverns containing enormous stalactites and stalagmites—proof that the caverns were once above water and have since sunk some hundreds of feet relative to sea-level, much as Atlantis itself is said to have done. Moreover, the unusual structure of the older stalactites shows clearly that the whole area underwent a gigantic geological upheaval in around 10,000 B.C. (almost exactly the date assigned by Plato to the final disappearance of Atlantis), as a result of which the local earth's crust became tilted by some fifteen degrees. Indeed, as a final curiosity, at least one of the caverns concerned has an almost spherical shape (the roof having now collapsed)—a phenomenon which is normally associated either with volcanic activity or with man-made underground explosions.

However, perhaps the most promising, not to say startling, revelations on the question of Atlantis come from a relatively unexpected quarter—Louis Charpentier's research into the origins of the Basque people, as reported in his book *Le Mystère Basque*. It is Charpentier's thesis—and a convincingly argued one—that the Basques represent an almost pure survival of Cro-Magnon man; that the ancient distribution of Cro-Magnon man correlates strongly with surviving dolmenic remains; and that it also correlates strongly with high percentages of blood-group 'O' in today's population. The Basques, it will be recalled, not only speak a language totally unrelated to any other European language, but also exhibit a uniquely high percentage of the Rhesus-negative gene and an unusually high percentage of 'O'-group blood (up to 75%). Apart from Iceland, the north-western British Isles and the Cotentin peninsula, only Crete, Sardinia and a small area of Tunisia around what used to be ancient Carthage exhibit comparable percentages of 'O'-group blood. In fact the areas where this group is most prevalent may be summed up as comprising the more mountainous Atlantic coasts of Europe and Africa and the shores and islands of the Mediterranean basin—within which its distribution corresponds to an uncanny degree with the areas listed by Plato as having been specifically colonised *by the Atlanteans.*

If this possible Cro-Magnon/Atlantean association is a valid one (and Charpentier adduces a great deal of evidence to suggest that it may be), then we have immediately to face two important facts. First, the percentage of 'O'-group blood has been found to be as high as 94% among exhumed mummies in the Canary Islands, and *as high as 100%* among various isolated Central and South American native populations—which would suggest a high probability that the Cro-Magnons/Atlanteans had a *western Atlantic* origin, if Charpentier's blood-group thesis is correct. This in turn would of course contradict the traditional (and probably biblically influenced) theory of the diffusion of culture throughout Europe from the Middle East—a theory which has in any case already started to become discredited in respect of the later megalith-culture, since its remains on the western seaboard of Europe often turn out to be older than their Middle Eastern counterparts,[29] as would be expected if our new theory is correct.

[29] Compare the researches of Dr. Colin Renfrew of Cambridge University.

Second, the main movement of Cro-Magnon man and his new technologies into Europe can be dated archaeologically as having taken place between about 13,000 and 8,000 B.C.—a period which comfortably encompasses Plato's controversial dating for the destruction of Atlantis (around 10,000 B.C.). It may well have been none other than the survivors of the Atlantean cataclysm, in short, who were responsible for the cave-art of Lascaux and Altamira, the dolmen-culture that was to lead to the later megaliths, and conceivably also the domestication of those very animals that are depicted in their major cave-paintings (notably cattle, bison, deer, reindeer, goat, sheep and horse). We may perhaps add to these achievements the original hybridisation of the wild plants which produced such crops as domestic wheat in Eurasia and maize in America— developments which are known to have taken place at about this time—citing in evidence the Mayan and Aztec traditions that the originators of maize were bearded fair- (or red-) skinned 'gods', the founders of Central American civilisation who appeared *from the east* at the dawn of time. In which case we should remember not only that the word 'Phoenician' is based directly on the Greek for 'red', but that the Minoans were always called The Red Ones by the Egyptians, while the ancient pharaohs themselves (but never their queens) were always depicted with red skins. An echo, perhaps, of the 'sons of the gods' in the biblical flood-account, who married the daughters of men when 'mankind began to increase *and spread all over the earth*'? (Gen. 6:1–3) Perhaps, in other words, Phoenicians (and thus the 'O'-group Carthaginians) and Minoans (the 'O'-group inhabitants of Crete) had, like the original Egyptians, an Atlantean origin[30]—a fact which would explain why the Egyptians always insisted that the land of the ancestors (or land of the dead) lay somewhere in the far west.

Meanwhile a western Atlantic origin for the Cro-Magnons/Atlanteans naturally presupposes a maritime technology of some sophistication—a technology reflected, perhaps, in the flood-epics of Noah and Gilgamesh. To such a technology we could conceivably trace back the later seafaring traditions of Phoenicians, Minoans, Egyptians and Carthaginians, not to mention those of ancient Tartessos (the biblical Tarshish) and the Basques themselves. And in this case would it be surprising to find 'Cretan' signs and 'Egyptian'

[30] The main Phoenician religious centre was the port of *Sidon*—a name actually attributed by the biblical account to a great-grandson of Noah, whose legend bears marked similarities to that of Atlantis. Compare also, once again, the name of Poseidon, described by Plato as the founder of Atlantis, and always associated by the Greeks with water and earthquakes.

cultural traditions among the Maya, a 'Phoenician' harbour off the coast of Bimini, a 'Phoenician' lighthouse (the Tower of Heracles) and a persistent Noah-legend (Christianised into that of St. James) in the Spanish province of Galicia, or a high percentage of 'O'-group blood in the Basque area, in Crete and around Carthage—not to mention an 'island' of it in Ethiopia, the homeland claimed by the first-known Egyptian pharaohs?[31]

Even ignoring, then, Edgar Cayce's persistent claims that large groups of (red-skinned) refugees from Atlantis settled, around 10,000 B.C., in Egypt, Central America (notably Yucatan) and the Pyrenees, there would appear to be a strong case for thorough ethnic and linguistic research into the possibility of an ancient link between the Basques on the one hand and the ancient ancestors of the Aztecs and the Maya on the other. That research may throw some interesting light on the following chain of supposedly unconnected circumstances:

(i) 10,000 B.C. marks a period of rapidly rising world sea-levels in the post-glacial era (Fairbridge: see pages 13 and 268–9).

(ii) 10,000 B.C. falls shortly after the earth's last recorded magnetic reversal, which could conceivably have been associated with a sudden, major shift of the earth's axis (Möner/Lanser/Hospers: see page 236).

(iii) In around 10,000 B.C. the area of the earth's crust around the Bahamas underwent a gigantic upheaval (Cousteau: see page 318).

(iv) Plato dates the disappearance of Atlantis beneath the sea at around 10,000 B.C., as did the sleeping Edgar Cayce (see pages 269 and 313).

(v) Between 13,000 and 8,000 B.C. Cro-Magnon man started to appear in large numbers on the western coasts of Eurasia and around the shores of the Mediterranean (see above).[32]

[31] One of the most famous and basic of Egyptian legends recounts how Osiris was shut up in a coffer by his treacherous brother Set and cast into the Nile. *Washed up at Byblos*, he was then hacked into small pieces and scattered throughout Egypt, whence he was reassembled and restored to life again by the goddess Isis.

Could this story, one wonders, represent an allegory of the original flight of the Atlantean founding fathers from their stricken homeland, bearing the accumulated wisdom of their civilisation in some sort of 'ark' (compare p. 260), and of an eventual landfall in Phoenicia? By way of the books with whose manufacture the very name of Byblos was subsequently to become identified, was that wisdom then scattered throughout the known world, whence it was destined eventually to he reassembled by the 'light of Egypt' into the resurrected body of knowledge and enlightenment which was once that of Atlantis?

[32] Cayce always claimed, in fact, that Atlantis underwent two periods of partial destruction even before its final disappearance in 10,000 B.C. The first of these, he

(vi) Around 10,000 B.C. wheat and maize appear to have been first domesticated in Eurasia and America respectively (Charpentier *et al*).

(vii) In 10,000 B.C. the Mediterranean Sea stretched as far as the present Cairo and the edge of the Giza plateau. Had the Great Pyramid been built at this time, direct water transport would have been possible all year round between the Moqattam quarries and the Pyramid's causeway (Pochan, based on published sedimentation rates for the Nile Delta: see note page 147).

(viii) It was in 10,490 B.C., according to Edgar Cayce, that the original construction of the Great Pyramid was begun (see pages 235–8).

(ix) In choosing some great stone symbol to act as 'guardian of the sacred places' (including the Great Pyramid) it would have been natural for the contemporary Egyptians to choose a form related to the then ruling sign of the zodiac. Between 4,000 and 2,000 B.C., for example, they would probably have decided on a colossal bull, while thereafter some kind of 'ram-god' would have been more appropriate (see section following). In fact, however, they chose to model their celebrated (and red-painted) Great Sphinx on the shape of the lion. *Leo was the ruling sign of the zodiac in 10,000 B.C.*[33]

The way in which the above facts dovetail into each other is little less than extraordinary, and suggests that the mystery of the disappearance of Atlantis could in some way supply the key to the whole complex of ancient knowledge to which this book is devoted.

Astrological Parallels

The term 'precession of the equinoxes' refers to the slow movement of the sun's apparent position at the vernal equinox (the ancient new year) backwards through the twelve signs of the zodiac—a process which has been calculated as taking over 25,900 years per zodiacal revolution. Consequently each 'zodiacal month'—i.e. each 'age' during which the vernal new year falls under any one of the twelve signs—lasts on average some 2,160 years at the present rate of

[32] *(continued)* insisted, occurred around 50,000 B.C., while the second caused an early wave of emigration around 28,000 B.C. It is interesting to note that some isolated Cro-Magnon remains in Europe appear to antedate by several millennia the more plentiful remains of the so-called Magdalenian period—*and that their earliest likely date appears to be around* 28,000 B.C.

[33] Period of the sign of Leo—*c.* 10,970 to 8,810 B.C. Numerous sources also state that the Egyptian Denderah zodiac (a fascinating, disc-shaped artefact now in Paris) takes Leo as its starting-sign.

precession. The Pyramid itself appears to give 25,826·4 years as the period of the zodiacal revolution at the time of its construction (see Appendix B) in terms of the sum of its base-diagonals.

Thus, since the zodiacal age of Taurus (the Bull) commenced in around 4,500 B.C., following the end of the age of Gemini (the Heavenly Twins), we may see the age of Aries (the Ram) as commencing around 2,300 B.C., and the age of Pisces (the Fishes) as commencing shortly before the birth of Jesus of Nazareth. The age due to begin in A.D. 2,010 may thus be seen as that of Aquarius (the Water-Carrier), and its successor, that of Capricorn (the Goat), will begin shortly after the year A.D. 4,000; to be followed in succession by those of Sagittarius (the Archer), Scorpio (the Scorpion), Libra (the Scales), Virgo (the Virgin) and Leo (the Lion). Following the age of Cancer (the Crab), the world will then re-enter Gemini, thus eventually beginning a new age of Taurus in around the year A.D. 21,000.

It is also perhaps worth noting that Pisces and Aquarius were originally closely linked—and were known to the Babylonians as the combined sign of the Fish-Man. Meanwhile Capricorn was likewise seen as the Goat-Man or Goat-Fish. Libra, for its part, was often referred to by the ancients as the 'claws of the Scorpion', while the original Virgo was referred to by the Babylonians as the 'ear of corn'.

Now it is a matter of historical fact that the age of Taurus, as defined above, corresponds chronologically to the ancient era of *bull-worship*, as known to us chiefly through the later bull-cults of ancient Crete, Assyria and Egypt. Again, the age of the Ram corresponds closely to the Old Testament era, which was characterised specifically by the cult of the sacrificial *lamb*. Meanwhile in Egypt the priest-rulers of the contemporary Middle Kingdom were likewise adepts of the cult of Amon, the ram-headed god. In both bull and ram, in fact, we find the contemporary symbols of human salvation synchronised with the progress of the zodiacal traversal. And the biblical 'golden calf' incident (representing a temporary reversion to the older bull-worship) suggests that the severest sanctions were liable to be visited by the adepts upon the celebration of any such symbol out of its duly-appointed cosmic time.

Furthermore the ancient Central American legends bear witness to a very early 'cult of the heavenly twins' (see page 311), which would appear to correspond to the ancient age of Gemini—and once again it was the twins or brothers in question who were regarded as the means and symbol of their people's salvation, rather after the style of the later Moses and Aaron. The notion of heavenly twins likewise finds an echo in Plato's account of Atlantis, in which he attributes to Poseidon five pairs of twin sons—the first ten kings of Atlantis. Meanwhile the redemptive aspect of the twins idea seems to be reflected in the ancient Greek legend of the brothers Castor and Pollux, of whom the latter insisted on forgoing the immortality to which he was entitled, so long as his mortal brother remained subject to mortality. The parallel with the notion of the voluntary Messianic 'descent into hell' for man's redemption (see 'The Antechamber', pages 110–17) is a clear one.

Under the terms of the zodiacal cycle, then, the advent of Jesus of Nazareth marks the end of the age of the Ram and the beginning of the age of Pisces, the Fishes. It is entirely appropriate, therefore, that the Lucan shepherds are reported as abandoning their 'sheep' in favour of the new saviour. Now, perhaps, we can see the reason for their apparently irresponsible act. Both ox and ram must now defer to the new dispensation, and the old Passover covenant sealed in the blood of a slaughtered lamb must be replaced by a new symbol. At the same time it is entirely appropriate astrologically that the accounts of Jesus' new salvationist initiative are full of references to fishes, catches of fish and fishers of men. The zodiacal fish, after all, were traditionally regarded as 'bound', and thus as prisoners.

That Christian water-baptism was itself originally intended as a Piscean symbol seems more than likely. And certainly it is symbolically apt that the Christian bishop traditionally supports himself with a crook (symbolic of Aries, and thus of the Old Testament) and is crowned with a fish-shaped mitre (symbolic of the Piscean New Testament dispensation). Meanwhile it is perhaps pure accident (unless it was a deliberate play on words) that the Greek original of his name—*episkopos*, meaning overseer—itself contains the root of the Latin word for fish. In fact an explicit link would not be too surprising, since the fish, or *vesica piscis*, is known to have been the geometrical secret sign by which the earliest Christians, in their astrological wisdom, habitually identified themselves and each other.

The coming age, as we have seen, is that of Aquarius, the water-carrier or fish-man. What salvationist or Messianic message is to be derived from this fact? Presumably the function of the 'water-carrier' is to carry the water (the water of mortality, perhaps) to the point where it is to be poured out, like the 'water of the womb', for the sake of the new birth of the human spirit—i.e. the release of the 'fishes' that are 'trapped within it'. In the sign of Aquarius, then, we seem to have a clear parallel to the Pyramid's message concerning the destiny of the future elect, while the closing date of the Aquarian age corresponds to an interesting degree with that of the King's Chamber as calculated above. And indeed there seem to be signs that Jesus of Nazareth was himself fully aware of this astrological link. Not only was the sole sign of the returning son of Man to be that 'of the prophet Jonah' (the biblical fish-man, and cognate with the Babylonian Oannes): the man who leads the disciples to the 'upper room', symbolising the exalted Final Age, is specifically described as 'a man carrying a pitcher of water'—a symbol which can only have been deliberate, since water carrying was strictly women's work. Both, in short, seem to be direct references to the zodiacal Aquarius.

Indeed, even the ensuing foot-washing incident reported by John seems to have been designed expressly as part of a baptismal rebirth-ritual—a ritual involving the 'Aquarian' *pouring* of water (compare John 13:5) rather than 'Piscean' immersion. The Last Supper takes place, it should be remembered, less than twenty-four hours before Jesus' death. And at this point the text suddenly has him declare that his disciples' continued fellowship with him must somehow be a function of their participation in this strange 'Aquarian' initiation ceremony (even though they cannot yet understand its full significance). In astrological terms, in fact, Jesus is apparently saying that if their association with him is to be resumed following his imminent death, then that association must become an 'Aquarian' one. They must all be physically reborn, in other words, to experience together the Golden Age of Aquarius, which Jesus, as the ceremonial 'water-carrier', would personally inaugurate. Only then could their fellowship be resumed, whether symbolised in the breaking of bread or the sharing of wine. In Jesus' own words at Matthew 26:29, 'Never again will I drink from the fruit of the vine *until I drink it new with you in the kingdom of my Father.*'

And at this point the possibility arises that *even the wine itself* may have an Aquarian significance. The proof-text in this case is of course the celebrated turning of water into wine reported in chapter 2 of John's gospel. For, whatever the historicity or otherwise of the story, one thing is clear—the alleged miracle carries a familiar symbolism. The wedding feast in question, with its strong Passover associations, naturally symbolises (as always in the gospel-teachings) the Messianic Banquet of the coming Kingdom of Heaven. Jesus (again, as always) is himself the bridegroom. And the central action of the story revolves around the filling of *six* stone purification-jars with water at his behest—water which, *when poured out* (again at his direction), is found to have 'turned into wine'. We may interpret the symbols, both numerological and otherwise, to mean that the aqueous or Piscean age initiated by Jesus is to be a *preparatory* age of purification leading directly to the glories of the expected Kingdom, *and that this is destined to commence with the advent of the Aquarian age.*

Nor should we forget Jesus' specific teaching at Matthew 13:47, where he likens the coming of the Kingdom to the lowering of a net into the sea, after which the good fish are put into pails and the net thrown away. Pails, it scarcely needs pointing out, are normally used for carrying water, so that there seem to be direct links here both with the Piscean and with the Aquarian symbolisms. The men and women of the Piscean age (we may interpret) will all be subjected to scrutiny with the coming of Aquarius—and only the best of them will be adjudged fit to 'occupy his pitcher', thus eventually being poured out by him on the immortal shore.

But if the Aquarian age is to be that of the new Messianic initiative, what significance are we to see in the succeeding age of the Goat and of the further ages which succeed it? If the text of Matthew 25 is any guide, we may perhaps see the age of the Goat (or Fish-Goat) as that of the 'goats' remaining after the elect have achieved their escape from the physical world—a further age of hell on earth, perhaps, corresponding to the destiny symbolised by the 'uncompleted' Dead End Passage (if our exegesis on page 146 is valid). Indeed, the goat notion has distinct affinities in the universal folk-memory with that of the horned, tailed and cloven-footed Devil himself. Whether the succeeding age of Sagittarius signifies an age of war or one of spiritual endeavour and escape is difficult to decide, but the age of Scorpio which follows it during the eighth millennium (compare the prophecies of Nostradamus) suggests an era during which the earth will be laid waste. Following this, perhaps, the remainder of mankind will in some way be 'judged' during the age of Libra (the Scales), and with the age of the Virgin (or of the 'ear of corn') a new beginning will be made, and the earth 'planted anew' . . .

If the above notions have any validity, then some important conclusions follow. The first is that history is indeed cyclic, as we earlier surmised, and as the Maya also long believed (see page 24), and that consequently the Pyramid's message may comprise a series of 'memories of the future' based on what *has happened before*. This would help to explain why the length of the year and the period of the equinoctial precession relative to that year are taken respectively by the Pyramid's architect as the length of the base-side and the sum of the base-diagonals. It would suggest that the Pyramid's designer wished to show that his predictions were 'based' on a cyclic view of history—a view, moreover, capable of being expressed in terms of astrological notions derived from the precession of the equinoxes. On this basis, we are faced with the interesting notion that history tends to repeat itself in some respects every twenty-six thousand years or so, which in turn would suggest that man may already have known a world comparable to our own in some at least of its accomplishments and tendencies in around, say, 102,000 B.C., 76,000 B.C., 50,000 B.C. and 24,000 B.C. In terms of current historical and archaeological theory the notion sounds wildly fantastic, of course, and yet, in view of the declarations of Edgar Cayce on such topics as Atlantis, one wonders . . .

Or alternatively we might perhaps more justly conceive of the progression of the ages in terms of a spiral. The march of evolution and history, in other words, displays a circular motion, but each revolution takes place at a different level (presumably a higher one) and is characterised by accomplishments of a different order. Indeed, the fact that the ancient Aztecs apparently regarded the conch-shell as symbolic of the succeeding ages would suggest that they subscribed to some such notion. Nor is the idea without its distinguished modern adherents: even Einstein is alleged to have subscribed to it. And the data set out in Appendix D suggests that the designer of the Great Pyramid may have adopted a similar view.

If history does tend to follow a spiral based on a 26,000-year cycle, then there must presumably be good and cogent reasons for the phenomenon. In order to establish the potential validity of such a theory, it would first of all be necessary to establish definite causal links between the vagaries of the earth's orbital cycle and the human drive towards cultural and technological evolution. Now the most obvious physical result of the 26,000-year precessional cycle must be a tendency for the historical pattern of terrestrial climatic conditions to repeat itself with a similar frequency. And indeed the geological record shows clearly that it has, to the extent that world sea-levels and sea-temperatures, for example, have tended in the past to oscillate at this frequency among others, in sympathy with independently-recorded precessional variations in solar radiation.[34] Further geological evidence going back to the beginning of the Pleistocene era, some two million years ago, confirms that the whole

[34] R.W. Fairbridge, 'The Changing level of the Sea', (*Scientific American,* May 1960).

period has been characterised by frequent and violent climatic changes, with consequent far-reaching effects on the spread and very survival of many types of plant and animal life. Thus it does not seem inherently improbable that man's own development may have been affected by these various changes.

None the less, the establishment of the possible *a priori* validity of a cyclic view of history, such as that suggested, would rely heavily on the discovery of specific links between climatic and environmental changes and the rate of evolution of human culture and civilisation. Nor does the idea seem an improbable one. If, at one end of the scale, a given human community were, for example, to be threatened by a sharp drop in average temperatures and/or the advance of the polar ice-sheet, then the community's continued survival might well depend either on the development of a sufficient degree of peaceful technology to keep themselves warm and provide for their continued sustenance, or on the development of a degree of military technology sufficient to allow the community to migrate to a warmer area and drive out or subdue its original inhabitants. Similarly, a rise in world temperatures, and a consequent drastic raising of world sea-levels, would once again give considerable impetus to the technologies of migration and conquest. Necessity, in other words, would as usual be the mother of invention.

Moreover; the fact that all the most ancient known civilisations of the present hypothetical cycle flourished within the range of latitude between approximately 20° and 40° north adds weight to the supposition that the drive to conquer has been a dominant factor in the development of human technology; and even the presumed location of the legendary Atlantis, at around 25° north, fits this notion perfectly (see map opposite). For this belt, it seems, has been the world's desirable area, the promised land, and to it its conquerors have been attracted like moths to the flame.

It is often stated that this particular band of latitude would naturally tend to produce civilisations because living conditions there are easier. Yet it is easily demonstrated that areas characterised by easy living-conditions do not, *ipso facto*, tend to produce 'high civilisations', and that living conditions in most of the areas which produced the ancient civilisations were in fact *far* from easy—indeed, in most cases, the civilisations in question arose out of a need to render unpromising swamp-areas habitable and to devise irrigation schemes efficient enough to permit the production of basic food requirements.

Meanwhile, the latitude-band in question lies well north of the

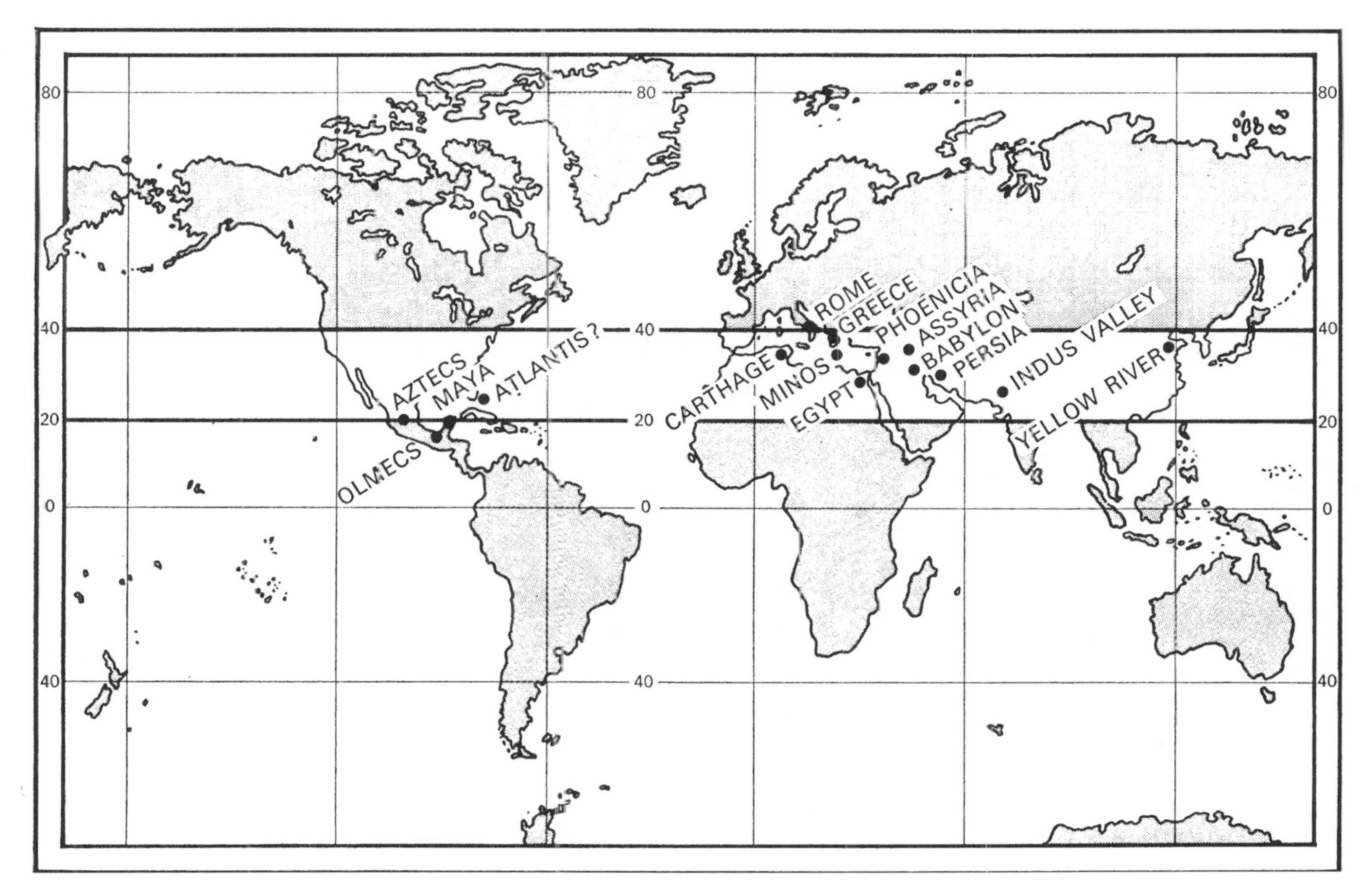

World-map showing the common latitude-band of ancient civilisations.

equator, and one would expect a similar band of ancient (B.C.) civilisations to the south of it. Yet neither in Australia nor in Africa did such civilisations apparently arise. Apart from the ancient precursors of the Incas in the Tiahuanaco area, only the Polynesian civilisation (of doubtful age) and the legendary land of Mu or Lemuria (of doubtful historicity, but possibly connected with the Tiahuanaco civilisation) would appear to fall within the band of latitude between 20° and 40° south. This fact tends to confirm that climatic conditions suitable for the development of civilisation do not of themselves *produce* civilisations, but rather tend to *attract* civilisations from elsewhere. In almost every case the various ancient civilisations appear to have been founded by parties of already advanced immigrants—whether we think in terms of the Olmecs, the Egyptians, the Yellow River civilisation, or indeed Polynesia. And if there was a lack of ancient civilisations south of the equator, then it may simply be a reflection of the presumed fact that man originally evolved in the northern hemisphere, and that the equatorial jungles tended to prevent the spread of early culture into the southern hemisphere, other than by means of water transport (compare page 53).

To return, however, to our postulated 26,000 year historical cycle, it seems quite possible that changing climatic conditions might result in the initial rise of human civilisations and their migration to certain areas—whatever other influences, real or imaginary, might be held to be at work. Meanwhile, at the other end of the scale, the eventual development of truly advanced technologies will tend to result in over-population, over-exploitation of natural resources, exceptionally severe types of individual and social neurosis arising from rapidly changing conditions and highly artificial environments, and the misuses of technology for selfish ends. Human powers for both good and evil must tend to be immeasurably increased, and the risk of the total self-destruction of the community—or the mutual destruction of rival communities—must therefore tend to be a high one. And thus, perhaps, with the destruction of the 'old world', a new 'cycle' of the spiral begins.

Consequently there appears to be nothing improbable in a cyclic view of world history. In fact the only real obstacle to the more general acceptance of such a view is the apparent lack of specific archaeological evidence to back it up—which may have more than a little to do with lack of knowledge of what to look for or where exactly to look for it. Whether, as Edgar Cayce has claimed, such evidence will in due course be found, only time will tell.

As for the astrological identifications of the signs of the zodiac which started us on this particular line of inquiry, it is far more likely that they *derive from* such a historically based knowledge of the future than that they directly *influence* it in any way. Similarly, it is much more likely that the almost equally 'occult' science of numerology originally derives from known features of the Great Pyramid's code, than that any particular number has any absolute 'magical' significance of its own. Both forms of semi-magical knowledge, in other words, may—like the Pyramid's own apparent precognitive aspects—be no more than inverted and mythologised versions of rationally derived knowledge acquired, and subsequently encoded, by the ancients.

* * *

Deep in the human subconscious, it seems, there lies a world-wide awareness, a premonition or even a memory, perhaps, that man has known better things and is one day destined to know them once again. Even in astrology, and in the supposedly unrelated mythology of Central America, we are repeatedly brought face to face with a Messianic symbolism similar to that which permeates the more familiar biblical and pyramidological records. Can one doubt that at the base of this vast column of mythological smoke there burns some sacred fire of truth?[35]

[35] Compare Acts of John 102, 100, 104: 'The Lord contrived all things symbolically and by a dispensation toward men, for their conversion and salvation . . . When the human nature is taken up, and the race which draweth near unto me and obeyeth my voice, he that now heareth me shall be united therewith, and shall no more be that which now he is, but above them, as I also am . . . If ye then abide in him and are builded up in him, ye shall possess your soul indestructible.' (Tr. M. R. James.)

10

The Sign of the Messiah

IN THE FOREGOING chapters we have seen how the awareness of a Messianic World-plan—whether it derives from some ancient folk-memory or from actual precognition—permeates much of the Ancient World's mythology. In particular we have observed its presence in the Hebrew, Egyptian and Mayan traditions, and have discovered clear evidence that it lies at the very root of the design of the Great Pyramid. Indeed, the further back in time we explore, the clearer man's awareness of the Plan appears to have been—for the Pyramid antedates by some millennia the available written records, and yet it is in the Pyramid that the Plan's details are revealed most clearly and explicitly. This in itself would seem to contribute a measure of confirmatory evidence for the supposition that the Plan was once known *in its entirety* by a race of men who were unable or unwilling to measure up to its standards in practice, and whose civilisation was subsequently—or perhaps consequently—overwhelmed by some gigantic cataclysm which left few traces, apart from the knowledge enshrined in the Great Pyramid and a number of obscure folk-memories and esoteric traditions. It may also add fuel to the notion that the same thing may in due time happen again, and that the 'test', 'hell' or 'river of fire' of the Messianic tradition is synonymous with the event symbolised by the biblical Flood—the cataclysmic event which will terminate just one more age in the vast cycles of cosmic time.

But if knowledge of the Messianic Plan was so deeply rooted—albeit subconsciously—in what we are pleased to call the Ancient World, it is no less securely rooted—even if often unrecognised—in the subconsciousness of man today. Increasingly, thinking men are prepared to admit to their awareness that the world is set on a more and more downhill course—spiritually, morally and even physically—and that none of the gods of modern man—whether politics, science, technology, wealth, comfort, security or obsessive entertainment—can save him from destruction unless they are subordinated to some form of higher, spiritual authority which is conspicuously lacking in the contemporary world. Increasingly, too, the young especially are overcome by dread and apprehension at the state of the world into which they have been brought, and are overwhelmed by a compelling urge to reject its values and 'escape' from it at all costs. Man, in short, is faced with an impasse—a Dead End Passage, to use the Pyramid's terminology—from which he is probably incapable of escaping by his own unaided efforts. Consequently he is increasingly prepared to turn to almost any 'saviour' that presents itself—so that, subconsciously, his mind is becoming steadily more attuned to that wavelength of receptivity which will eventually ensure maximum effect for the expected Messianic return.

Meanwhile, man's scientific advance is (whether he realises it or not) ever more urgently preparing the ground for acceptance of the Messianic teachings. Steadily, scientific man is coming to realise that the observable physical world is in reality very different from what he has for so long assumed—that indeed it represents only a minute part of existence—while at the same time the more he knows the more he becomes aware of the acute limitations of his knowledge.

The stage, then, is set for the Messianic initiative. In the words of the ancient 'Hymn of the Robe of Glory' (from the *Acts of Thomas*) the 'king and queen of the east' (symbolic of the solar divinity) are about to dispatch their son to rescue from Egypt the pearl of great price guarded by the serpent. The sunrise of the new age will bring with it the Messianic figure, an emissary from the world of pure spirit, who will rescue the precious jewel of man's soul from the terrestrial snakepit. Putting off his princely raiment, the redeemer will put on Egyptian clothes 'lest I should seem strange, as one that had come *from without* to recover the pearl, and lest the Egyptians should arouse the serpent against me'. We may interpret this sentence to mean that the Great Initiate will come *as an ordinary man* whose message will be expressed in terms acceptable from the point of view of contemporary science and knowledge. The story then concludes: 'And I caught away the pearl and turned back to bear it unto my fathers. And I stripped off the filthy garment and left it in their land, and directed my way forthwith to the light of my fatherland in the East.' And there, we may assume, his mortal body having been left behind him, the robe of glory will be waiting for him, and amid great rejoicing the redeemer will once more put it on.

Ultimately, however, what will eventually do more than anything else to make men's minds ripe for the Messianic initiative is scientific 'proof'—proof produced by the fulfilment of predictive statements based on known and incontrovertible evidence. It is here that the Great Pyramid, in association with the surviving biblical texts, seems to have a vital rôle to play. For if the Pyramid's predictions are borne out by historical fact in the years to come, then our interpretation of its symbolism will be validated and the truth of its message confirmed. And this in turn, in association with modern scientific knowledge, will help to permit human acceptance of the Great Initiate's teachings. Man's destiny will then be unmistakably revealed to him in all its breadth and splendour, and each human being will be faced with the ultimate choice of fulfilling or ignoring the conditions necessary for its fulfilment.[1]

But when will these final events stir into motion, and how will man recognise the coming Messiah and know that he is at hand? The Pyramid, as we have seen, is quite specific in its dating, and its Granite Leaf suggests that the sign of the Messiah will have something in common with the shape of the rainbow—which is also the sign of the divine covenant in the Noah story, and here too is apparently symbolic of the abolition of universal death. We should note, however, that in both cases the 'bow' in question is facing not to the right or left, nor yet to the east or west, but *upwards*. Could there be any significance in this?

One does not have far to seek for the answer: for a bow by itself is not a very useful weapon—before it becomes effective it needs to have an arrow fitted to it. And that symbolic arrow, we must deduce, stands for nothing less than the souls of men—those same 'unbreakable reed arrows' which the god Atlaua, having plucked them from the waters of mortality over which he is master, will fire upwards with such force that they will pierce the heart of the cosmos, never again to return to earth. Indeed, it is *men's souls* which, like Atlaua, will then 'arise and come forth like the Quetzal bird', until, with the birdlike soul of Quetzalcoatl himself, they become one with the morning and evening star, the inseparable herald and attendant of the solar divinity.[2]

[1] See Appendix J: The Messianic Plan—a Tentative Summary.

[2] See pp. 111–12 and 312–15, and compare the bow-and-arrow symbolism of traditional Shinto, Zen Buddhist and South American Indian rites. Note, too, the words of the Hindus' *Mundaka Upanishad*: 'Take the great bow of the Upanishads and place in

What, then, is needed to permit this ultimate process to take place? Clearly, man needs to be prepared *to fit his own soul to the Messianic bow*—a marriage highly reminiscent of Jesus' constant bridegroom symbolism. One is also reminded of his celebrated call: 'Come to me all whose work is hard, whose load is heavy: and I will give you relief. Bend your necks to my yoke, and learn from me, for I am gentle and humble-hearted, and your souls will find relief. For my yoke is good to bear, my load is light.' The yoke in question is of course none other than the traditional ox-bow or ox-yoke—the point being that the fitting of the neck of the arrow (i.e. the human soul) to the Messianic yoke or bow will result not in the perpetuation of the soul's eternal animal enslavement, but in its being fired or loosed into the perfect freedom of eternal spirituality.

But not all men are willing to undergo this complete uprooting and 'launching into orbit', and first of all it is necessary for one man to show the way. And that is perhaps the outstanding symbolism of Jesus of Nazareth's crucifixion: for the Christian cross itself is a symbol for the Messianic bow—a crossbow, if you will—which the priest-Messiah

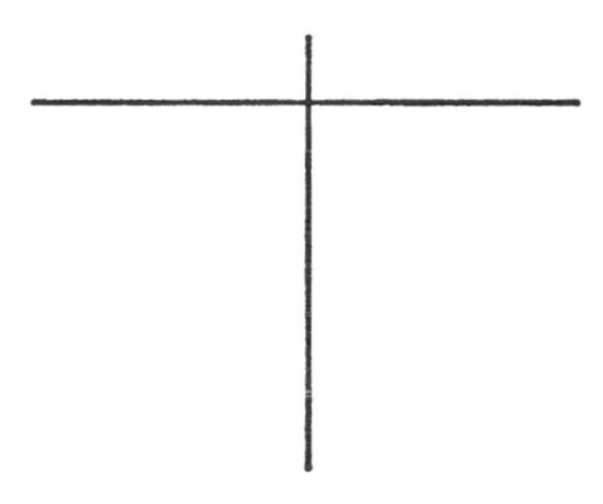

himself was the first to 'bend' with the weight of his own body—

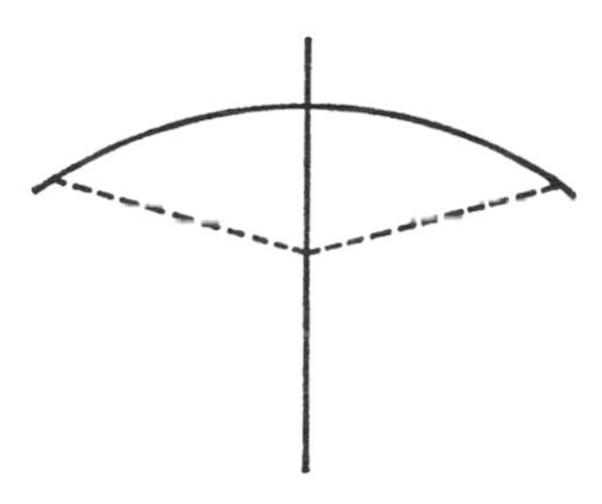

[2] (*continued*) it an arrow sharp with devotion. Draw the bow with concentration on him and hit the centre of the mark, the same everlasting spirit. The bow is the sacred OM, and the arrow is our own soul. Brahman is the mark of the arrow, the aim of the soul. Even as an arrow becomes one with its mark, let the watchful soul be one in him' (Tr. J. Mascaró).

So gigantic is the tension of this mighty bow that it is destined eventually to achieve, in symbol, the launching of the souls of all men into eternity.

Presumably, then, if the cross indeed symbolises the Messianic bow, it is the Messiah himself who will draw it. But at this point it is also worth noting that the Egyptian quadrant already referred to (page 6) is itself the outline of a drawn bow. And if we pause to consider what feature is drawing the bow in question the answer is precisely what one would expect—the Great Pyramid itself, the ancient Messianic pillar of witness set at the centre of the world.

Moreover the noon-reflections of the Pyramid's polished white limestone casing, used for centuries by the Egyptians as a form of solar calendar, also had a clear bow-symbolism; for, as reference to Davidson and Aldersmith's research will reveal, the shape of these reflections from midwinter until (significantly) midsummer corresponds strikingly to the straightening of a drawn bow and the projection of an arrow due southwards (the direction which throughout the Great Pyramid's symbolic passages constantly indicates the direction of progress through time), culminating at the summer solstice *in the cross-like shape of a star* (see opposite).

The correspondence between these facts and the Mayan symbolism already referred to is most striking. One is again reminded strongly of the arrows of Atlaua and of the ascending flock of birds representing the feathered soul of Quetzalcoatl (itself reminiscent of the feathered flights of the arrows) which are at length transformed for ever into the planet Venus, the constant heavenly companion of the divine sun. And meanwhile the Pyramid itself confirms, in symbol, the identity of the arrows in question—for its design incorporates, in side elevation, two surprising and unmistakable upward-pointing arrowheads (see page 17) in the gables of the Queen's Chamber and the topmost Construction Chamber, together with a third over the entrance passage (see page 181) with its barbs set at the spirit-angle of 51°51′14·3″ *to the vertical* (unlike the Pyramid's sides which are set at the same angle *to the horizontal*). All three gables thus seem specifically to symbolise giant upward steps in the development of the human soul.

Further, we should note that the seventh angel of St. John's Revelation (chapter 10) is likewise described as being 'wrapped in cloud, *with the rainbow round his head*'; and it is also this angel whose appearance heralds the fulfilment of the 'hidden purpose of God ... as he promised to his servants the prophets'. This figure clearly demands identification with the Great Initiate, the One-who-is-to-

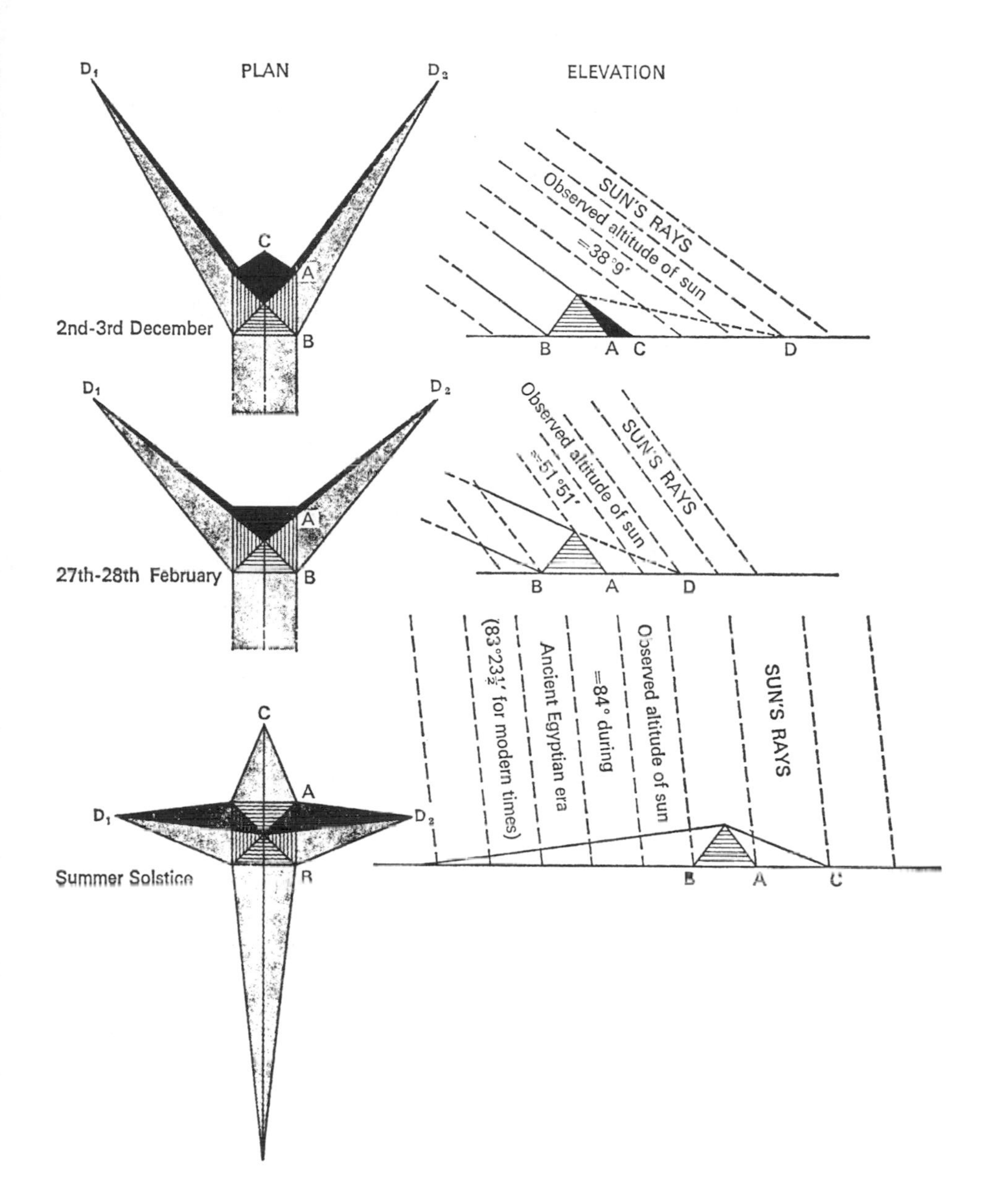

Noon-reflections of the Great Pyramid's original polished limestone casing (capstone assumed for diagrammatic simplicity).

come; while it also reminds us of the equally rainbow-crowned Egyptian goddess of the foundation, Sefkhet-Aabut, who is related both to the Egyptian Hathor (the mythical instigator of the primeval Deluge and Flood, and thus of the post-diluvian age) and to the Pleiades. The latter (alternatively known specifically as the

Atlantides—the daughters of Atlas) were known in ancient Euphratean astronomy as The Foundation, as well as being intimately bound up with the establishment of the astronomical datum for the Great Pyramid's entire chronology (see page 54).

In the Messianic sign of the bow, in short, we are reminded both of the inception and of the eventual fulfilment of the Messianic Plan: it speaks to us not merely of the expected future Messianic visitation, but of the very Alpha and Omega, the Flood and the Exodus, the Eden and the Heavenly City, of the whole great plan for the evolution of mankind.

The various Messianic cross, bow and bridegroom symbolisms are in essence all one, their combined message being that man's soul must be yoked to the Messianic initiative if he is to achieve his true destiny, which has been prepared for him from the beginning of the world. Perhaps it is specifically in this light that we should understand Jesus' statement at John 12:32, '*I shall draw all men to myself*, when I am lifted up from the earth'. The evangelist himself, who clearly feels that the statement needs some interpreting, sees it as an allusion to the Crucifixion—which it clearly is—but the symbolism of the bow and arrow seems to be even more applicable.

Again and again, it seems, the deep symbolism of the Messianic Plan comes down to us through the centuries and works on us, almost without our knowledge, at the subconscious level, so that when the literally crucial time comes we may be able at last to reach out and grasp that great Reality which is our undoubted birthright from eternity. Miraculous it may seem to us now, but when that day comes our translation will seem—*must* seem—as natural as breathing itself, as involuntary as the first breath of the newborn child. The alternative is stillbirth and extinction.

And so it is for the sign of the Messianic bow, flashing like the lightning from horizon to horizon, that the world is now being prepared. When that sign comes, it will be the final signal, hoisted upon the mountains, that the long-awaited Messiah and Great Initiate is at hand, and that the last great act in the present cycle of the human drama is about to begin.[3]

[3] The bow in question will almost certainly be a technological phenomenon quite explicable in terms of the science of the day—as it clearly was not for Ezekiel's generation, if the latter's description of the extraterrestrial, rainbow-like phenomenon in 1:4 to 2:15 is anything to go by. The Messiah too will presumably be a flesh-and-blood figure, wielding non-magic powers, possibly with an instrument like the Egyptian *ankh*. We should do well to remember that, as Arthur C. Clarke's 'Third Law' puts it, 'Any sufficiently advanced technology is indistinguishable from magic.'

'O lovers, O lovers, it is time to abandon the world;
The drum of departure reaches my spiritual ear from heaven.
Behold, the driver has risen and made ready the file of camels,
And begged us to acquit him of blame: why, O travellers, are you asleep?
These sounds before and behind are the din of departure and of the camel-bells;
With each moment a soul and a spirit is setting off into the Void.
From these stars like inverted candles, from these blue awnings of the sky
There has come forth a wondrous people, that the mysteries may be revealed.
. . . O soul, seek the Beloved, O friend, seek the Friend,
O watchman, be wakeful: it behoves not a watchman to sleep.
On every side is clamour and tumult, in every street are torches and candles,
For tonight the teeming world gives birth to the world everlasting.
Thou wert dust and art a spirit, thou wert ignorant and art wise.'

From *Shamsi Tabriz*, by the Sufi mystic
Jalal'ud-Din Rumi (1207–73)
(Translated by R. A. Nicholson)

Appendices

APPENDIX A

The Great Pyramid's Units of Measure

1. When the metric system was devised at the time of the French Revolution, the basic unit of measurement (the metre) was taken to represent a ten-millionth part of the mean distance from either of the Earth's poles to its centre, *measured round the earth's surface*—i.e. one ten-millionth of the distance from pole to equator.

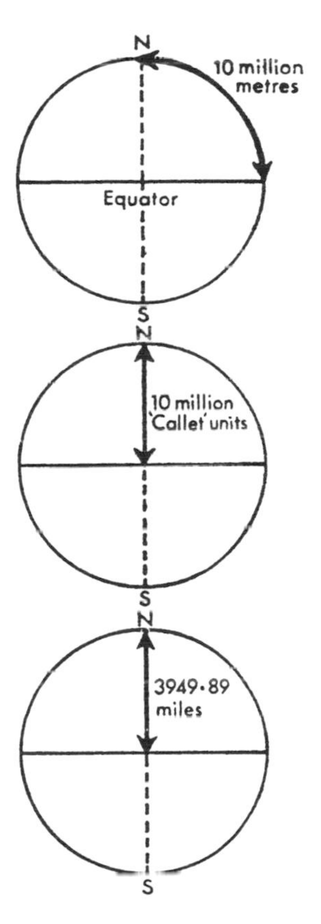

2. This distance is almost impossible to ascertain accurately, however, since the earth is not a true sphere. As the mathematician Callet observed in 1795, a more scientific standard of measure would have been the ten-millionth part of the direct distance from the pole to the centre of the earth.

3. In the course of the geodetic research undertaken during International Geophysical Year (1957–8), observation of the orbits of earth-satellites established that the Earth's mean polar radius was 3949·89 miles—a distance equal to 250,265,030·4 standard inches.

4. The 'ideal' metric unit equal to a ten-millionth of this distance would therefore equal 25·0265 standard inches.

5. But this very measurement is found three times (correct to four decimal places) inside the Great Pyramid . . .

(a) Inclined N.-S. distance between north wall of Grand Gallery and north wall of Well-Shaft (i.e. length of initial section of Gallery floor)
= 25·0265 standard inches.

(b) Horizontal distance between E.-W. axis of Queen's Chamber and E.-W. axis of Niche = 25·0265 standard inches.

(c) Distance from east end of Granite Leaf (embedded in wall) to centre of Boss or Seal
= 25·0265 standard inches.

. . . and twice in its exterior design, where it represents the distance from the top and the bottom of the 35th course to its vital aroura-defining axis.

(See p. 350)
(See 12, 13, 14 below)

6. Published surveys reveal that the side of the Pyramid's designed base-square, defined by the still surviving foundation-sockets on the Pyramid's south side, measures 9140·7+ standard inches. This distance equals 365·242 of the units of measure referred to in 5 above (actual value = 9140·73"), while the designed base-length of each of the Pyramid's concave sides measures 365·256, and the 'constructed' line AbB 365·259, of these same units. *These are, respectively, the precise number of days in the Earth's mean solar tropical, sidereal and anomalistic years.*

(See AB on diagram p. 354)
(AEFB on diagram p. 354)

7. The conclusion is inescapable that a unit of measure equal to 25·0265 standard inches was one of those used by the Pyramid's architect. This unit seems to correspond to the Long or Sacred Cubit known to have been used by the ancient Egyptians in the design of their larger religious monuments. There is also evidence that a cubit of similar length was formerly used by both Persians and Hebrews.

8. A further measurement found at least four times in the Pyramid's Antechamber represents *an exact twenty-fifth subdivision* of the presumed Sacred Cubit, i.e. 1·00106 standard inches.

(See points 46, 47, pp. 111–12)

9. Therefore the smaller unit too seems to demand recognition as a basic measure used by the

architect, especially as it is the *only* measurement in the Pyramid which represents an *exact* subdivision of the Sacred Cubit. While an obvious name for it would be the 'Sacred Digit', students of the Great Pyramid generally refer to this unit as the Pyramid or Primitive Inch.

(See lower diagram p. 125 and compare listed data)

10. In confirmation of points 6 and 9 above, a measurement of exactly 365·242 Primitive Inches occurs twice in the King's Chamber Complex, while both the Antechamber and the King's Chamber display measurements of 116·26 Primitive Inches—the diameter of a circle of circumference 365·242 Primitive Inches.

11. A further unit of measure known to have been used by the ancient Egyptians is the so-called Royal Cubit, of which actual measuring-sticks still survive. These define the length of the Royal Cubit as 20·63 standard inches. This shorter cubit too was used in the Great Pyramid's design, and the widths of its major passageways and the basic dimensions of both King's and Queen's Chambers are direct functions of it. Examination of these permits a further refinement of the above figure, giving the Royal Cubit as 20·6284+ standard inches, or 20·60659+ Primitive Inches. The Royal Cubit (RC) was normally subdivided by the ancient Egyptians into one hundred *n*.

(a) Width of major passageways = 2 Royal Cubits.

(b) Queen's Chamber floor measures 10 RC × 11 RC.

(c) King's Chamber floor measures 10 RC × 20 RC.

12. The basic unit of Egyptian area is known to have been the aroura. According to Herodotus, 'the aroura is a square of a hundred (Royal) cubits'. The side-length of the square was thus 2060·659+ Primitive Inches, and the total area of the aroura was some 4,246,317 square Primitive Inches.

13. But a square of 4,246,317 square Primitive Inches is equal in area to a rectangle measuring 3652·4235+ Primitive Inches by 1162·6025+ Primitive Inches . . .

14. . . . which is itself equal in area to a parallelogram similarly having as its base-side length 3652·4235+ P″ and as its height 1162·6025+ P″.

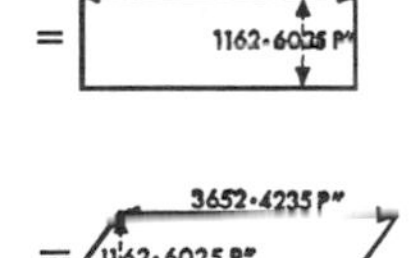

15. The full-design Pyramid itself clearly exhibits, in the vertical cross-section of each of its four faces, two such aroura-parallelograms. These are defined by the Pyramid's base-line, its sloping sides, the fifth-scale inset triangle let into each of its faces, and its 35th-course axis—which itself lies exactly one Sacred Cubit from the top and bottom of that course.

(AFED and EGBH on diagram p. 350)
(DEH on diagram p. 350)

16. As a check on point 15, the base of the Pyramid's side, as illustrated on p. 350, is equal to AD plus HB plus DH
= 3652·4235+ P″ + 3652·4235+ P″
+ (25 × 3652·4235+ P″)/5
= (10 + 10 + 5) × 365·24235+ P″
= 25 × 365·24235+ P″
= 365·24235+ Sacred Cubits
This is the known side-length of the Pyramid's base between the surviving foundation-sockets (south side)—see point 6.

(DEH is fifth-scale triangle: base therefore equals 1/5 that of Pyramid)

17. Meanwhile, the vital dimensions of both aroura-rectangle and aroura-parallelogram (3652·4235+ P″ and 1162·6025+ P″) are respectively the circumference and the diameter of a circle whose area is exactly one quarter-aroura, as is demonstrated below, since 3652·4235+ P″ = 1162·6025+ P″ × π

Clearly, in the four circles depicted . . .
. . . total area = $\pi r^2 + \pi r^2 + \pi r^2 + \pi r^2$
$= 4\pi r^2$.

But if, in each of the above circles, radius = r, then r = 1162·6025+ P″/2

Consequently, in rectangle ABCD,
AB = π × 1162·6025+ P″
= 2π × [1162·6025+ P″/2]
= $2\pi r$.
And BC = 2 × [1162·6025+ P″/2]
= 2r.
Thus, area of rectangle = AB × BC
= $2\pi r \times 2r$
$4\pi r^2$

∴ total area of all four circles
= total area of rectangle
= 1 aroura.

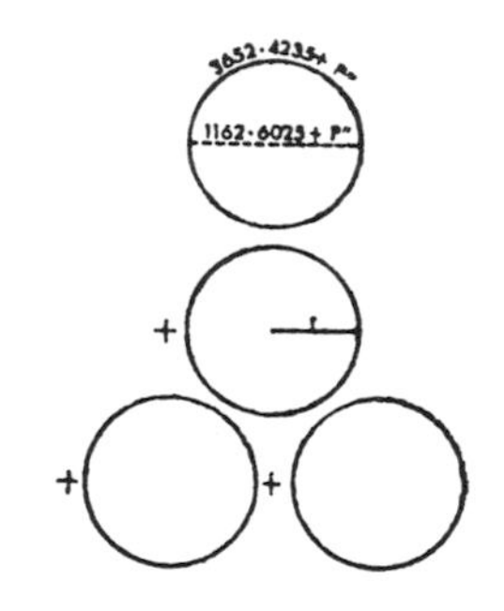

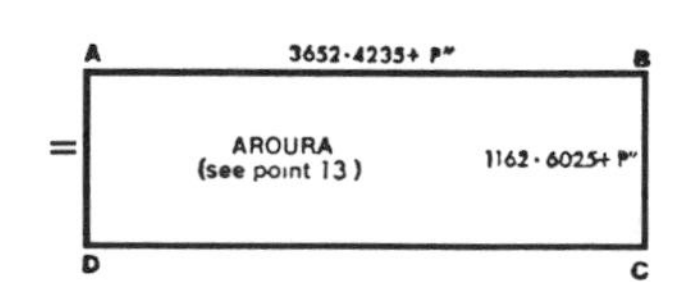

Therefore the area of a circle having as its circumference the number of Primitive Inches equal to ten times the number of days in the Earth's mean solar tropical year equals one quarter-aroura.

18. That the ancient Egyptians were aware of this correlation is revealed by the writer Horapollon, who stated: 'To represent the current year they depict the fourth part of an aroura.'

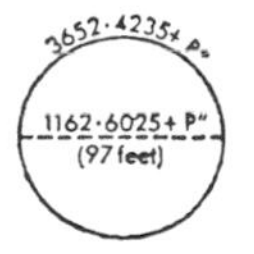

Sarsen the 'year-circle' or quarter-aroura

19. Meanwhile, at Stonehenge, the stone lintels of the sarsen-ring were carefully carved to fit the shape of the circle they enclosed. The published diameter of this circle is 97 feet, or 1164 standard inches. But the diameter of the Egyptian year-circle or quarter-aroura (1162·6025+ P") is itself equivalent to 1163·8 standard inches. *The conclusion is therefore unavoidable that the circle enclosed by the sarsen-ring at Stonehenge is none other than the Egyptian year-circle.*

20. To conclude, the evidence is overwhelming that the Royal Cubit (RC), Primitive Inch (P") and Sacred Cubit (SC) are direct functions of each other in terms of the Egyptian year-circle, of which an actual example still exists today at Stonehenge. And all four are apparently derived from the Earth's dimensions and from its orbital and rotational frequency.

21. Indeed, the relationship between Sacred Cubit and Royal Cubit is expressible mathematically as follows:

1 Sacred Cubit =
$10^3\sqrt{\pi}/(4 \times 365{\cdot}24235+)$ Royal Cubits.

22. That the builders of the Great Pyramid were fully aware of these relationships is exemplified not merely by the Pyramid's overall design, but in particular by the design of the Antechamber. Here, as the lower diagram on page 125 illustrates, the area of a circle inscribed in the 116·26 P"-long Antechamber (a circle whose circumference therefore measures 365·242 P") is equal to the area of the square ABCD (defined by the granite portions of the Antechamber's floor and east wainscot). Since the side-length of

this square is 5 Royal Cubits exactly, the area of both the Primitive-Inch-based circle and the Royal-Cubit-based square is 25 square Royal Cubits, or one four-hundredth of an aroura. One function at least of the Antechamber, in other words, is to define the relationship between Royal Cubit and Primitive Inch.

The Antechamber, in short, contains the geometric key to the Pyramid's whole system of basic measures. It would therefore be no less than fitting if, as is suggested on page 120, the Antechamber were also found to contain the key to the monument's symbolic message.

APPENDIX B

The Great Pyramid's Geometric Symphony

1 Astronomical Links with the Building's Dimensions

The precise dimensions of the Great Pyramid *as built*, and the alignments and disposition of the building's foundation-sockets, are to be found set out to very fine tolerances in a number of published surveys (e.g. Piazzi-Smyth, Petrie, Edgar, Rutherford, Cole).

The various surveyors disagree to some extent, however, in their estimates of the precise shape and size of the *full-design* Pyramid. These estimates are necessarily tentative extrapolations from the existing data, since nobody can be sure about the precise relationship between the reduced-design Pyramid *as built* and the original foundation-sockets—which can still be seen cut into the rock of the Giza Plateau, *well beyond the limits of the original building* (see diagram page 354). A particular cause of uncertainty is the fact that the positioning of the foundation-sockets is itself somewhat 'eccentric' (a fact well known ever since Petrie's survey of 1881).

Rutherford's view (fully explained in his multivolume *Pyramidology*) is what may be described as the traditional one—namely that the full-design Pyramid was intended to be seen as having a perfectly square base-platform. Each side would be indented towards the centre to the extent of 35·762+ P″, however, to conform with the parallel, observed concavity of the existing core-masonry (see diagram page 354).

Rutherford's view (supported by a number of professional surveyors) would thus see the south-east socket (being the 'furthest out') as marking the full-design structure's actual south-east corner, whilst it and the south-west socket between them indicated the length (9140+ standard inches) of its southern base-side. The outer corners of the remaining sockets (both of them interior to the resulting square) would then have had the sole function of marking the line of the building's diagonals, using the procedure known

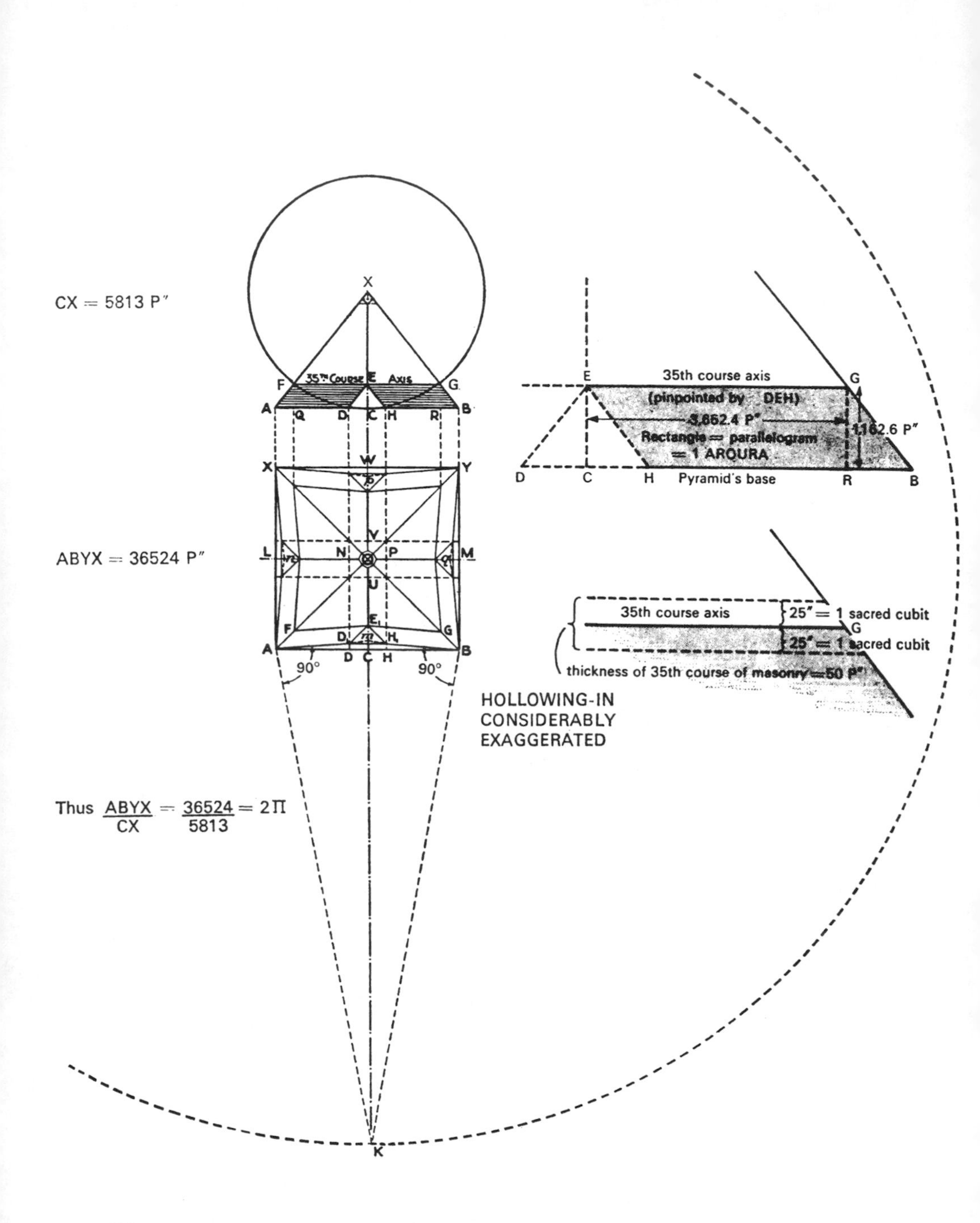

Diagram showing the integrated relationships between the full-design Pyramid's geometry, the Aroura, the Sacred Cubit, the value π, and known astronomical data.

as 'stretching the cord'—immemorially regarded as the vital preliminary to the commencement of large-scale building operations. With the building's centre thus precisely defined, the base-limits could then be measured—and the first courses of masonry laid—outwards from that centre.

Rutherford's view of the full-design Pyramid would thus give a base-square of side 9131·05+ P" (365·242 + Sacred Cubits) and a square base-perimeter of 36524·2+ P". On this basis, with an observed angle of slope of 51° 51′ 14·3″, the building's designed height would work out at just over 5813 P".

Perhaps the outstanding example of the 'opposite' point of view is the survey by J. H. Cole, also a professional surveyor, as set out in his somewhat neglected *Determination of the Exact Size and Orientation of the Great Pyramid* (Cairo: Government Press, 1925). According to Cole's survey, the base-sides of the Pyramid *as built* would have varied in their alignments by up to 3½ minutes of arc from the true square, which would itself now be aligned 2½ minutes of arc west of true north (a result of polar precession). At the same time the four sides' respective lengths would have varied by up to 8 inches.

As L. C. Stecchini points out in his comprehensive appendix to Peter Tompkins's *Secrets of the Great Pyramid*, Cole's actual figures would have some interesting consequences. In particular, the geometry of the Pyramid's north face would have been a direct function of the φ-ratio, while that of the west face would have demonstrated the quantity π. Comparison with Appendix D would thus suggest for the Pyramid's north/south aspect a symbolism connected with the dynamic 'spiral' development of the human soul, while the 'divine' dimension would be represented by the 'static' π-based geometry of the Pyramid's east-west aspect. And certainly such notions would accord entirely with the symbolic read-out undertaken in chapter 3 of this book.

Cole's figures for the base of the Pyramid *as built* are as follows (shown here with Primitive-Inch conversion):

West side	230,357 mm	9059·5766 P"
North side	230,251 mm	9055·4078 P"
East side	230,391 mm	9060·9137 P"
South side	230,454 mm	9063·3914 P"
Total perimeter of base	921,453 mm	36239·2895 P"

Cole himself suggests an average tolerance here of some 31 mm (1¼ ins.) per side. Extrapolating from Cole's figures for the original Pyramid on the basis of the observed concavity of 35·762 + P" (see diagram page 354), the base-square of the *full-design* Pyramid would thus have the following dimensions:

West side	9131·1021 P"
North side	9126·9333 P"

East side	9132·4392 P″
South side	9134·9169 P″
Total perimeter of base	36525·3915 P″

Taking Cole's figures as a basis, then, the full-design Pyramid's base-perimeter would be *within one-and-quarter inches or so* of the figure suggested by Rutherford (36524·2 P″). Bearing in mind the fact that the difference is well within Cole's own suggested tolerances, and that the total distance involved is *well over half-a-mile*, it is clear that the two figures can for all practical purposes be regarded as identical, as well as providing impressive checks on each other's accuracy.

Both views of the full-design Pyramid, in other words, lead to the conclusion that the base-perimeter was intended to be directly related to the length of the mean solar tropical year (365·242 days). And from this fact the data listed below can be directly derived. As to whether we should regard the full-design Pyramid (with Rutherford) as standing on a perfectly square base, or (with Cole) on a slightly distorted one, this must remain a matter for personal preference, since the actual construction was never undertaken. Yet perhaps this fact itself is significant in the context. Had the full-design Pyramid ever been completed, then *either* Rutherford's Pyramid *or* that based on Cole's measurements would have been ruled out of account. As it is, however, *both* possibilities exist: the architect, so to speak, has succeeded in having his cake and eating it too. The fact that both perimeter-measurements are for all practical purposes identical suggests that both Rutherford's version *and* Cole's were present in the architect's mind, the one being merely another version of the other. And in this case it is clear that the *basic* design must have been the simple square as proposed by Rutherford.

Interpreted symbolically, then, we may perhaps see the basic design (founded on π and the perfect square) as representing, among other things, the divine or eternal 'in a state of rest'. At this point, however, a tension or 'distortion' comes into the picture. A basic polarisation ensues, whereby the human soul (represented by φ and the distorted Pyramid's north/south aspect) separates from the divine matrix (represented by π and retained in the 'distorted' Pyramid only in its east/west aspect). This evolutionary downturn is reflected in the reduced dimensions of the still-distorted Pyramid as actually built. A process of dynamic growth must now occur, in order to permit the reunion of soul with its divine origin, and thus the re-attainment of the primeval 'state of rest'. And the internal message of the Pyramid seems to be dedicated (as chapter 3 reveals) to the specific detailing of the stages of that growth-process.

Returning to the Pyramid's vital statistics, the following list of possible geophysical, geometric and astronomical correlations is based on Rutherford's figures for the basic, or perfect Pyramid. None of the references is of course *specific*: the supposed 'links' are derived simply from experimental application of the published data involving scales of 1:1, 1:10,

1:25, 1:100 and 1:10,000,000. Their validity therefore hangs on one's personal assessment of the likelihood that all the various figures involved could have come together accidentally. And it would seem, *a priori*, that the likelihood must be a very remote one. In the diagram above:

1. AB = 365·242 Sacred Cubits (SC)
   AD_1H_1B = 365·256 SC
   AmB = 365·259 SC
   But . . .
   365·242 days = *length of mean solar tropical year*
   365·256 days = *length of sidereal year*
   365·259 days = *length of anomalistic year*

2. Pyramid's Sacred Cubit = 25 P″
   = 250,000,000 P″ /10,000,000
   = 3949·89 statute miles/10,000,000
   But . . .
   3949·89 statute miles = earth's mean polar radius (confirmed 1957 by satellite observations)

3. VOU/OK = 1826·21235/470860·606
   = 0·003878414
   and WOC/OK = 9131·061625/470860·606
   = 0·0193924
   But . . .
   Known maximum and minimum values for eccentricity of earth's orbit = 0·004 and 0·019 respectively.

4. OK = 7·4393819 statute miles
   = 185,984,540 miles/25,000,000
   But . . .
   185,984,540 miles is a very close approximation for the diameter of the earth's orbit, giving a mean distance of the earth from the sun's centre of 92,992,270 miles.

5. AY + BX = 25,826.4 P″
   But . . .
   25,826·4 years is an acceptable figure for the duration of the cycle of the precession of the equinoxes (period subject to slow variation —present value not yet precisely determined).

6. AXYB (plan)/CX (section) = 36524·2/5813
   = 6·28319
   But . . .
   6·28319 = 2 × 3·14159+
   = 2π

From the foregoing we have little choice but to conclude that the basic, full-design Pyramid represents not merely the divine or eternal 'in a state of rest': it also stands for the planet earth itself, in harmony with its cosmic environment. The distortion and diminution observable in the actual Pyramid *as built* therefore seem to speak of a spiritual falling-away reflected in a present cosmic disharmony. As the following pages show, the measure of that disharmony is then reiterated within the passageways themselves.

2 The Astronomically-based Geometry of The Great Pyramid's Base, and its Links with the Building's own Internal Measurements

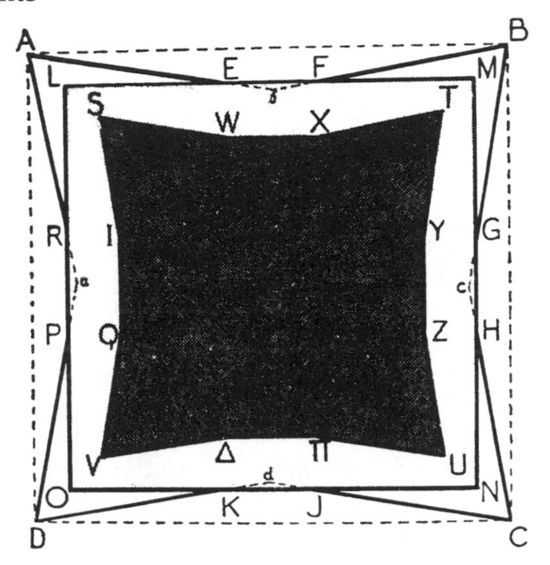

LMNO – Base as actually built
SWXTYZUπΔVQI – Base of core masonry
AEFBGHCJKDPR – Base of full-design pyramid
(relative casing-thickness and concavity greatly exaggerated).

On the above plan of the Pyramid's base, as has been pointed out in section 1,

AB	= 365·242 SC	= number of days in Mean Solar Tropical Year (equinoctial)
AEFB	= 365·256 SC	= Number of days in Sidereal Year (earth's solar revolution)
AbB	= 365·259 SC	= number of days in Anomalistic Year (earth's orbital revolution)

Identical ratios are of course obtainable from the full-circuit measurements expressed in Primitive Inches (= 1·00106 standard inches), which give 36524·2, 36525·6 and 36525·9 P″ respectively. Indeed, the same ratios necessarily apply *whatever the units of measurement used*.

Meanwhile,

35·762+ P″ = depth of side-concavity
= height of Great Step
= distance of Well-Shaft axis beyond north wall of Grand Gallery
= height of Subterranean Passage.

Now side of outer (designed) circuit-square	= 9131·05 P″
	= 365·242 Sacred Cubits
Therefore, perimeter of outer circuit-square	= 36524·2 P″
But side of *inner* (built) circuit-square	= 9059·5 P″
	= 9131·05 – (2 × 35·762+) P″
Therefore, perimeter of *inner* circuit-square	= 36238 P″
	= 36524·2 – (8 × 35·762+) P″
	= 36524·2 – 286·102+ P″

Thus **286·102+ P″** = difference between inner and outer circuit squares, *but also*
= half designed base-length of capstone (see Appendix H)
= distance of passage-axis to east of Pyramid's axis
= mean height of Grand Gallery roof above Ascending Passage roof
= floor-distance (east side) between lower Well-Shaft opening and bottom end of Descending Passage.

The significant distances 35·762+ P″ and 286·102+ P″ (its multiple) are thus found both outside and inside the Pyramid, and appear to have been determined initially by astronomical considerations.

APPENDIX C

The Great Pyramid's Dimensions and the Egyptian King-List Data

All the figures on page 357 (omitting fractions—see Tolerances, page 41) are derived either directly or indirectly *from the chronologies of the ancient dynasties* reported by Herodotus and Manetho, and later edited by various other commentators. Details are given in the table on page 358.

Herodotus, in his chronology of the Egyptian dynasties, refers to:

8 'gods' (Hephaistos, Helios, Agathodaimon, Kronos, Osiris and Isis, Typhon (Set), Horus, Ares)
12 further 'gods' (Anubis, Heracles, Apollo, Ammon, Tithoes, Sosos, Zeus and five others)
4 dynasties of 'demigods' and
30 human dynasties.

Details of the number of years in each reign were subsequently listed by the Egyptian priest Manetho (third century B.C.), whose work survives in the varying editions and reports of Africanus, Eusebius, Syncellus and others. The resulting figures are tabulated by Davidson and Aldersmith on page 77 of *The Great Pyramid* (1925), and from them all the data listed on page 358 may be extracted.

Some of the figures given may of course bear purely accidental relationships to the Pyramid's dimensions, but the similarities seem to be far too numerous to permit the assumption that *all* of them are coincidental. If this were the case, after all, then a similar number of coincidences could be expected in respect of the Pyramid's dimensions expressed in metres or feet—yet there is a total lack of evidence to this effect.

Consequently it is virtually certain that the king-lists themselves were devised or re-modelled on the basis of the known measurements of a

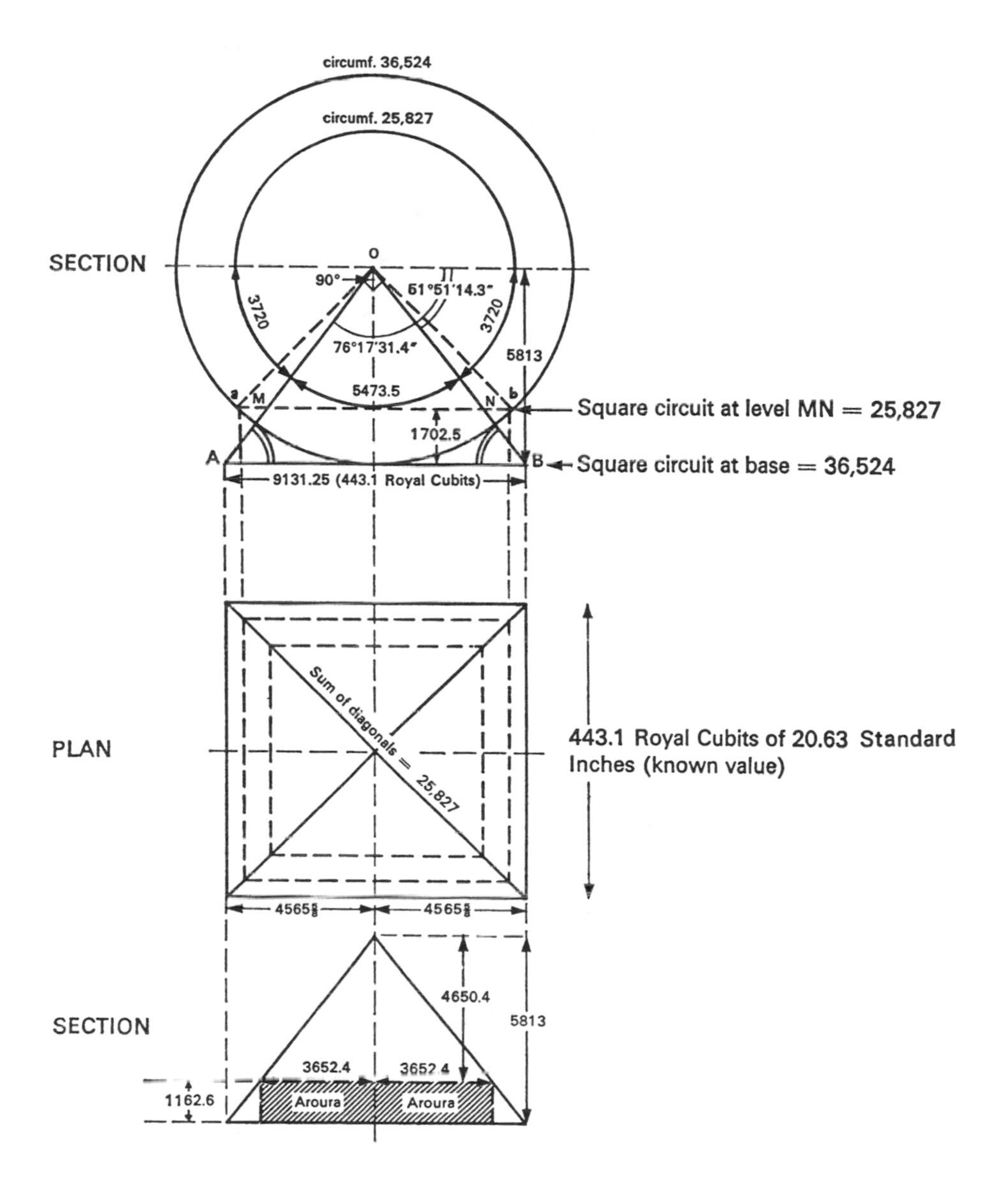

Diagram showing how all the Great Pyramid's basic external measurements are preserved in the Ancient Egyptian fictitious king-lists (all measurements in P″ unless otherwise stated

Source	*Figure*	*King-list attribution*	*Pyramidal application (full-design)*
Syncellus	36,525	Total years, divine and human dynasties to Amasis II (end of 30th dynasty)	Base perimeter in P″[1]
Manetho	5,813	Total years, dynasty of Manes	Height in P″
Manetho	29,220	Total years, gods and kings	Circuit at 35th course axis in P″
Let aMNb subtend at O an angle of 90° in outer circle (centre O) circumference 36,524 P″.			
Manetho Africanus	25,827	Total years, divine dynasties	Circuit at level MN in P″ Sum of base-diagonals in P″ Circumference of inner circle, centre O, to touch MN
Manetho	1,702	Total years, dynasty 3 of demigods	Height of level MN from base in P″
Africanus	5,474	Total years, human dynasties 1 to 31	Length in P″ of arc subtending at O the Pyramid's apex-angle (76° 17′ 13·4″) in inner circle (centre O, circumference 25,827 P″)
Castor	3,720	Total years, human dynasties 1 to 18	Length in P″ or arc subtending at O the Pyramid's angle of slope (51° 51′ 14·3″) in inner circle (circumference 25,827 P″)
Eusebius	4,565	Total years, human dynasties 1 to 31	Half base-side in P″
Syncellus	443	Total in years of the 15 generations of the Cynic (Sothic) cycle after the divine dynasties	Length of base-side *in Royal Cubits*
Africanus	5,151	Total years, first 26 human dynasties	Side of square equal in area to Pyramid's cross-section, in P″

Standard Pyramid such as that shown, and that the Pyramid in question *was the Great Pyramid.*

Experimental application of the king-list data therefore clearly suggests:

that the unit of measurement = 1 Primitive Inch,

that the Primitive Inch = 1·001+ standard inches (in consequence of the independently quoted base-side length of 443 Royal Cubits of known length),

that the angle of slope with the horizontal = 51° 51′ 14·3″,

that the apex angle = 76° 17′ 31·4″,

that the square base-circuit = 36,524+ Primitive Inches, or independently,
= 1772+ Royal Cubits,

that the height from base to apex = 5813 P″,

[1] Also (a) quoted by Syncellus as the number of years in the Egyptians' 'fabled period of precession of the equinoxes'—apparently in erroneous substitution for 25,827 (see fourth entry in table),

(b) traditionally the number of treatises written by the 'god' Thoth.

(c) ten times the number of P″ in the year-circle or quarter-aroura.

that the square circuit at 1702 P″ above the base = 25,827 P″ (both figures quoted independently),

that the square-circuit at 1162 P″ above the base = 29,220 P″,

that the Pyramid's vertical section was equal in area to a square of length of side 5151 P″ (= 25 quarter-arouras).

These data therefore not only indicate beyond all reasonable doubt that the Standard Pyramid of Measures referred to in the ancient Egyptian texts (and apparently used as a basis for the remodelling of the dynastic chronologies) was indeed the Great Pyramid, but also confirm gratifyingly the accuracy of the reconstruction of the Pyramid's original dimensions made from geometrical and trigonometrical observations arrived at on the site. Even if the Great Pyramid no longer existed, in other words, it would be possible, by experimental geometrical application of the figures contained in the ancient king-lists, to reconstruct it to its intended full-design with a fine degree of accuracy.

APPENDIX D

The Great Pyramid's Dimensions and the Phi-Ratio

1. The quantity π is not the only important mathematical ratio inherent in the Great Pyramid's design. Examination shows that the design also incorporates an extremely close approximation to the value φ (*phi*)—better known, perhaps, in terms of the Golden Section.

2. The ratio of a rectangle's length to its width can, of course, range from 1 (in the case of a square) to infinity (in the case of a theoretical rectangle of infinite length). But it has long been realised that certain shapes of rectangle seem to the human eye to be aesthetically more satisfactory than others. Indeed, given a large range of rectangular shapes to choose from, most people, it is said, will tend to choose as most satisfactory one whose length bears to its width the ratio $(\sqrt{5} + 1)/2$, or 1·61803+, the same ratio, in fact, as the sum of both dimensions will then bear to the length alone. The resulting proportions are thus both mathematically and aesthetically elegant, and it is their governing ratio of 1·61803+, exhibited in the Athenian Parthenon and a number of other buildings of classical antiquity, which is known as the Golden Section or the irrational number φ.

3. In the Great Pyramid (assuming Rutherford's 'regular' full-design), the ratio between the slope-height and half-base (i.e., the secant of the angle of slope) is 1·61899 (correct to five decimal places)—or within ·001 of the φ-ratio.

4. It needs to be pointed out, however, that this appearance of the φ-ratio is *an inevitable consequence* of the architect's choice of a four-sided pyramid incorporating the 'π-angle' of 51° 51′ 14·3″ as its angle of slope. None the less it is true to say that, whether by accident or design, the *Great Pyramid embodies a direct geometrical expression of the approximate relationship between the quantities* π (3·14159+) *and* φ (1·61803+).[1]

[1] Approximation is, of course, inevitable, since both numbers are irrational. Other approximate expressions of the $\pi:\varphi$ relationship are $\pi/4 = 1/\sqrt{\varphi}$ and $5\pi/6 = \varphi^2$.

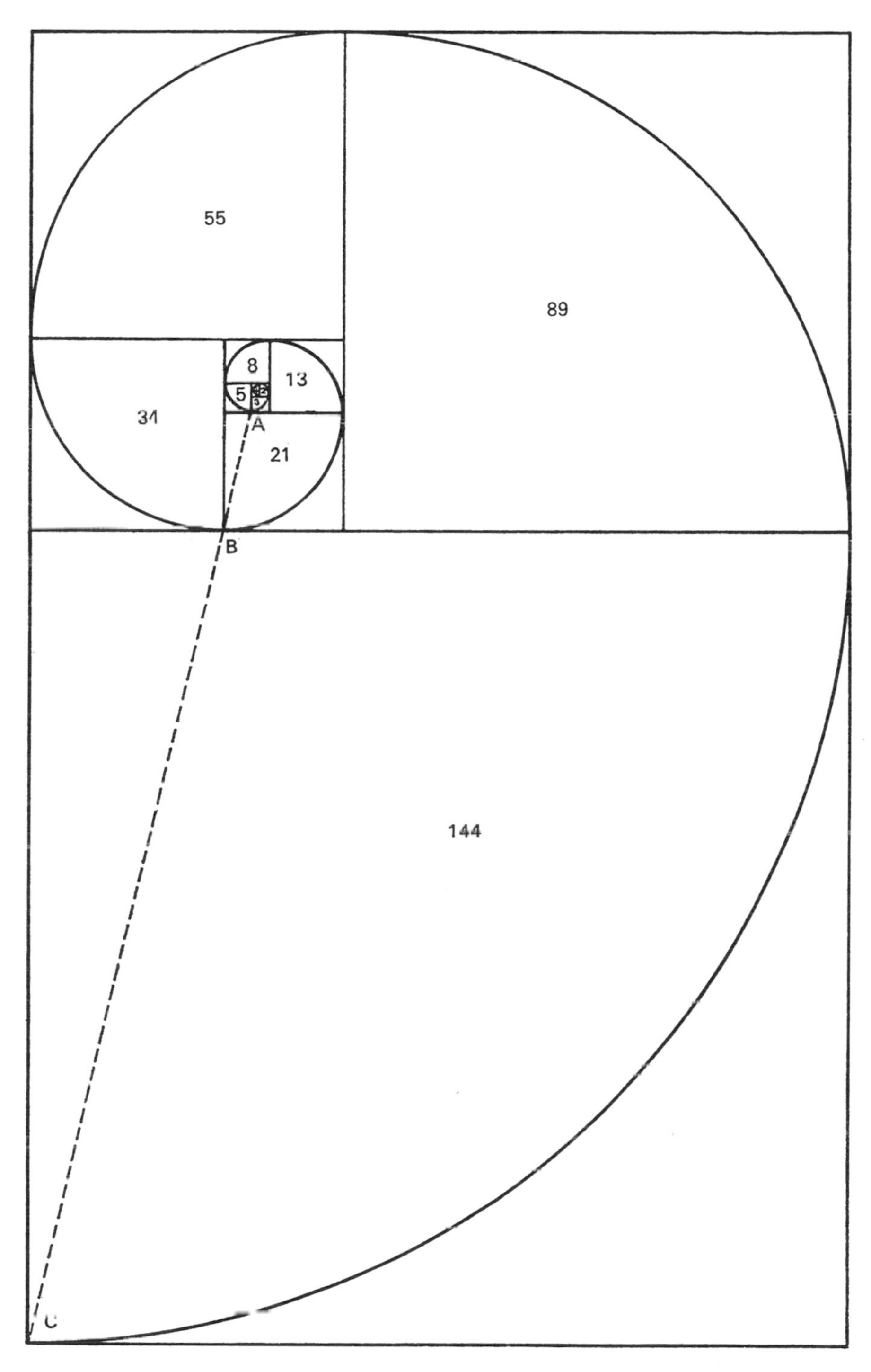

Logarithmic Spiral based on the Fibonacci Series. The radius of each quarter-turn of the spiral is determined by the length of side of the square in which it is inscribed, and the value of this for each succeeding square is in turn determined by the Fibonacci Series.

5. The number φ has several remarkable mathematical properties. Its square, for example, is equal to itself plus one, while its reciprocal equals itself minus one. But perhaps its most intriguing feature is its link with the so-called Fibonacci Series.

6. The Fibonacci Series is the sequence of numbers starting 0, 1, 1, 2, 3, 5, 8, 13, 21, 34, 55, 89, 144 . . ., in which each number equals the sum of its two predecessors. It is found with surprising frequency in nature—notably in patterns of plant growth, in flower-petal arrangements and fir-cone design, in the laws of Mendelian heredity and in the ratios between planetary orbits—and, when plotted diagramatically, it provides the mathematical formula for the construction of a logarithmic spiral which is also found in nature—notably in the shell of the Nautilus (see diagram page 361).

7. Like the ratio between the adjacent sides of a rectangle, the ratio between each member of the Fibonacci Series and its successor varies initially between one and infinity; but the further up the series one goes, the more the ratio tends to settle down towards a more-or-less constant figure. Investigation reveals that this 'central' figure is none other than φ, or 1·61803+ .

8. The ratios between the lower Fibonacci numbers quoted in 6. (above) are as follows:

∞; 1; 2; 1·5; 1·6; 1·$\dot{6}$; 1·625; 1·615+ ; 1·619+ ; 1·6176+ ; 1·61818+ ; and 1·6179+. It will thus be seen that the ratios between the succeeding Fibonacci numbers alternately 'box in' the φ-ratio from above and below. When plotted as a graph, the result is a kind of 'two-dimensional spiral', as shown in the diagram opposite.

9. John Ivimy, in his book *The Sphinx and the Megaliths*, claims that if one assumes a unit of measurement of 4 Royal Cubits, the Great Pyramid's slope-height in fact measures 89, and its half-base 55, such units. And consequently, he points out, there seems to be a direct link between the Pyramid and the Fibonacci Series, since 55 and 89 are successive numbers in that series.

None the less it should be pointed out

that these measurements apply only to the original Pyramid *as built*, and not to its full-design counterpart;

that, even then, neither measurement quoted is exact, being some 4″ and 3″ too great respectively;

that the φ-ratio resulting from the dimensions quoted by Ivimy is 1·61818+ instead of the true figure for the secant of the Pyramid's angle of slope, which is 1·61899+ .

In fairness it should be said, however, that Ivimy takes the Pyramid's angle of slope to be 51° 50′ instead of its true value (51° 51′ 14·3″). And, in any case, it must be admitted that the 'near-miss' is an extraordinarily close

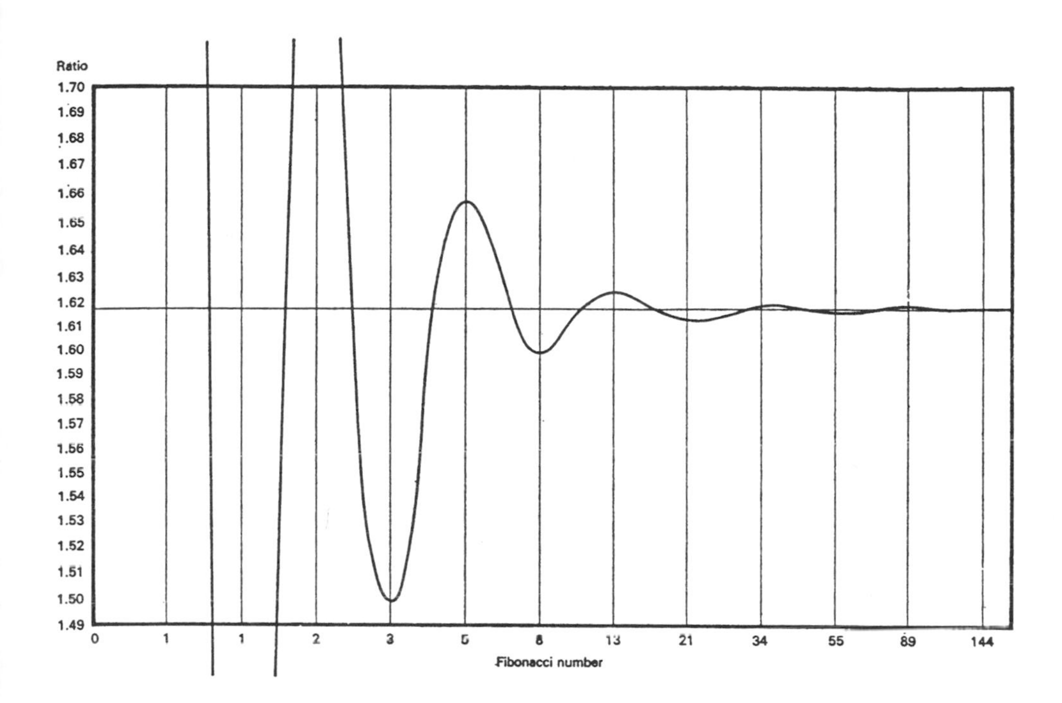

Graph of the ratios between successive Fibonacci numbers.

one, amounting to a factor of scarcely more than 8 in 10,000 as far as the Pyramid's slope angle secant is concerned.

10. From the above, then, the following facts seem to have emerged:

that the Great Pyramid's design demonstrates not only the ratio π, but also an extremely close approximation for the ratio φ;
that the Pyramid's demonstration of the φ-ratio is a direct function of its π-ratio geometry;
that the dimensions of the original, reduced-design Pyramid may have had a direct and intended link with the Fibonacci Series;
that the geometric form characteristic of the Fibonacci Series is the spiral;
that graphical representation of the ratios between successive Fibonacci number shows a 'two-dimensional spiral' centred around—and tending ever more closely towards—the value φ.

What, then, if any, could be the combined significance of these facts?

11. In terms of the Pyramid's hypothetical code (see chapter 2), π stands for the divine or the eternal. In view of this, might not the fact that φ is here presented as a *function* of π indicate that it in turn is intended to represent some particular *aspect* of the divine or eternal?

If this possibility is accepted, then the next point to observe is that the full-design Pyramid seems to be designed to exhibit π rather more than φ, while the reduced original building seems to be intended, via its overt reference to the Fibonacci series, to attest to some dynamic process of spiral development leading inexorably 'inwards' towards the quantity φ. Stecchini's speculations (see Appendix B) may be of relevance here, and suggest a *specific* φ-reference in the original design.

At this point we should recall that the Pyramid's designer seems to have seen human history—and with it, presumably, the very development of the human soul—as basically cyclic or spiral in form (see 'Astrological Parallels' page 322). And bearing this in mind, we can perhaps translate the symbolism of the above (if valid) to mean that the present reduced-design Pyramid denotes in some way the 'imperfect', reincarnating human soul spiralling on its inexorable karmic way towards union with the eternal or divine, as symbolised by the full-design Pyramid's essentially π-based geometry. At which point it is worth noting that the French archaeologist and philosopher Schwaller de Lubicz is on record as affirming specifically that the Egyptians saw the φ-ratio as a symbol of reproduction and of the eternal, on-going creative function—which is a notion by no means unrelated to the above.

12. But is there any supporting evidence for this admittedly speculative view? Here it is the Fibonacci series itself which, taken in conjunction with the Pyramid's own number-code, offers one or two of the more interesting possibilities.

We have suggested above that the reduced-design Pyramid's apparent reference to the Fibonacci Series might symbolise the still-imperfect reincarnating soul. But reference to the Fibonacci Series itself, as set out on page 362 above, shows not only that 8 and 13 (Pyramid-code for rebirth and soul) are both numbers in the series—and successive ones at that—but also *that 13 is itself the eighth item*. Could this, perhaps, be seen as a possible link between the symbolism of the spiral and the notion of the reincarnating soul? Could it, indeed, have been one of the designer's reasons for allocating the meanings rebirth and 'soul' to the numbers 8 and 13 in the first place?

13. Again, it will be observed that, in the diagram on page 361, the completion of the first, second and third revolutions of the spiral at A, B and C is associated with the numbers 3, 21 and 144 respectively—in turn, Pyramid-code for perfection, utter spiritual perfection and the elect (men of men). Could this in some way be taken as a kind of 'shorthand' for the path leading to the eventual flowering of the human soul—a shorthand once again made possible by the original allocation of appropriate meanings to those numbers in the Pyramid's arithmetical code?

Again, given that it is the *third* revolution of the spiral that is associated with the number 144, one wonders what connections there might be with the persistent Messianic notion that the 'third day' is characteristically the 'day of the elect'.

14. In view of the approximate nature of the various mathematical correlations involved, it is difficult to draw any firm conclusions regarding the symbolic significance of the ratio φ in the Great Pyramid. There may, indeed, not be any. And yet the possibility remains that the Pyramid's designer may have seen in the φ-ratio and the Fibonacci-spiral a means of shedding further light on the evolution and destiny of the human soul. And it is not entirely beyond the bounds of possibility that the characteristics of the Fibonacci Series may even have influenced his choice of numbers in constructing the Pyramid's basic code.

APPENDIX E

The Great Pyramid, Stonehenge, Chartres Cathedral, Carnac—and the Limits of Coincidence

Coincidences that are not exact are, by definition, not coincidences: at best they are similarities. And whereas a multiple series of 'coincidences' argues—as in the case of the Great Pyramid's astronomical links—the probability of a single causal link, a multiple series of similarities can do no more than suggest the *possibility* of one. In short, in exploring such series of similarities for their supposed significance, one is venturing into the sphere of inspired speculation—usually a fairly pointless exercise unless backed up by corroborative facts, since the imagined links do not carry their own self-verification in the same way as exact coincidences do.

None the less, there has of recent years been a rapid increase in books attributing 'esoteric' significance to a number of ancient texts, buildings and traditions, and tracing supposed links between them on the grounds of information that is often plainly inaccurate. On this basis, similarities are then treated as 'coincidences', and adduced as proof of the author's original 'esoteric' thesis—however valid that thesis may or may not be in its own right. The result may seem impressive to those not equipped with the facts, and the 'mind-blowing' exercise involved may itself not be without some value to the reader; but the conclusions drawn are too often misleading, occasionally quite nonsensical, and always liable to lead to eventual disillusionment in the reader. The Great Pyramid in particular has tended to suffer from this treatment.

Thus, to take a few examples, the Pyramid's π-angle, or angle of slope (51° 51′ 14·3″) is almost (but not quite) the same as the angle of latitude of Stonehenge (51° 10·8″ N), whose famous bluestone-circle was apparently quarried in the megalith-littered Presely Mountains in Pembrokeshire, South Wales from a site (Carn Meini) which in fact lies on latitude 51° 57½′ N. Meanwhile the azimuth of the midsummer sunrise from Stonehenge—the basis of the monument's general alignment—is N. 51° 12′ E.[1] (i.e. within 1·2 minutes of arc of its latitude), while the Stonehenge outer circle can be used to construct a triangle which has almost (but not quite) the dimensions of the Great Pyramid's cross-section.[2] At the same time the layout of Stonehenge's Aubrey-holes *seems* to have been planned around an isosceles triangle having a slope-angle of 51° 19′ and a height of 100 Egyptian Royal Cubits, while the dimensions of its sarsen-ring are apparently related to the same circle by the geometry of the seven-pointed star.[3] But this same angle of 51° 19′ is also the angle of slope (according to Ivimy) of the Third Pyramid of Giza, and is nearly (but not quite) one seventh of a circle of 360°.[4] This in turn means that it is the angle subtended at the centre of Stonehenge by an arc of any 9 of its 56 Aubrey-holes. Moreover it is also the angle whose secant is defined by the Fibonacci-ratio 8/5, and thus the base-angle of any 5:4:4 isosceles triangle. And as Louis Charpentier points out in his remarkable *Les Mystères de la Cathédrale de Chartres*, this triangle is one of those which can be formed using what he calls the 'Druids' cord'—a cord of thirteen equal sections separated by 12 equally-spaced knots:[5]

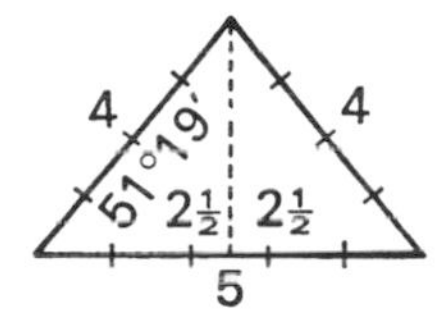

This triangle in turn, Charpentier suggests, may have been used in constructing the seven-pointed star which apparently is basic not only to the layout of Stonehenge but also to the design of Chartres cathedral. But then the latter itself seems to be based (according to Charpentier) on a measure of ·82 metres—almost (but not quite) identical to the 'megalithic yard' established by Prof. Alexander Thom as the basic measure of most western European megalithic stone-circles.[6] Again, the side of what Charpentier calls the cathedral's *table carrée* is 23·192 metres—almost (but not quite) one

[1] Sunrise here being defined as the moment when the sun's lower limb sits tangent on the horizon—the definition which seems to have been usual among the megalith-builders.

[2] See John Michell's *The View over Atlantis.*

[3] See John Ivimy's *The Sphinx and the Megaliths.*

[4] Actual value—51° 25′ 43″ approximately.

[5] 13 × 12—'the soul of man'?

[6] 32·28″ instead of 32·64″ (see Thom's *Megaliths of Ancient Britain*).

hundredth of the length of side of the Great Pyramid,[7] while the *proportions* of the cross-section of its choir apparently bear close similarities to those of the Pyramid's King's Chamber. Meanwhile the interior width of the choir is almost (but not quite) a direct function of the cathedral's terrestrial latitude, while the various levels of its nave correspond almost (but not quite) to the frequency-ratios between the notes of the musical scale. Charpentier seems not to notice, incidentally, that the cathedral's orientation (E. 46° 54′ N.) is almost (but not quite) that of the extreme midwinter *moonrise* (E. 47° 34′ N.) from the latitude of Chartres.

But that is not all. The celebrated Pythagorean 3:4:5 triangle (exemplified in the Great Pyramid's King's Chamber) can also be made with twelve of the segments of the Druid's cord:

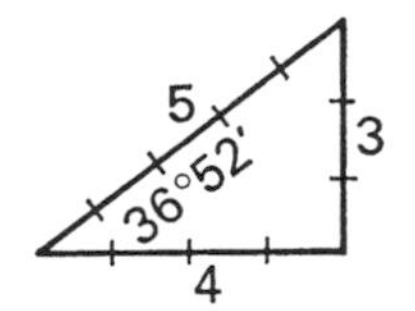

But as Professor Thom has pointed out, the 3:4:5 triangle is also basic to the geometry of many of the megalithic stone-circles, as well as to one in particular of the megalithic structures at Carnac in Brittany. Rectangular in shape, this structure forms an enclosure orientated accurately to the four cardinal points, and its sides measure 3 and 4 megalithic yards respectively. Its diagonal therefore measures five megalithic yards, and points in the direction E. 36° 52′ N., *the bearing at the latitude of Carnac of the midsummer sunrise*. But then, why in any case does this famous megalithic monument—apparently as old as any in Egypt—bear the same name as the even more famous Egyptian temple of Karnak at Luxor, likewise characterised by great avenues of stone pillars . . .?

One could of course continue almost *ad infinitum* with this chain of 'links'—or rather 'pseudo-links'—taking in such subjects as ley-lines and extra-terrestrial visitors on the way. And indeed, if all these facts could be shown to be significant, one would possibly be forced to the conclusion that they had all been 'planned' by some former advanced civilisation, either for our benefit or for 'magical' or scientific purposes unknown to us. Unless, of course, it could be shown that these facts tend to be *necessary* functions of each other—in which case they might convince us that all knowledge and all phenomena are somehow one and interconnected, as indeed they ultimately must be.

[7] 9130·477 standard inches instead of 9140·7284. But oddly enough, almost exactly the right number of *Primitive* inches in the Pyramid's base side. If it could be shown, in other words, that this measurement was originally made by the cathedral's builders in what they thought were Primitive Inches, which had in fact by that time contracted to the present value of the *standard* inch, there might be some significance in Charpentier's interesting observation.

But facts such as these cannot reasonably be taken to be significant until it can be shown that they are more than mere similarities. Either we must find *exact correspondences*, or we need reliable independent evidence on the reality of the supposed links. It would help, for example, if we were to find evidence that there was once a Proto-Stonehenge with definite Egyptian connections in Pembrokeshire at latitude 51° 51′ 14·3″ N.—in the vicinity of Llandisilio, perhaps (whose name is oddly reminiscent of the name Atlantis). But failing evidence of this kind, much of the information adduced is largely worthless—and, on these terms, only a handful of the above similarities would appear to qualify as inherently significant.[8]

We owe a great deal to the historical speculations of Ivimy, Michell, Tomas, von Däniken, Kolosimo, Charpentier, T. C. Lethbridge and others. In the same way we owe much to the great futuristic science-fictionists such as Clarke and Asimov. All of them have served to broaden considerably the conceptual horizons of humanity. But, ultimately, it is hard facts—accurate information of the kind amassed by Thom and Rutherford, and rigorous scientific research such as that undertaken by the scientists of N.A.S.A.—which alone will bring these efforts to fruition in the world-wide realisation that man's past is not what he has hitherto supposed it to be, while his future achievements may surpass his wildest imaginings.

[8] One of the most impressive of those listed is perhaps the extraordinarily close link between the azimuth of the midsummer sunrise from Stonehenge and the monument's terrestrial latitude. This is, after all, the only latitude in the northern hemisphere where the two angles are equal—just as the latitude of Carnac is the only one in the northern hemisphere where the hypotenuse of a 3:4:5 triangle having an east-west base of 4 units and a north-south height of 3 units points directly at the midsummer sunrise.

APPENDIX F

The Great Pyramid's Principal Interior Measurements

The data which follow, all in Primitive Inches, are based throughout on the figures supplied by Rutherford for the full-design Pyramid. Readers surprised by the extraordinary degree of precision claimed would do well to consider the following facts:

(i) During the last century or so the Pyramid has been measured almost *ad nauseam*, both inside and out, not merely by amateur archaeologists, but by numerous professional surveyors, some of whom have spent literally months on the site using the most up-to-date equipment available—parts of it designed specifically with the Great Pyramid in mind. Due and precise allowance having been made for the effects of observed temperature-variation on the instruments themselves, for subsidence-distortion, and for wear and exfoliation of the ancient stone, the result has been a series of figures already of outstanding accuracy, and expressed in each case to clearly defined tolerances.

Rutherford's own figures for the Pyramid as designed are, almost without exception, well within the stated tolerances of the most authoritative surveys (on which, indeed, they are largely based) and are thus, by definition, necessarily as valid as any other figures which conform to those tolerances.

(ii) One of the procedures used in the measurement of the Pyramid involves the averaging of a large number of different measurements of the same feature. The resulting average figure tends, not unnaturally, to ignore preconceived notions of what a 'neat' answer should be, and often displays many digits to the right of the decimal point. At this juncture the temptation is great to round the figures up or down. Yet a moment's thought reveals the potential danger of such a policy—for how is the researcher to know that he is not moving *away* from the correct figure? Consequently the resulting figures need initially to be accepted as they stand.

(iii) Certain of the measurements thus arrived at turn out to correspond to exact numbers of Egyptian Royal Cubits—a measurement whose value is already known to very fine tolerances. Examination of these measurements in the Pyramid permits even further refinement of the accepted figure (see Appendix A)—and their exact expression in Primitive Inches necessarily involves, once again, several digits to the right of the decimal point.

(iv) Some of the measurements involved (including the Royal Cubit itself) turn out to be clear and direct functions of the distance 365·242 P″ and the quantity π (see, once again, Appendix A). The fact that π can theoretically be calculated to an infinite number of decimal places once again produces figures in which fractions figure prominently.

(v) Finally, in side-elevation, the Pyramid and its passageways present a clear geometric figure composed largely of straight lines and based on known angles and levels (compare page 372), these too having been meticulously surveyed on many occasions. Trigonometrical calculation therefore makes it possible finally to check many of the 'raw' measurements resulting from (i) to (iv) above *against each other*, thus exposing even the slightest inaccuracy. For this, absolutely precise data are essential, and any prior rounding up or down would invalidate the results.

(vi) Rutherford's final figures, as listed below, are almost alone in passing this crucial test with flying colours, in that they 'fit' each other trigonometrically to make a perfect and self-consistent system—as any correct assessment of the Pyramid's measurements ultimately must do. There seems to be no alternative, therefore, but to accept Rutherford's figures as they stand, as representing the best available assessment of the Pyramid's intended dimensions.

A List of the More Significant of the Great Pyramid's Passage Measurements

(all figures in Primitive Inches, and based on the data supplied by Rutherford)

Descending Passage

Vertical distance, beginning of floor to beginning of roof	37·995
Vertical height of passage	52·7452
Perpendicular height of passage	47·2842
Width	41·2132 (2 RC)
Entrance to foot of Scored Lines	481·7457

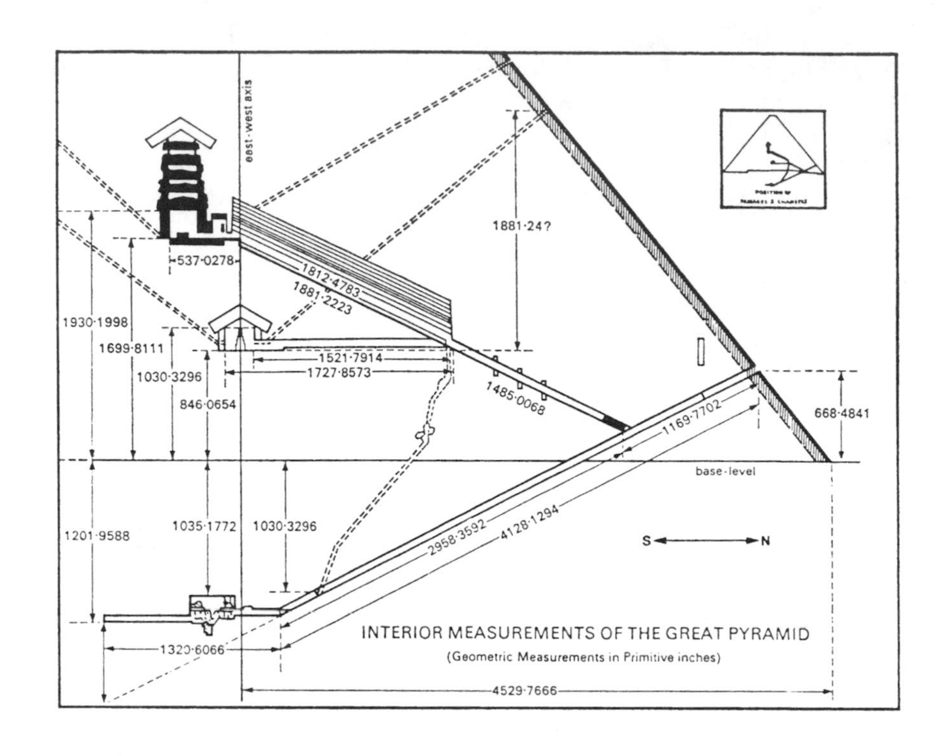

Scored Lines to intersection with Ascending Passage floor-line — 628·5079

Scored Lines to point vertically below beginning of Ascending Passage floor — 688·0245

Entrance to intersection with Subt. Passage floor-line — 4128·1294

Entrance to end of Dead End Passage — 5448·736

Subterranean Chamber Passage

Height — 35·762766

Width — 33·5204

Centreline floor-length from intersection with Descending Passage floor-line — 352·2933

Distance to Recess from intersection with Descending Passage floor-line — 220·3984

Distance across Recess from north to south — 72·35187

Recess to furthest point southwards of passage floor — 59·5430

Great Subterranean Chamber

Length east to west — 553·1598

Width north to south — 322·7711

Upper part of pit, depth	67·5946
Lower part of pit, depth	41·2132 (2 RC)
Total depth of pit	108·8078
Dead End Passage	
Mean height	29·8412
Mean width	29·8412
Length	645·5422
Ascending Passage	
Vertical height	52·7452
Perpendicular height	47·2842
Width	41·2132 (2 RC)
Floor-line length from intersection with Descending Passage floor	1544·5234
Beginning of floor from intersection with Descending Passage floor	59·5166
Beginning of floor to level of Queen's Chamber Passage	1451·4952
Top end of passage from level of Queen's Chamber Passage	33·5116
Possible original length of Granite Plug	206·0659 (10 RC)
Queen's Chamber Passage	
Height of low first portion	46·988
Height of high second portion	67·5946
Height of downward step at beginning of first portion	5·32075
Height of downward step at beginning of second portion	20·6066 (1 RC)
Width throughout	41·2132 (2 RC)
Virtual beginning of passage (under Grand Gallery north wall) to Well-Shaft-opening axis	35·7628
Distance across Well-Shaft opening	26·7021
Virtual length of low first portion from Grand Gallery N. wall	1305·2246
Actual length of floor of low first portion from first downward step	1282·81285
Length of high second portion	216·5668

Queen's Chamber

Length east to west	226·67252 (11 RC)
Width north to south	206·06593 (10 RC)
Height of north and south walls	184·26425
Height of east and west walls to gable apex (inner)	243·75036
Height of gable apex (inner) above north and south walls	59·48611
Height of niche	184·26425
Width of niche at bottom (north to south)	61·81978 (3 RC)
Width of niche at top (north to south)	20·60659 (1 RC)
Depth of niche (east to west)	41·21319 (2 RC)
East-west distance from Queen's Chamber Passage axis to N.-S. niche-axis	41·21319 (2 RC)
North-south distance from Queen's Chamber axis to E.-W. niche-axis	25·0000 (1 SC)

Grand Gallery

Extreme height	338·8473
Extreme height above Ascending Passage roofline	286·10213
Width between ramps	41·2139 (2 RC)
Width of roof	41·2139 (2 RC)
Width over top of ramps	82·42637 (4 RC)
Length of floor-line from north to south wall	1881·2223
Length of roof (approx.)	1836·0000
Height of bottom section of east and west walls	89·80568
Floor length to downward step into Queen's Chamber Passage	25·0000 (1 SC)
Floor-line length to intersection of Grand Gallery floor-line and Queen's Chamber Passage roof-line (approx.)	94·0000
Floor-line length to Great Step	1812·47832
Produced floor-length under Great Step to Hidden Step	68·74398
Height of Great Step riser	35·762766
Distance across top of Great Step (north to south)	61·62660
Distance across top of Great Step (east to west)	82·42637 (4 RC)

King's Chamber Passage—First Low Section

Height	41·21319 (2 RC)
Width	41·21319 (2 RC)
Length	52·02874

Antechamber

Height (1/100 × sum of height and base length of Pyramid)	149·44071
Floor-width	41·21319 (2 RC)
Width above east wainscot	53·23458
Width above west wainscot	65·25603
Length (365·24235/π)	116·26025
Length of limestone part of floor	13·22729
Length of granite part of floor	103·03296 (5 RC)
Height of east wainscot	103·03296 (5 RC)
Height of west wainscot	111·8034
Height of limestone portion of south wall (approx.)	12·0000
Height of Granite Leaf above floor	41·21319 (2 RC)
Dressed-down thickness of Granite Leaf	15·75
Thickness of boss or seal on north face of Granite Leaf	1·0000
Height of boss or seal at junction with Granite Leaf (approx.)	5·0000
Width of surface of boss or seal (approx.)	5·0000
Height of bottom edge of boss above bottom of upper slab of Leaf	5·0000
East-west distance from boss axis to Granite Leaf axis	1·0000
Distance from face of boss to north wall of chamber	20·60659 (1 RC)
Distance from axis of boss to east end of Granite Leaf (in wall)	25·0000 (1 SC)
Height of bottom of boss from floor (approx.)	74·0000

King's Chamber Passage—Second Low Section

Height	41·21319 (2 RC)
Width	41·21319 (2 RC)
Length	101·04629
(Total length of both Low Sections	153·07503)

King's Chamber

Length east to west ($2 \times 365{\cdot}24235/\sqrt{\pi}$)	412·13186 (20 RC)
Width north to south ($365{\cdot}24235/\sqrt{\pi}$)	206·06593 (10 RC)
Height ($\sqrt{5} \times 365{\cdot}24235/2\sqrt{\pi}$)	230·38871
Floor diagonal ($\sqrt{5} \times 365{\cdot}24235/\sqrt{\pi}$)	460·77743
Diagonal of east and west walls ($3 \times 365{\cdot}24235/2\sqrt{\pi}$)	309·09889 (15 RC)

Diagonal of north and south walls ($\sqrt{21} \times 365{\cdot}24235/2\sqrt{\pi}$)	472·15636
Cubic diagonal ($5 \times 365{\cdot}24235/2\sqrt{\pi}$) (see also Rutherford pages 1010–1011)	515·16482 (25 RC)
Length of coffer (N.-S.)	89·80568
Width of coffer	38·69843
Height of coffer (sum of coffer's height, width and length = 1/5 King's Chamber's height, width and length)	41·21319 (2 RC)
Thickness of coffer's sides (mean)	6·0000
Thickness of coffer's base (approx.)	7·0000
Combined distance to north and south walls at either end of coffer ($365{\cdot}24235/\pi$)	116·26025
Distance from north wall to end of coffer ($365{\cdot}24235/2\pi$)	58·13013
Distance from north wall to interior of coffer (N. end) (approx.)	64·13013
Distance of coffer's N.-S. axis to west of King's Chamber Passage axis (assumed)	286·10213
Distance from mid-point of Antechamber to S. wall of King's Chamber	365·242

APPENDIX G

The Great Pyramid: Core-Masonry Data

The course-thicknesses shown on the graph are mean values based on Petrie's 1881 measurements at the north-east and south-west corners. Detailed inspection of individual courses suggests an average accuracy of around ± ·6 P″ for Petrie's figures, while few courses are themselves absolutely constant in thickness.

Allowing for this degree of built-in error, it will be seen from the graph

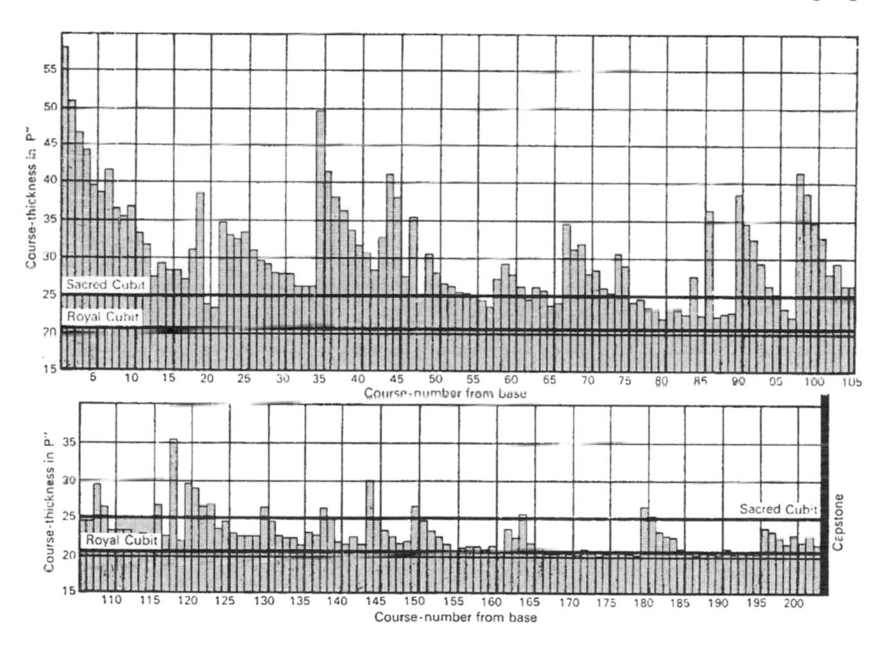

The Great Pyramid's core-masonry: graph of course-thicknesses.

that a number of surprisingly regular 'curves' result, interspersed with a number of sudden 'peaks'—features which may conceivably have an objective, exterior significance in some field of specialist inquiry yet to be identified. Internally, however, it is noteworthy that factorisation of the course-numbers of the 26 unmistakably peak courses produces as factors the numbers 1, 2, 3, 4, 5, 6, 7, 8, 9, 10, 11, 12, 13, 19, 29, 41 and 67 (all of them features of the Pyramid's geometric and/or arithmetical codes as set out in chapter 2), plus the further numbers 17, 23, 37, 43 and 59. Since it would presumably have been perfectly possible for the designer to have chosen exclusively *prime* numbers for his peak courses—or numbers divisible only by 2, 3 and 5, say—it seems reasonable to see in his choice of peak courses a deliberate indication of the essential signals comprising his internal code and a hint that factorisation might play a rôle in its application. In which case, of course, the last five numbers listed may also be in some way symbolically significant.

To take a case in point, the fact that 17 is one-ninth of 153 might suggest for it a meaning such as *enlightenment.* In this way 153 would in the first instance signify the utter perfection of enlightenment (9 × 17), and thus, by extension, the enlightened themselves.

Except in this connection, the number 17 nevertheless appears not to have been used in the Pyramid's passage-symbolism. On the other hand, it does appear to be relevant to the symbolism of the core-masonry itself. Without its capstone, the present Pyramid has 203 courses: it therefore appears to stand for the death of spiritual perfection (29 × 7). With the addition of the final capstone, however, the masonry courses will necessarily total 204, thus apparently symbolising (in view of the above) utter terrestrial enlightenment (3 × 4 × 17) or the enlightenment of mankind (17 × 12). It has to be admitted that these facts, if numerologically valid, are extraordinarily apt in the light of our overall interpretation in chapter 3.

APPENDIX H

The Great Pyramid's Internal Geometric Definition of its Missing Capstone

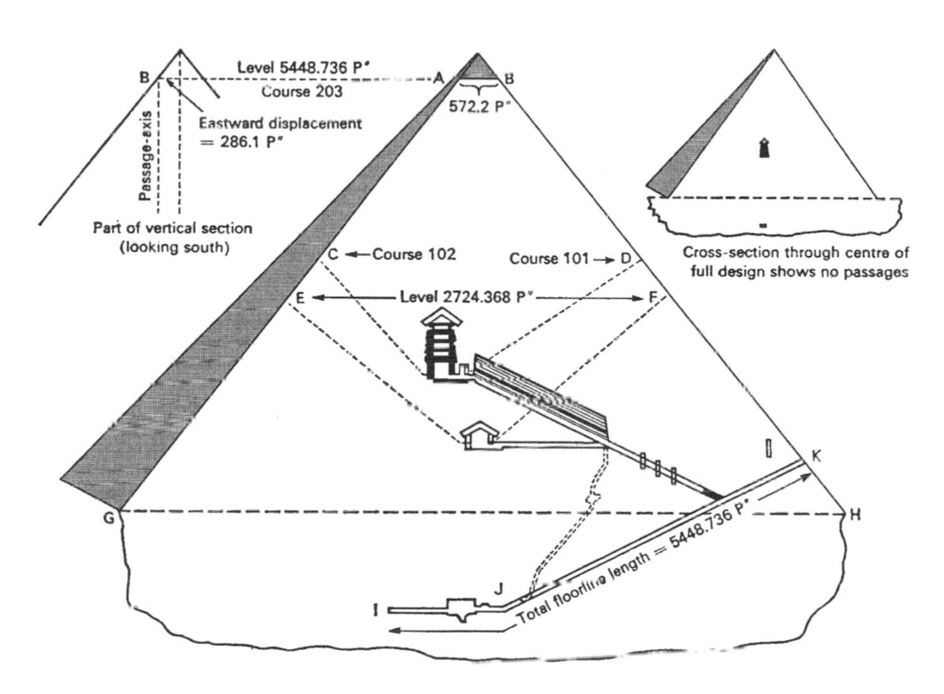

Vertical section through the Great Pyramid at the passage-axis, 286·1 P″ east of building's N-S axis, looking west.

In the main diagram above:

(i) AB represents the top of a cross-section taken through the full-design Pyramid at the Passage-Axis, 286·1 P″ to the east of the building's own north-south axis (see left-hand diagram, and compare other applications of this measurement in Appendix B).

AB therefore measures 2 × 286·1 P″, or 572·2 P″.

(ii) Meanwhile height of AB above base GH
= 5448·736 P″ geometrically.

(iii) But total floor-line length of lower passageways also
= 5448·736 P″ (compare Appendix F).

(iv) Combined heights of Queen's Chamber air-shaft outlets E and F in core-masonry (see chapter 5)
= 2724·368 P″ + 2724·368 P″[1]
= 5448·736 P″.

(v) Combined course levels of King's Chamber air-shaft outlets C and D in core-masonry (see chapter 3: 'The King's Chamber')
= 102 courses + 101 courses
= 203 courses.

(vi) But authoritative sources (e.g. Rutherford) put the original height of the Pyramid's present 203rd course (now only vestigial) at 5448·736 P″.

(vii) Taking (ii), (iii), (iv) and (vi) together, it becomes clear that the Pyramid's architect wished to emphasise that level AB was of some overriding importance, and that he therefore deliberately included the above four references to it in his design.

(viii) It seems logical to conclude that this level was intended to be seen as that of the Pyramid's planned 'temporary summit'—which, in the full design, would have taken the form of a square platform of side 572·2 P″ and perimeter 2288·8 P″ (or 8 × 286·1 P″).

(ix) In the reduced-design Pyramid as built, the perimeter of the summit-platform (like that of each of the Pyramid's courses) would thus have measured 286·1 P″ *less* than in the full design, or 2002·7 P″ (7 × 286·1 P″).

(x) The above facts would necessarily produce the following data for the intended full-design capstone:
Base-side = 572·2 P″ (or 2 × 286·1 P″)
Base-perimeter = 2288·8 P″ (or 8 × 286·1 P″)
Height = 364·2765 P″
Slope = 51° 51′ 14·3″
Height of base above Pyramid's base = 5448·736 P″ (see code)
Number of courses above King's Chamber floor = 153 (see code)

(xi) In the event it is clear that no such capstone was ever installed by the

[1] Exact vertical measurement still to be confirmed by on-site inspection.

builders. Whether it was in fact cut and transported, we do not yet know for certain. But we do know that 'the stone which the builders rejected' would have been too big for the Pyramid as built (compare (viii) and (ix) above), and its final installation as the 'chief cornerstone' must therefore depend upon the completion of the world-symbolising Pyramid to its full, or perfect, dimensions. The fact that the weight of the capstone described would have been over a thousand tons might also have had something to do with its non-installation—though engineering feats of this order are known to have been successfully undertaken by the builders of the great terrace at Baalbek in the Lebanon.

* * *

If the Great Pyramid's designed capstone was never added, then, what was? It is my own belief that the capstone was replaced by some kind of false apex. This was to satisfy the natural expectations of the reigning pharaoh, who saw the building as his cenotaph and would not have tolerated its being left incomplete. Based on the 203rd course, this false apex was, I surmise, constructed of small blocks of inferior local limestone, and was probably dry-built—i.e. uncemented—these features being designed expressly to ensure its eventual destruction by sandstorm erosion and earth movements without noticeable damage to the main structure.

As evidence for this piece of designed obsolescence I would cite the fact that the neighbouring pyramids of Khafra and Menkura—both of them clearly modelled on the Great Pyramid's design—were indeed completed to their apexes, a fact which would have been unlikely had the great Khufu recently set a new trend in 'apexless' pyramids. Again, the successive reports of the seventeenth-and eighteenth-century observers Greaves, Melton, Davison and Abbé do indeed reveal a severely dilapidated and rapidly disappearing superstructure of small blocks above the present summit-platform, and if the classical historian Diodorus Siculus is to be believed, this process of decay had already set in by the first century B.C., since he describes the building as being already apexless to the extent of some eight feet in his own day.

Yet the same authority insists that it was otherwise in perfect condition—a fact which confirms beyond all reasonable doubt that no capstone as such was even installed or cemented in place, for the removal of such a capstone would have meant climbing to the summit, which in turn would have involved some breaking-up of the Pyramid's steep, smooth casing—as would, of course, the dislodging of the massive capstone itself. Evidently no such spoliation had occurred, and we can discount the stone's removal by natural means on the evidence of the same lack of damage to the casing. And in any case the neighbouring Second Pyramid—clearly inferior in construction to its immediate predecessor—succeeded in retaining its apex-stone through at least two earthquakes until quite recent times, when most of its casing was stripped by the mosque-builders (like that of the Great Pyramid) for its building stone. It seems logical, then, to assume that the Great Pyramid itself would likewise have retained its capstone—had it ever been given one—at least until its casing started to be removed in the ninth century of our era. But evidently this was not the case, and it therefore seems likely that the false apex theory, or something very like it, is the right one.

We are left, however, with the question of what is symbolised by the capstone's designed height of $364\frac{1}{4}$ P″. True, 364 factorises as $7 \times 4 \times 13$, which would seem (in the light of the hypothetical code proposed in chapter 2, and supplemented in chapter 4) to signify the spiritual perfection of the terrestrial soul. This reading in itself seems to fit the Pyramid's general symbolism very aptly.

On the other hand, $364\frac{1}{4}$ P″ seems to bear so obvious a relationship to the pyramidally ubiquitous quantity $365\frac{1}{4}$ as to suggest a deliberate reference, *via the addition of an extra inch*. What, then, could the reference be? Once again invoking the terms of the reconstructed code, that reference turns out to be a symbolic statement of the most remarkable appositeness.

For our reading must now suggest that man, having at last established a perfect world-order (the completed Pyramid), is destined to use it as a springboard to create a culminating age ($365\frac{1}{4}$ P″) involving the realisation of full divinity (1P″). That awesome achievement, meanwhile, must necessarily be reflected in an extension of the lines of the world-symbolising Pyramid *past* its apex and into a new dimension entirely. With man's final apotheosis, in other words, even the physical world—even, perhaps, the universe itself as known by man—becomes strangely transmuted and even reversed. And so there at last comes into realisation the ancient vision of 'a new heaven and new earth.'

APPENDIX J

The Messianic Plan—A Tentative Summary

The Garden of Eden What we call, for want of a better term, men's souls have in some sense existed for ever. At some unimaginably distant point in the remote past they became involved in the earth-planes during the culminating stages of terrestrial evolution. But in the process they became fatally *The Fall* enmeshed in the pleasures and pains of the physical world. *The prisoners* Not only are they now unable to escape, but men have to *The blind* a large extent ceased to be aware that escape is even possible, or that they have any claim to independent *spiritual* existence at all. Consequently their souls are constantly drawn back to the physical world—birth after birth, *The dead* and consequently death after death. There are a few who *The hungry* manage, over many existences, to 'drag themselves up by their own boot-straps' to the spiritual level where only one more incarnation during the Final Age will ensure their escape into immortality—the few who actually find the 'narrow way' and who 'take the kingdom of heaven by *The lame* storm'—but for the many the broad highway of mortality remains the norm. From this condition escape is, of course, paramount, for the physical 'host-world' cannot last for ever, and its eventual destruction must inevitably result in the destruction of all its parasites as well, and consequently *The Messianic Plan* something akin to the 'death' of the soul. Hence the Messianic Plan, whose purpose is to drive a motorway through the mountains of death and despair and to open up to man the fertile uplands of immortality. It was the mission of *Jesus of Nazareth* Jesus of Nazareth, a man who deliberately personified the second in the Messianic trinity of 'prophet, priest and

Charity
Hope
Faith
The Second Coming
The Resurrection
The harvest
Forgiveness
The Judgement
The Exodus
The Promised Land
Salvation

king', to symbolise the plan in his own life and death, and to initiate the process of putting it into effect. This involved (i) preparing the soil, by encouraging man to make a fresh start on freeing himself from the bonds of mortal self-gratification at the expense of others (the moral teachings); (ii) planting the seed, which is the new knowledge and realisation in man (and it *was* new to many Jews of his day) that he is a spiritual creature shackled to a mortal body, but with rights to a divine and eternal inheritance; and (iii) fertilising the crop—i.e. giving his followers faith in their ability to make this giant evolutionary step, by himself showing that he, physically a mere man, could achieve it. This demonstration will come to full fruition with his return at the end of the age, at the beginning of the third 'day' of a thousand years after his death. All men will be alive to witness this event, and will thus have final 'proof' at last—a proof that will catalyse, if they wish it, their ultimate return to the Real world, which the world of pure spirit (far from condemning man) has been trying, in its 'mercy', to engineer now since the beginning of what we call time. There will, however, be some who reject the opportunity, and who will prefer to condemn their own souls to 'death'. But, for their part, the 'sons of God' will, by a process of rigorous self-perfection, be led by a succession of Messianic figures into the 'way of Truth', which at length will lead them to the gateway of 'a new heaven and new earth'. Then at last the great Exodus will begin and man, conquering the last death, will emerge victorious into the Promised Land of the spirit. The Messianic Plan for human evolution will have been fulfilled.

Select Bibliography

The following list is confined to indicating various sources to which reference has been made, together with some recommended works for further reading. It makes no claim to balance or homogeneity—the works listed range from the abstruse to the popular. However, it should not automatically be assumed that validity of outlook bears any relation to stiffness of cover, or that scholastic soundness is in any way incompatible with directness of language or with popular appeal. Even the historical figure who, in the event, plays a central rôle in this book is far more widely celebrated as a daring popular teacher than as a safe academic authority.

Readers interested in pursuing further any particular branch of inquiry, and requiring rather more exhaustive bibliographical details, will find this need amply supplied by the works asterisked.

Works consulted

Berlitz, C., *Mysteries from Forgotten Worlds* (Souvenir, 1972)

Blavatsky, H. P., *The Secret Doctrine* (Theosophical University Press, 1888; 1966)

Blumrich, J. F., *The Spaceships of Ezekiel* (Corgi, 1974)

Brod, M., *Paganism—Christianity—Judaism* (University of Alabama Press)

Budge, Sir E. A. Wallis, *The Book of the Dead* (Theban Recension: Routledge, 1960)

Burland, Nicholson and Osborne, *Mythology of the Americas* (Hamlyn, 1970)

Carter, M. E., *Edgar Cayce on Atlantis* (Coronet, Paperback Library, 1968)

Cayce, E. E., *Edgar Cayce on Reincarnation* (Coronet, Paperback Library, 1968)

Cayce, H. L. (Ed.), *Edgar Cayce on Reincarnation* (Coronet, Paperback Library, 1967); *Venture Inward* (Coronet, Paperback Library, 1966)

Cerminara, G., *Many Mansions* (Morrow, 1950)

Charpentier, L., *Les Mystères de la Cathédrale de Chartres* (Laffont, 1971); *Le Mystère Basque* (Laffont, 1975)

Cruden, A., *Cruden's Complete Concordance to the Old and New Testaments* (Revised edition: Lutterworth, 1954)
Däniken, E. von, (Tr. M. Heron), *Chariots of the Gods?* (Corgi, 1969) and *The Gold of the Gods* (Souvenir, 1973)
Davidson and Aldersmith, *The Great Pyramid: Its Divine Message*, Vol. I (Williams and Norgate, 1925)
de Sabato, M., *Confidences d'un Voyant* (Hachette, 1971)
* Desroches-Noblecourt, C., *Tutankhamen (The Connoisseur* and Michael Joseph, 1969)
Dixon, J., *My Life and Prophecies* (Muller, 1969)
* Downing, B. H., *The Bible and Flying Saucers* (Sphere, 1974)
Drake, W. Raymond, *Gods and Spacemen in the Ancient East* (Sphere, 1973); *Gods and Spacemen in the Ancient West* (Sphere, 1974)
Edgar, J. and M., *The Great Pyramid Passages and Chambers*, Vols, I and II (Bone and Hulley, 1925)
Eleiade, M. (Tr. P. Mairet), *Myths, Dreams and Mysteries* (Collins/Fontana, 1968)
* Every, G., *Christian Mythology* (Hamlyn, 1970)
Fairbridge, R. W., 'The Changing Level of the Sea' (*Scientific American*, May 1960, Vol. 202, No. 5)
Frazer, J. G., *The Golden Bough: A Study in Magic and Religion* (abridged, Macmillan, 1941)
Galanopoulos, A. G., and Bacon, E., *Atlantis: The Truth Behind the Legend* (Nelson, 1969)
Gardiner, Sir A. H., *Egyptian Grammar* (3rd Ed.: Oxford University Press, 1957)
Gauquelin, M., *Astrology and Science* (Mayflower, 1972)
Glass, J., *The Story of Fulfilled Prophecy* (Cassell, 1969)
* Hawkins, G. S., *Stonehenge Decoded* (Souvenir, 1966)
Head and Cranston (Ed.), *Reincarnation in World Thought* (Julian Press, 1967)
* Honoré, P., *In Quest of the White God* (Futura, 1975)
* Humphreys, C., *Buddhism* (Pelican, 1951)
Ivimy, J., *The Sphinx and the Megaliths* (Turnstone, 1974)
* James, M. R. (Tr.), *The Apocryphal New Testament* (Oxford, 1924)
Johnson, Raynor C., *The Imprisoned Splendour* (Hodder, 1953)
Keller, W., *The Bible as History* (Hodder, 1956)
Kingsland, W., *The Great Pyramid in Fact and in Theory* (Rider, 1932)
Kolosimo, P., *Not of this World* (Sphere, 1971)
Langer, J., *Nine Gates* (Clarke, 1961)
Larousse, *World Mythology* (Hamlyn, 1965)
Lethbridge, T. C., *The Legend of the Sons of God* (Routledge & Kegan Paul, 1972)
MacKinnon, D., and others, *Objections to Christian Belief* (Constable, 1963)
Mascaró, J. (Tr.), *The Bhagavad Gita* (Penguin, 1962)
Michell, J., *The View over Atlantis* (Sphere, Abacus, 1973)

New English Bible, The (Oxford University Press, Cambridge University Press, 1970)
Nicoll, M., *The New Man* (Robinson & Watkins, 1967)
Petrie, Sir W. M. F., *The Pyramids and Temples of Gizeh* (1881)
Plato (Tr. H. D. P. Lee), *Timaeus* and *Critias* (Penguin, 1971)
Pochan, A., *L'Enigme de la Grande Pyramide* (Laffont, 1971)
* Powell Davies, A., *The Meaning of the Dead Sea Scrolls* (Mentor, 1956)
Rampa, T. L., *Chapters of Life* (Corgi, 1967)
Red Sea Pilot, The (Admiralty, 1926)
Reed's *Nautical Almanac*
Roof, S., *Journeys on the Razor-Edged Path* (Hodder, 1960)
Rutherford, A., *Pyramidology*, Vols. I–IV (Institute of Pyramidology, 1957 onwards)
Saddhatissa, H., *The Buddha's Way* (Allen & Unwin, 1971)
Santesson, H. S., *Understanding Mu* (Coronet, Paperback Library, 1970)
* Schonfield, H. J., *The Passover Plot* (Hutchinson, 1965); **Those Incredible Christians* (Bantam, 1969); **The Pentecost Revolution* (Macdonald, 1974); *The Authentic New Testament* (Dobson, 1955)
* Sen, K. M., *Hinduism* (Pelican, 1961)
Stevenson, I., *Twenty Cases Suggestive of Reincarnation* (American Society for Psychical Research, 1966)
Sugrue, T., *There is a River* [The story of Edgar Cayce] (Dell, 1967)
Thom, A., *Megalithic Sites in Britain* (Oxford University Press, 1967)
* Thompson, J. E., *The Rise and Fall of Maya Civilisation* (University of Oklahoma Press, 1954)
Tomas, A., *Atlantis: from Legend to Discovery* (Hale, 1972); *We Are Not The First* (Sphere, 1972); *Beyond the Time Barrier* (Sphere, 1974)
* Tompkins, P., *Secrets of the Great Pyramid* (Allen Lane, 1973)
* Vaillant, G. C., *Aztecs of Mexico* (Pelican, 1950)
Velikovsky, I., *Oedipus and Ikhnaton* (Sidgwick & Jackson, 1960)
Weatherhead, L. D., *The Christian Agnostic* (Hodder, 1965)
Wiesel, E., *Souls on Fire* (Weidenfeld & Nicolson, 1972)
Young, R., *Analytical Concordance to the Holy Bible* (8th Ed.: Lutterworth, 1939)

Suggestions for further reading (including some fictional 'food for thought')

Barjavel, R. (Tr. C. L. Markham), *The Ice People* (Mayflower, 1972)
Beesley, R. P., *The Path of Esoteric Truthfulness* (White Lodge Publications)
Benavides, R., *Dramatic Prophecies of the Great Pyramid* (Editores Mexicanos Unidos—tr. of 11th Ed., 1970)
Cayce, H. L. (Ed.), *The Edgar Cayce Reader* (Coronet, Paperback Library, 1969)
Clarke, Arthur C., *Childhood's End* (Pan, 1954); *2001: A Space Odyssey* (Arrow, 1968); *Profiles of the Future* (Pan, 1964)
Challoner, H. K., *The Wheel of Rebirth* (Theosophical Publishing House, 1969)

* Cross, C., *Who was Jesus?* (Hodder, 1970)
David-Neel, A., *Magic and Mystery in Tibet* (Corgi, 1971)
Govinda, A., *The Way of the White Clouds* (Rider, 1973)
Grant, J., *The Winged Pharaoh* (Sphere, 1973)
Hoyle, F., *The Black Cloud* (Penguin, 1960)
Landsburg, A. and S., *In Search of Ancient Mysteries* (Corgi, 1974)
Lehmann, J. (Tr. M. Heron), *The Jesus Report* (Souvenir, 1972)
Lewis, C. S., *Out of the Silent Planet* (Pan, 1952); *Voyage to Venus* (Pan, 1953)
Lindsey, H., *The Late Great Planet Earth* (Zondervan Press, 1970)
* Lunan, D., *Man and the Stars* (Souvenir, 1974)
Michell, J., *City of Revelation* (Abacus, 1973)
Needleman, J., *The New Religions* (Allen Lane, 1972)
Pauwels L., and Bergier, J., *The Morning of the Magicians* (Mayflower, 1971); *Eternal Man* (Mayflower, 1973); *Impossible Possibilities* (Mayflower, 1974)
Rampa, T. L., *The Third Eye*, et alia (Corgi, 1956 onwards)
Spangler, D., *Links with Space* (Findhorn Publications); *Revelation, the Birth of a New Age* (Findhorn Publications)
Vacca, R., *The Coming Dark Age* (Panther, 1974)
* Watson, L., *Supernature* (Hodder, 1974); *The Romeo Error* (Hodder, 1974)
* Wilson, C., *The Occult* (Mayflower, 1973)
Yerby, F., *Judas, My Brother* (Mayflower, 1971)

Index